OBSERVED CONFIDENCE LEVELS
Theory and Application

OBSERVED CONFIDENCE LEVELS
Theory and Application

Alan M. Polansky
Northern Illinois University
Dekalb, U.S.A.

CRC Press
Taylor & Francis Group
Boca Raton London New York

CRC Press is an imprint of the
Taylor & Francis Group, an **informa** business

A CHAPMAN & HALL BOOK

CRC Press
Taylor & Francis Group
6000 Broken Sound Parkway NW, Suite 300
Boca Raton, FL 33487-2742

First issued in paperback 2019

ISBN-13: 978-1-58488-802-4 (hbk)
ISBN-13: 978-0-367-38842-3 (pbk)

Library of Congress Cataloging-in-Publication Data

Polansky, Alan M.
  Observed confidence levels : theory and application / Alan M. Polansky.
     p. cm.
  Includes bibliographical references and index.
  ISBN 978-1-58488-802-4 (alk. paper)
  1. Observed confidence levels (Statistics) 2. Asymptotic expansions. I. Title.

QA277.5.P65 2007
519.5--dc22   2007032838

Visit the Taylor & Francis Web site at
http://www.taylorandfrancis.com

and the CRC Press Web site at
http://www.crcpress.com

TO

CATHERINE AND ALTON

AND

ADLER, AGATHA, HOLMES, KINSEY, LISA, MILHOUSE, AND RALPH

Contents

Preface

Observed confidence levels provide an approach to solving multiple testing problems without the need for specifying a sequence of null and alternative hypotheses or adjusting the experimentwise error rate. The method is motivated by the problem of regions, where the practitioner is attempting to determine which of a set of subsets of a parameter space a parameter belongs to on the basis of observed data generated by the probability model of interest. Observed confidence levels measure the amount of confidence there is that the parameter is within each of the regions of interest. In application, the use of observed confidence levels is similar to the use of posterior probabilities from the Bayesian framework, but the method lies within the frequentist framework and does not require the specification of prior probabilities.

This book describes the basic development, theory, and application of observed confidence levels for a variety of common problems in statistical inference. The core of the theoretical development is contained in Chapters 2–4 where the theory and application of observed confidence levels are developed for general scalar parameters, vector parameters, and linear models. Chapter 5 considers nonparametric problems that are often associated with smoothing methods: nonparametric density estimation and regression. Further applications are briefly described in Chapter 6, including applications in generalized linear models, classical nonparametric statistics, multivariate analysis, and survival analysis. Some comparisons of the method of observed confidence levels to methods such as hypothesis testing, multiple comparisons, and Bayesian posterior probabilities are presented in Chapter 7. Asymptotic expansion theory lies at the heart of the theoretical development in this book. The Appendix provides some background into this theory.

The focus of the book is the modern nonparametric setting where many of the calculations are based on bootstrap estimates, usually within the smooth function model popularized in this setting by Peter Hall. It is this framework that allows for relatively simple solutions to many interesting problems, and also allows for substantial theoretical development. The use of observed confidence levels is certainly not constrained to these types of applications. Some examples in the book, therefore, demonstrate the use of observed confidence levels in classic parametric applications. Results from classical nonparametric statistics are also briefly discussed in Chapter 6, with obvious extensions to other problems in that area.

There are a myriad of techniques that have been developed to handle the multiple testing problem, a problem that has generated polarizing views and substantial controversy. While controversy often breeds radical new ideas and solutions, it can also overshadow fundamental ideas, especially for newly proposed methods. In this atmosphere it is often majority rule, and not the merits of new methods that decide their fate. Therefore, this book does not intend to make the case that observed confidence levels are necessarily superior to these methods, or that observed confidence levels offer the practicing statistician a tool that will always solve every problem without fail. It is my opinion that observed confidence levels do provide a new and simple method for solving many problems. I have talked with many practitioners of statistics who have expressed interest and delight at the possibilities that these methods hold. Therefore, I am willing to let the examples presented in this book stand on their own and let the readers decide for themselves what their final opinion is of these methods.

This book could serve as the basis for a graduate course in statistics that emphasizes both theory and applications. Exercises have been included at the end of the chapters and the appendix. The most reasonable approach to using this book in a course would be to cover the chapters in sequential order, as written. Chapter 6 could certainly be optional as the theoretical development in that chapter is very light. Nonetheless, applied exercises have been included at the end of this chapter if one wishes to cover it.

I have endeavored to include as many real data sets as possible in the book. Many of these will be familiar to some practitioners, and will hopefully provide a novel way of looking at some standard problems. In some cases I have used references to suggest a type of problem, but have simulated new data when the original data had some unrelated problems such as very small sample sizes or obvious outlying values. In these cases I have attempted to retain the flavor of the original application with the simulated data. Nearly all of the data used in the exercises are simulated, but are based on ideas from real applications. This allowed me to easily control the flow and character of the problems. Again, I have attempted to retain a flavor of the original applications. Though several people from other disciplines helped me craft many of these problems, any obvious errors in magnitudes of effects, units or applications are entirely my own.

The applied problems in the book are solved using this R statistical computing environment. This software is one of the few statistical software packages that is flexible enough to allow for simple application of bootstrap problems. The fact that R is freely available for download to everyone, including students, only convinces me further that this package is the wave of the future, and is certainly used by many practitioners and researchers today. To aid researchers and students in applying this software package to the problems presented in the book, most of the chapters include a section that demonstrates how many of the examples were solved using R. I am not an expert in this language, and there are certainly more efficient methods for solving some of the problems that are presented in this book. However, the included code should provide novice and intermediate users a starting point for solving their own problems.

Many of the data sets and R code used in the book are available on my professional website. At the time of publication this address is

http://www.math.niu.edu/~polansky/oclbook

This website also includes errata for the book, links to relevant references, and news about current applications of observed confidence levels in the literature.

I was introduced to the problem of regions by Michael Newton during a visit to the University of Wisconsin. In my subsequent research there were many who provided invaluable assistance. Among these are Brad Efron, Rob Tibshirani, and Peter Hall who always responded quickly and thoughtfully to my emailed questions about their research. Brad Efron also provided me with the data from the cholesterol study used in the regression chapter. I would also like to thank Dr. Stephen Rappaport and Dr. Peter Egeghy for making the naphthalene data used in the multiple parameter chapter available for analysis in the example. Support for collection of the naphthalene data was provided by a grant from the U.S. Air Force and by NIEHS through Training Grant T32ES07018 and Project Grant P42ES05948. Many other colleagues have contributed, either directly or indirectly, to the writing of this book. They include Sanjib Basu, Nader Ebrahimi, Mike Ernst, Dick Gunst, Bala Hosmane, Rama Lingham, Jeff Reynolds, and Bill Schucany. My Ph.D. student Cathy Poliak provided special assistance to this project by developing most of the results in Chapter 4 as part of her dissertation. Special thanks are due to Tom O'Gorman who provided valuable insights into the research project and also provided valuable advice on writing and publishing this book. Chiyo Kazuki Bryhan also provided special assistance by translating an article that was written in Japanese. The local Tex and Mac guru in the Department of Mathematical Sciences at Northern Illinois University, Eric Behr, provided extensive help and advice that helped keep my G5 and MacBook running smoothly throughout this project. This book would not have been possible without the help and support of the people from Taylor and Francis who worked on the manuscript, including Karen Simon, Katy Smith and Samantha White. I especially wish to thank my editor Bob Stern whose belief in this project made it possible for this book to become a reality. Of course there is always family, so I wish to thank Mom, Dad, Gary, Dale, Karin, Mike, Jackie, Frank, and their families for always being there. Special thanks are also afforded to my wife Catherine Check, who not only provided the emotional support required to complete this project, but also read large parts of the manuscript and provided consultation for the examples within her areas of interest which include chemistry, physics, geology, ecology, and astronomy. Catherine also had infinite patience with my work schedule as the final copy of the manuscript was prepared. My son Alton, who was born midway through this project, also provided valuable emotional support by just being there when I needed him. Each evening as I fed him and put him to bed he always provided me with a smile or a laugh that made it easy to keep working.

Alan M. Polansky
Creston, IL, USA
May 2007

CHAPTER 1

Introduction

1.1 Introduction

Observed confidence levels provide useful information about the relative truth of hypotheses in multiple testing problems. In place of the repeated tests of hypothesis usually associated with multiple comparison techniques, observed confidence levels provide a level of confidence for each of the hypotheses. This level of confidence measures the amount of confidence there is, based on the observed data, that each of the hypotheses are true. This results in a relatively simple method for assessing the truth of a sequence of hypotheses. This chapter will introduce the basic concepts of observed confidence levels. Section 1.2 focuses on a problem in statistical inference which provides the basic framework for the development of observed confidence levels. The section also introduces the basic definition of observed confidence levels. Section 1.3 presents several applied examples and poses questions where observed confidence levels can provide useful answers. These examples serve as a preview for the types of problems that will be discussed throughout the book. Section 1.4 provides a preview of the material that is covered in the remaining chapters.

1.2 The Problem of Regions

The development of observed confidence levels begins by constructing a formal framework for the problem. This framework is based on the *problem of regions*, which was first formally proposed by Efron and Tibshirani (1998). An expanded version of the paper is provided in the technical report by Efron and Tibshirani (1996), and is available for download from Rob Tibshirani's website. The problem is constructed as follows.

Let $\mathbf{X}$ be a d-dimensional random vector following a d-dimensional distribution function F. We will assume that F is a member of a collection, or family, of distributions given by $\mathcal{F}$. The collection $\mathcal{F}$ may correspond to a parametric family such as the collection of all d-variate normal distributions, or may be nonparametric, such as the collection of all continuous distributions with finite mean. Let $\boldsymbol{\theta}$ be the parameter of interest and assume that $\boldsymbol{\theta}$ is a functional parameter of F. That is, define a function $T : \mathcal{F} \to \mathbb{R}^p$ such that $\boldsymbol{\theta} = T(F)$.

1

The parameter space for $\boldsymbol{\theta}$ under the collection $\mathcal{F}$ is then defined as

$$\Theta = \{\boldsymbol{\theta} : \boldsymbol{\theta} = T(F), F \in \mathcal{F}\}.$$

Note that this definition depends on the collection $\mathcal{F}$. Therefore a single parameter, such as a mean, may have different parameter spaces depending on the collection $\mathcal{F}$ that is being used. For example, suppose that $\boldsymbol{\theta}$ is the mean parameter defined by

$$\boldsymbol{\theta} = \int_{\mathbb{R}} \mathbf{t} dF(\mathbf{t}). \tag{1.1}$$

If $\mathcal{F}$ is the collection of all continuous univariate distributions with finite mean, then $\Theta = \mathbb{R}$, whereas if $\mathcal{F}$ is the collection of univariate gamma distributions then $\Theta = \mathbb{R}^+ = (0, \infty)$.

For a given parameter space Θ, let $\{\Theta_i\}_{i=1}^{\infty}$ be a countable sequence of subsets, or regions, of Θ such that

$$\bigcup_{i=1}^{\infty} \Theta_i = \Theta.$$

For simplicity each region will usually be assumed to be connected, though many of the methods presented in this book can be applied to the case where Θ_i is made up of disjoint regions as well. Further, in many of the practical examples in this book the sequence will be finite, but there are practical examples that require countable sequences. There will also be examples where the sequence $\{\Theta_i\}_{i=1}^{\infty}$ forms a partition of Θ in the sense that $\Theta_i \cap \Theta_j = \emptyset$ for all $i \neq j$. In the general case, the possibility that the regions can overlap will be allowed. In many practical problems the subsets technically overlap on their boundaries, but the sequence can often be thought of as a partition from a practical viewpoint.

The statistical interest in such a sequence of regions arises from the structure of a specific inferential problem. Typically the regions correspond to competing models for the distribution of the random vector $\mathbf{X}$ and one is interested in determining which of the models is most reasonable based on the observed data vector $\mathbf{X}$. Therefore the problem of regions is concerned with determining which of the regions in the sequence $\{\Theta_i\}_{i=1}^{\infty}$ that $\boldsymbol{\theta}$ belongs to based on the observed data $\mathbf{X}$.

An obvious simple solution to this problem would be to estimate $\boldsymbol{\theta}$ based on $\mathbf{X}$ and conclude that $\boldsymbol{\theta}$ is in the region Θ_i whenever the estimate $\hat{\boldsymbol{\theta}}$ is in the region Θ_i. For most reasonable parameters and collections of distribution functions a reasonable estimate of $\boldsymbol{\theta}$ can be developed using the plug-in principle.

To apply the plug-in principle, suppose a reasonable estimate of F can be computed using the observed data vector $\mathbf{X}$. That is, there is an estimate $\hat{F} = V(\mathbf{X})$ for some function $V : \mathbb{R}^d \to \mathcal{F}$. The existence of such an estimate certainly depends on the structure of the problem. As an example, suppose

$\mathbf{X}' = (X_1, X_2, \ldots, X_d)$ and consider the collection of distributions given by

$$\mathcal{G}_{\mathrm{IID}} = \left\{ G(\mathbf{x}) : G(\mathbf{x}) = \prod_{i=1}^{d} F(x_i), F \in \mathcal{F} \right\}, \tag{1.2}$$

where $\mathcal{F}$ is taken to be the collection of all univariate distribution functions and $\mathbf{x}' = (x_1, \ldots, x_n)$. If $G \in \mathcal{G}_{\mathrm{IID}}$, then $X_1, \ldots, X_d$ is a sequence of independent and identically distributed random variables following the distribution F. In this case a reasonable estimate of F is given by the empirical distribution function

$$\hat{F}_d(t) = V(t; \mathbf{X}) = d^{-1} \sum_{i=1}^{d} \delta(t; [X_i, \infty)), \tag{1.3}$$

for all $t \in \mathbb{R}$ where δ is the indicator function defined by

$$\delta(t; A) = \begin{cases} 1 & t \in A, \\ 0 & t \notin A. \end{cases}$$

The empirical distribution function is a point-wise unbiased and consistent estimator of F. In addition the Glivenko-Cantelli Theorem implies that

$$\sup_{t \in \mathbb{R}} |\hat{F}_d(t) - F(t)| \xrightarrow{a.s.} 0,$$

as $d \to \infty$. See Section 2.1 of Serfling (1980) for further details on these results. Many other collections of distributions allow for simple estimators of F as well. For example consider a parametric family

$$\mathcal{F}_{\boldsymbol{\lambda}} = \{F(\mathbf{x}) : F(\mathbf{x}) = F_{\boldsymbol{\lambda}}(\mathbf{x}), \boldsymbol{\lambda} \in \Lambda\},$$

where $F_{\boldsymbol{\lambda}}$ is a parametric function of $\boldsymbol{\lambda}$ and Λ is the parameter space of $\boldsymbol{\lambda}$. Suppose $\hat{\boldsymbol{\lambda}}$ is an estimator of $\boldsymbol{\lambda}$ based on the sample $\mathbf{X}$ using maximum likelihood or method of moments estimation, for example. Then F can be estimated using $\hat{F}(\mathbf{x}) = F_{\hat{\boldsymbol{\lambda}}}(\mathbf{x})$.

Assuming that a reasonable estimator of F exists, a simple estimator of $\boldsymbol{\theta}$ can be computed as $\hat{\boldsymbol{\theta}} = T(\hat{F})$. Such an estimate is often known as a *plug-in estimate*. See Putter and van Zwet (1996) for further theoretical details on such estimates. For example, suppose $G \in \mathcal{G}_{\mathrm{IID}}$ with F defined as in Equation (1.2) and θ is the mean parameter of the univariate distribution F defined in Equation (1.1). Then a plug-in estimate of the mean parameter θ can be derived as

$$\hat{\theta} = \int_{\mathbb{R}} t d\hat{F}_d(t) = d^{-1} \sum_{i=1}^{d} X_i, \tag{1.4}$$

where the integral is been taken to be a Lebesgue integral. In the case where $F \in \mathcal{F}_{\boldsymbol{\lambda}}$, it is usually the case that $\boldsymbol{\theta}$ is a function of $\boldsymbol{\lambda}$, that is $\boldsymbol{\theta} = m(\boldsymbol{\lambda})$ for some function $m : \Lambda \to \Theta$. In this case using the plug-in estimate $\hat{\boldsymbol{\theta}} = T(\hat{F})$ is equivalent to using $\hat{\boldsymbol{\theta}} = m(\hat{\boldsymbol{\lambda}})$.

The problem with simply concluding that $\boldsymbol{\theta} \in \Theta_i$ whenever $\hat{\boldsymbol{\theta}} \in \Theta_i$ is that $\hat{\boldsymbol{\theta}}$ is subject to sample variability, that is, $\hat{\boldsymbol{\theta}}$ is itself a random variable. Therefore, even though we may observe $\hat{\boldsymbol{\theta}} \in \Theta_i$, it may actually be true that $\boldsymbol{\theta} \in \Theta_j$ for some $i \neq j$ where $\Theta_i \cap \Theta_j = \emptyset$, and that $\hat{\boldsymbol{\theta}} \in \Theta_i$ was observed simply due to chance. If such an outcome were rare, then the method may be acceptable. However, if such an outcome occurred relatively often, then the method would not be useful. Therefore, it is clear that the inherent variability in $\hat{\boldsymbol{\theta}}$ must be accounted for in order to develop a useful solution to the problem of regions.

Multiple comparison techniques are based on solving the problem of regions by using a sequence of hypothesis tests. Adjustments to the testing technique helps control the overall significance level of the sequence of tests. Modern techniques have been developed by Stefansson, Kim and Hsu (1988) and Finner and Strassburger (2002). Some general references that address issues concerned with multiple comparison techniques include Hochberg and Tamhane (1987), Miller (1981) and Westfall and Young (1993). A general discussioni of multiple comparison techniques can also be found in Section 7.3. Some practitioners find the results of these procedures difficult to interpret as the number of required tests can sometimes be quite large.

An alternate approach to multiple testing techniques was formally introduced by Efron and Tibshirani (1998). This approach computes a measure of confidence for each of the regions. This measure reflects the amount of confidence there is that $\boldsymbol{\theta}$ lies within the region based on the observed sample $\mathbf{X}$. The method used for computing the observed confidence levels studied in this book is based on the methodology of Polansky (2003a,b).

Let $\boldsymbol{C}(\alpha, \boldsymbol{\omega}; \mathbf{X}) \subset \Theta$ be a $100\alpha\%$ confidence region for $\boldsymbol{\theta}$ based on the sample $\mathbf{X}$. That is, $\boldsymbol{C}(\alpha, \boldsymbol{\omega}; \mathbf{X}) \subset \Theta$ is a function of the sample $\mathbf{X}$ with the property that

$$P[\boldsymbol{\theta} \in \boldsymbol{C}(\alpha, \boldsymbol{\omega}; \mathbf{X})] = \alpha.$$

The value α is called the *confidence level* or *confidence coefficient* of the region. The vector $\boldsymbol{\omega} \in \Omega_\alpha \subset \mathbb{R}^q$ is called the *shape parameter vector* as it contains a set of parameters that control the shape and orientation of the confidence region, but do not have an effect on the confidence coefficient. Even though Ω_α is usually a function of α, the subscript α will often be omitted to simplify mathematical expressions. In later chapters the vector $\boldsymbol{\omega}$ will also be dropped from the notation for a confidence region when it does not have a direct role in the current development. Now suppose that there exist sequences $\{\alpha_i\}_{i=1}^\infty \in [0,1]$ and $\{\boldsymbol{\omega}_i\}_{i=1}^\infty \in \Omega_\alpha$ such that $\boldsymbol{C}(\alpha_i, \boldsymbol{\omega}_i; \mathbf{X}) = \Theta_i$ for $i = 1, 2, \ldots$, conditional on $\mathbf{X}$. Then the sequence of confidence coefficients are defined to be the observed confidence levels for $\{\Theta_i\}_{i=1}^\infty$. In particular, α_i is defined to be the observed confidence level of the region Θ_i. That is, the region Θ_i corresponds to a $100\alpha_i\%$ confidence region for $\boldsymbol{\theta}$ based on the observed data. This measure is similar to the measure suggested by Efron and Gong (1983), Felsenstein (1985) and Efron, Holloran, and Holmes (1996). It is also similar in application to the methods of Efron and Tibshirani (1998), though the formal definition of the measure differs slightly from the definition used above.

See Efron and Tibshirani (1998) and Section 7.4 for further details on this definition. Some example applications of observed confidence levels are given in Section 1.3.

1.2.1 Normal Mean Example

To demonstrate this idea, consider a simple example where $X_1, \ldots, X_n$ is a set of independent and identically distributed random variables from a normal distribution with mean θ and variance σ^2. Let $\hat{\theta}$ and $\hat{\sigma}$ be the usual sample mean and variance computed on $X_1, \ldots, X_n$. A confidence interval for the mean that is based on the assumption that the population is normal is based on percentiles from the t-distribution and has the form

$$C(\alpha, \boldsymbol{\omega}; \mathbf{X}) = (\hat{\theta} - t_{n-1;1-\omega_L} n^{-1/2} \hat{\sigma}, \hat{\theta} - t_{n-1;1-\omega_U} n^{-1/2} \hat{\sigma}), \qquad (1.5)$$

where $t_{\nu;\xi}$ is the ξth percentile of a t-distribution with ν degrees of freedom. In order for the confidence interval in Equation (1.5) to have a confidence level equal to $100\alpha\%$ we take $\boldsymbol{\omega}' = (\omega_L, \omega_U)$ to be the shape parameter vector where

$$\Omega_\alpha = \{\boldsymbol{\omega} : \omega_U - \omega_L = \alpha, \omega_L \in [0, 1], \omega_U \in [0, 1]\},$$

for $\alpha \in (0, 1)$. Note that selecting $\boldsymbol{\omega} \in \Omega_\alpha$ not only insures that the confidence level is $100\alpha\%$, but also allows for several orientations and shapes of the interval. For example, a symmetric two-tailed interval can be constructed by selecting $\omega_L = (1 - \alpha)/2$ and $\omega_U = (1 + \alpha)/2$. An upper one-tailed interval is constructed by setting $\omega_L = 0$ and $\omega_U = \alpha$. A lower one-tailed interval uses $\omega_L = 1 - \alpha$ and $\omega_U = 1$.

Now consider the problem computing observed confidence levels for the normal mean for a sequence of interval regions of the form $\Theta_i = [t_i, t_{i+1}]$ where $-\infty < t_i < t_{i+1} < \infty$ for $i \in \mathbb{N}$. Setting $\Theta_i = C(\alpha, \boldsymbol{\omega}; \mathbf{X})$ where the confidence interval used for this calculation is the one given in Equation (1.5) yields

$$\hat{\theta} - t_{n-1;1-\omega_L} n^{-1/2} \hat{\sigma} = t_i, \qquad (1.6)$$

and

$$\hat{\theta} - t_{n-1;1-\omega_U} n^{-1/2} \hat{\sigma} = t_{i+1}. \qquad (1.7)$$

Solving Equations (1.6) and (1.7) for ω_L and ω_U yields

$$\omega_L = 1 - T_{n-1} \left[\frac{n^{1/2}(\hat{\theta} - t_i)}{\hat{\sigma}} \right],$$

and

$$\omega_U = 1 - T_{n-1} \left[\frac{n^{1/2}(\hat{\theta} - t_{i+1})}{\hat{\sigma}} \right],$$

where T_{n-1} is the distribution function of a t-distribution with $n - 1$ degrees of freedom. Because $\omega \in \Omega_\alpha$ if and only if $\omega_U - \omega_L = \alpha$ it follows that the observed confidence level for the region Θ_i is given by

$$T_{n-1}\left[\frac{n^{1/2}(\hat{\theta} - t_i)}{\hat{\sigma}}\right] - T_{n-1}\left[\frac{n^{1/2}(\hat{\theta} - t_{i+1})}{\hat{\sigma}}\right].$$

Further details on this example will be given in Section 2.2.1.

1.3 Some Example Applications

1.3.1 A Problem from Biostatistics

In most countries new drugs are required to pass through several levels of clinical trials before they are approved for sale to the general public. In the United States the governing body overseeing this process is the Food and Drug Administration (FDA). In some cases treatments already exist for a particular application and a drug company may wish to introduce a new treatment to the market that contains the same active ingredients, but may be delivered in a different form, or may be manufactured in a new way. In this case the drug companies wish to show that the new treatment is *bioequivalent* to the approved treatment. That is, the effect of the new treatment is the same as the old treatment.

Efron and Tibshirani (1993) consider a bioequivalence study of three hormone supplement medical patches. The first patch has already been approved by the FDA. The second patch is manufactured at a new facility, and is designed to have an identical effect on hormone levels as the approved patch. The third patch is a placebo. The study consists of having eight patients wear the patches in random order, each for a specified period of time. The blood level of the hormone is measured after each patch is worn. Box-plots of the observed hormone levels for the eight patients after wearing each of the three patches is given in Figure 1.1. The data are given in Table 1.1.

Let A_i, N_i, and P_i be the blood hormone levels in Patient i after wearing the approved, new, and placebo patches, respectively. Of interest in the bioequivalence problem is the parameter

$$\theta = \frac{E(N_i - A_i)}{E(A_i - P_i)}.$$

It is assumed that the parameter θ is the same for each patient so that θ does not depend on i. According to Efron and Tibshirani (1993), the FDA requirement for bioequivalence is that a 90% confidence interval for θ lie in the range $[-0.2, 0.2]$. An alternative approach would consist of computing observed confidence levels for $\Theta_1 = (-\infty, -0.2)$, $\Theta_2 = [-0.2, 0.2]$, and $\Theta_2 = (0.2, \infty)$. Bioequivalence would then be accepted if the observed confidence

Table 1.1 *Data from the small bioequivalence study of Efron and Tibshirani (1993). The observations are the blood level of a hormone after wearing each of the patches.*

Subject	Placebo (P_i)	Approved (A_i)	New (N_i)
1	9243	17649	16449
2	9671	12013	14614
3	11792	19979	172174
4	13357	21816	23798
5	9055	13850	12560
6	6290	9806	10157
7	12412	17208	16570
8	18806	29044	26325

level for Θ_2 exceeds 0.90. A plug-in estimator of θ is

$$\hat{\theta} = \frac{\sum_{i=1}^{n}(N_i - A_i)}{\sum_{i=1}^{n}(A_i - P_i)},$$

where n is the number of patients in the study. For the data given in Table 1.1, $\hat{\theta} = -0.071 \in \Theta_2$. How much confidence is there, based on the observed data, that θ is really in Θ_2? Using the methods developed in Chapter 2 for single parameter problems such as this, it is shown that a good estimate of this confidence level is given by 0.8611, or that there is approximately 86.11% confidence that the patches are bioequivalent.

1.3.2 A Problem from Statistical Quality Control

The application of statistics in industrial manufacturing processes is generally concerned with two major features of the process. The first is to be able to establish that the manufacturing process is stable and produces items in a consistent way. Once this stability is established, the second concern is the quality of the manufactured items. In particular, a manufacturing process is said to be *capable* is the process consistently produces items within the specifications set by the engineers.

Consider a simple situation where the items produced by a manufacturing process have a single measurement associated with them that insures the quality of the product. We will denote this measurement, known as a *quality characteristic*, as X. Suppose that design engineers have studied the product and have concluded that the quality of the product is acceptable if $L \leq X \leq U$, where L and U are constants known as the lower and upper *specification limits* of the quality characteristic, respectively. To assess the capability of

Figure 1.1 *Box-plots of the observed hormone levels for the eight patients after wearing each of the three patches in the bioequivalence study. The data originates from Efron and Tibshirani (1993).*

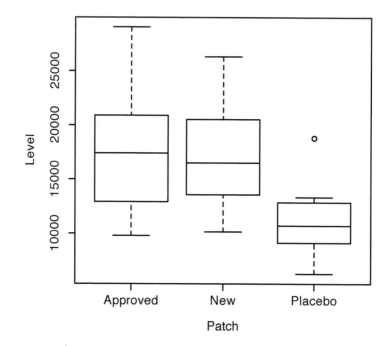

the process that manufactures these items, Kane (1986) suggested a *process capability index* of the form

$$C_{pk} = \min \left\{ \frac{U - \mu}{3\sigma}, \frac{\mu - L}{3\sigma} \right\},$$

where $\mu = E(X)$ and $\sigma^2 = V(X)$. This index has been developed so that larger values of the index indicate manufacturing processes that have a greater capability of producing items within specifications. Many other indices have been developed. See Polansky and Kirmani (2003) and Kotz and Johnson (1993) for a detailed discussion of this problem. It is usually the case that the parameters μ and σ are unknown. In this case one can observe n items from the manufacturing process to obtain information about μ and σ. Let $X_1, \ldots, X_n$ denote the resulting measurements from a set of observed items and let $\hat{\mu}$ and $\hat{\sigma}$ be the usual sample mean and sample standard deviation computed on these measurements. An estimate of the C_{pk} process capability index can then be computed as

$$\hat{C}_{pk} = \min \left\{ \frac{U - \hat{\mu}}{3\hat{\sigma}}, \frac{\hat{\mu} - L}{3\hat{\sigma}} \right\}.$$

Table 1.2 *Summary of the observed process data obtained from the four potential suppliers of piston rings.*

Supplier	1	2	3	4
Sample Size	50	75	70	75
Estimated μ	2.7048	2.7019	2.6979	2.6972
Estimated σ	0.0034	0.0055	0.0046	0.0038
Estimated C_{pk}	1.5392	1.1273	1.3333	1.5526

Problems in process capability analysis often involve the comparison of several process capability indices from the processes of competing suppliers or production methods. For example, Chou (1994) considers a process capability problem where there are four potential suppliers of piston rings for automobile engines. The edge width of a piston ring after the preliminary disk grind is a crucial quality characteristic in automobile engine manufacturing. Suppose that the automotive engineers have set the lower and upper specification limits of this quality characteristic to be $L = 2.6795$mm and $U = 2.7205$mm, respectively. Samples of output from each of the suppliers is observed in order to obtain information about the capability of each of the suppliers manufacturing processes. Histograms of the observed edge widths from each supplier, along with the manufacturer's specification limits are given in Figure 1.2. A summary of the observed process data is given in Table 1.2. It is clear from Table 1.2 that Supplier 4 has the largest estimated process capability. Given this process data, how much confidence is there in concluding that Supplier 4 has the most capable process? Using the methods that are developed in Chapter 3 for multiple parameter problems, it will be shown that a good estimate of this confidence level is 50.66%. That is, based on the observed process data, there is 50.66% confidence that Supplier 4 has the most capable process.

1.3.3 A Problem from Biostatistics

A study presented by Potthoff and Roy (1964) concerns the analysis of growth curves of the distance (mm) from the center of the pituitary gland to the pterygomaxillary fissure in young children. Sixteen boys and eleven girls participated in the study, which was conducted at the University of North Carolina Dental School. The distance measurement was taken at ages 8, 10, 12, and 14 for each individual in the study. The original intent of the study was to compare the growth curves of males and females. The individual growth curves for each of the sixteen male subjects are displayed in Figure 1.3 and the individual growth curves for the female subjects are presented in Figure 1.4. The average growth curve for each gender is displayed in Figure 1.5. The data are also analyzed by Davis (2002) and are displayed in Table 1.3.

Figure 1.2 *Frequency histograms of the observed process data from the four potential suppliers of piston rings. The dashed vertical lines represent the specification limits of the manufacturer.*

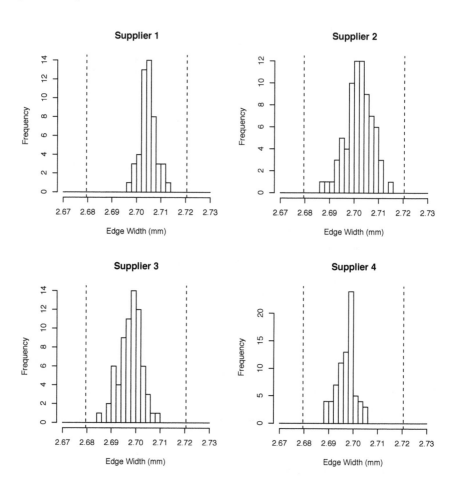

Represent the set of four measurements from the ith male subject in the study as $\mathbf{X}_i$, a random vector in $\mathbb{R}^4$. Assume that $\mathbf{X}_1, \ldots, \mathbf{X}_{16}$ is a set of independent and identically distributed random vectors from a distribution F_M with mean vector $\boldsymbol{\theta}_M = E(\mathbf{X}_i)$. Represent the set of four measurements from the ith female subject in the study as $\mathbf{Y}_i$, a random vector in $\mathbb{R}^4$. Assume that $\mathbf{Y}_1, \ldots, \mathbf{Y}_{11}$ is a set of independent and identically distributed random vectors from a distribution F_F with mean vector $\boldsymbol{\theta}_F = E(\mathbf{Y}_i)$. Let the parameter of interest be the mean vector of the joint distribution of F_M and F_F given by $\boldsymbol{\theta}' = (\boldsymbol{\theta}'_M, \boldsymbol{\theta}'_F) \in \mathbb{R}^8$.

Figure 1.3 *Observed measurements from the sixteen boys in the dental study. The measurements of each subject are connected by a solid line.*

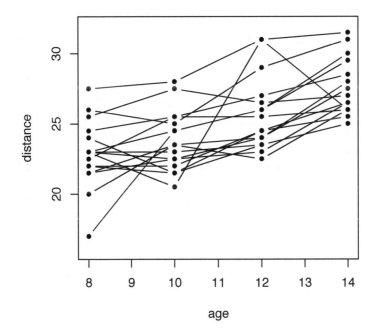

Many questions may arise as to what trend there may be in the mean vector $\theta' = (\theta_1, \ldots, \theta_8)$. For example, how much confidence is there that the mean distance is monotonically nondecreasing in both genders, that is that $\theta_1 \leq \theta_2 \leq \theta_3 \leq \theta_4$ and $\theta_5 \leq \theta_6 \leq \theta_7 \leq \theta_8$? How much confidence is there that the mean distances for females is always less than that of the males, that is $\theta_5 < \theta_1$, $\theta_6 < \theta_2$, $\theta_7 < \theta_3$, $\theta_8 < \theta_4$? The results from Chapter 3 will demonstrate how to obtain reliable observed confidence levels for these parameter regions.

1.3.4 A Problem from Ecology

The Whooping Crane (*Grus americana*) is an endangered North American bird that has been the subject of protection efforts by the National Audubon Society and the governments of the United States and Canada. The single wild population of this bird breeds in the Wood Buffalo National Park in The Northwest Territories of Canada. The population migrates to the Aransas National Wildlife Refuge on the Gulf coast of Texas in the United States during the winter. An annual winter census of Whooping Cranes at the Aransas National Wildlife Refuge has taken place since 1938. The observed number of birds from this census is plotted in Figure 1.6. The data can be found in

Figure 1.4 *Observed measurements from the eleven girls in the dental study. The measurements of each subject are connected by a solid line.*

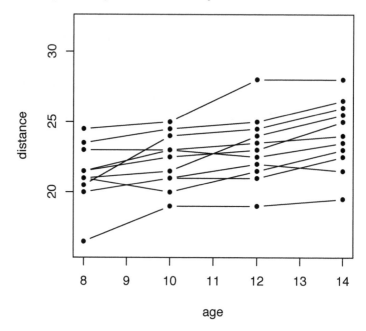

Canadian Wildlife Service and U. S. Fish and Wildlife Service (2005). This data includes adjustments as descibed in Boyce (1987).

On interest in this type of data is the probability of extinction of the species. Extinction is this situation is defined as the population size reaching a specified lower threshold v. This threshold size is usually $v = 1$, where it is obvious that a closed, sexually reproducing population will cease to exist with probability one. Such a threshold could be set larger in terms of wildlife management policy. For example, a threshold $v > 1$ may be established as a lower bound for a population to avoid the effects of inbreeding. A threshold $v > 1$ may also be established as a bound for which some regulatory protective action may be taken. Dennis, Munholland, and Scott (1991) propose estimating the extinction probability of species in a single population by modeling the log of the population size as a Wiener process with drift. In this context the extinction probability is given by

$$\theta = \begin{cases} 1 & \beta \le 0, \\ \exp(-2\beta d/\tau^2) & \beta > 0, \end{cases} \tag{1.8}$$

where β and τ^2 are the infinitesimal mean and variance of the Wiener process, and $d = \log(x_n/v)$ where x_n is the initial population size, which in this case will be the last observed population count. Dennis, Munholland, and Scott

Figure 1.5 *Average measurements for each gender in the dental study. The average measurements of the female subjects are connected by a solid line. The average measurements of the male subjects are connected by a dashed line.*

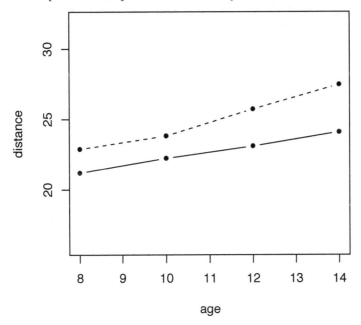

(1991) argue that the parameters in Equation (1.8) can be estimated as follows. Suppose the population is observed at times $t_0 < t_2 < \cdots < t_n$ and let x_i be the observed population size at time t_i for $i = 0, \ldots, n$. Then the extinction probability can be estimated as

$$\hat{\theta} = \begin{cases} 1 & \hat{\beta} \leq 0 \\ \exp(-2\hat{\beta}d/\hat{\tau}^2) & \hat{\beta} > 0, \end{cases}$$

where

$$\hat{\beta} = \frac{\sum_{i=1}^{n} \log(x_i/x_{i-1})}{\sum_{i=1}^{n}(t_i - t_{i-1})}, \tag{1.9}$$

and

$$\hat{\tau}^2 = n^{-1} \sum_{i=1}^{n}(t_i - t_{i-1})^{-1}[\log(x_i/x_{i-1}) - \hat{\beta}(t_i - t_{i-1})]^2, \tag{1.10}$$

where the sum in Equation (1.10) can be divided by $n-1$ instead of n to obtain an unbiased estimator. Dennis, Munholland, and Scott (1991) consider wildlife management thresholds of $v = 10$ and 100. Applying the estimators given in Equations (1.9)–(1.10) to the data in Figure 1.6 yields $\hat{\beta} = 0.0364$ and $\hat{\tau}^2 = 0.0168$. The last observed population count is $x_n = 194$. Hence $\hat{\theta} =$

Table 1.3 *The data of Potthoff and Roy (1964) showing the distance (mm) from the center of the pituitary gland to the pterygomaxillary fissure in young children. Sixteen boys and eleven girls participated in the study which was conducted at the University of North Carolina Dental School. The distance measurement was taken at ages 8, 10, 12, and 14 for each individual in the study.*

Gender	Age 8	Age 10	Age 12	Age 14
Male	26.0	25.0	29.0	31.0
Male	21.5	22.5	23.0	26.5
Male	23.0	22.5	24.0	27.5
Male	25.5	27.5	26.5	27.0
Male	20.0	23.5	22.5	26.0
Male	24.5	25.5	27.0	28.5
Male	22.0	22.0	24.5	26.5
Male	24.0	21.5	24.5	25.5
Male	23.0	20.5	31.0	26.0
Male	27.5	28.0	31.0	31.5
Male	23.0	23.0	23.5	25.0
Male	21.5	23.5	24.0	28.0
Male	17.0	24.5	26.0	29.5
Male	22.5	25.5	25.5	26.0
Male	23.0	24.5	26.0	30.0
Male	22.0	21.5	23.5	25.0
Female	21.0	20.0	21.5	23.0
Female	21.0	21.5	24.0	25.5
Female	20.5	24.0	24.5	26.0
Female	23.5	24.5	25.0	26.5
Female	21.5	23.0	22.5	23.5
Female	20.0	21.0	21.0	22.5
Female	21.5	22.5	23.0	25.0
Female	23.0	23.0	23.5	24.0
Female	20.0	21.0	22.0	21.5
Female	16.5	19.0	19.0	19.5
Female	24.5	25.0	28.0	28.0

3.31×10^{-6} for $v = 10$ and $\hat{\theta} = 0.0699$ for $v = 100$. Given these estimates, how much confidence is there based on the observed data, that there is less than a 10% chance of crossing each threshold? How much confidence is there that there is less than a 5% chance of crossing the threshold? Observed confidence levels for these problems are considered in Chapter 2.

Figure 1.6 *The number of Whooping Cranes observed during the winter season at the Aransas National Wildlife Refuge on the Gulf coast of Texas.*

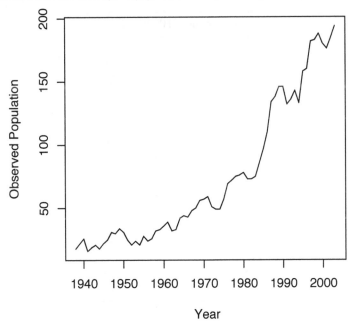

1.3.5 A Problem from Biostatistics

A study presented by Woolson (1987) concerns a group of hyperactive adolescent males who participated in an experiment to determine which of four treatments (Labled *A–D*) for hyperactivity is most effective. There were sixteen subjects in the study, which were grouped into blocks of size 4 according to several factors relevant to the treatment of hyperactivity including age, length of illness, and severity of illness. Within each group the subjects were randomly assigned to one of the four treatments for hyperactivity, resulting in a randomized complete block design for the study. The subjects in the study were rated using a structured schedule for the assessment of hyperactivity. The observed data from this study are given in Table 1.4. The data are plotted in Figure 1.7. A standard analysis of variance procedure performed on the data indicates that there is a significant difference between at least two of the four treatments with a *p*-value less than 0.001. Woolson (1987) performs Tukey's multiple comparison technique on the data and finds that treatments *A* and *D* are not significantly different from one another, and that treatments *B* and *C* are not significantly different from one another. However, treatments *A* and *D* are significantly more effective than either treatment *B* or *C*. All comparisons used the 5% significance level. Hence it appears that either treatment *A*

Table 1.4 *Observed hyperactivity scores for the patients receiving four treatments, labeled A–D, in four blocks. The data originate in Woolson (1987).*

Treatment	Block	Score
A	1	41
	2	48
	3	53
	4	56
B	1	61
	2	68
	3	70
	4	72
C	1	62
	2	62
	3	66
	4	70
D	1	43
	2	48
	3	53
	4	52

or D would be recommended. A more direct analysis of this problem would determine how much confidence there is, based on the observed data, that each of the treatments is most effective in treating hyperactivity. Chapter 4 considers methods for computing observed confidence levels for linear model and regression problems. The methods developed in that chapter can be used to show that there is approximately 70% confidence that treatment D is most effective, 30% confidence that treatment A is most effective, and virtually no confidence that either treatment B or C is most effective.

1.3.6 A Problem from Astronomy

Chondrite meteorites are mainly composed of stone with an age of approximately 4.55 billion years, which puts their formation at about the same time as the solar system. As such they are considered samples of matter from the time when the solar system was formed. Most meteorites classified as chondrites contain small spherical bodies called *condrules* which are mineral deposits formed by rapid cooling. These condrules are composed of various metals such as aluminum, iron and magnesium silicate. See Snow (1987). The composition of chondrites is considered an important indicator as to the origin of the meteorite. Ahrens (1965) reports on the content of magnesium oxide and silica dioxide in a sample of twenty-two chondrite meteorites. The data

Figure 1.7 *Observed hyperactivity scores for the patients receiving four treatments in four blocks. The four treatments are labeled A (solid line), B (dashed line), C (dotted line) and D (dash-dot line). The data originate in Woolson (1987).*

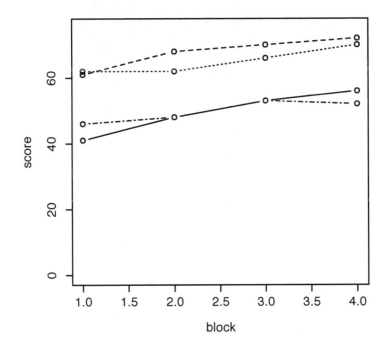

are given in Table 1.5. The data are also considered by Good and Gaskins (1980) and Scott (1992). A kernel density estimate calculated on the percentage of silicon dioxide in the meteorites using a bandwidth estimated using a two-stage plug-in procedure is presented in Figure 1.8. The density estimate has two modes which indicates two possible subpopulations in this data. How much confidence is there that the data do not come from a unimodal density? Methods for computing such observed confidence levels is addressed in Chapter 5.

1.3.7 A Problem from Geophysics

Azzalini and Bowman (1990) considers data based on 299 eruptions of the Old Faithful geyser in Yellowstone National Park in the United States between August 1, 1985, and August 15, 1985. The data include two observations for each eruption corresponding to the waiting time from the last eruption to the current one in minutes, and the duration of the current eruption in minutes. A scatterplot of these variables along with a local constant regression line

Table 1.5 *The percent of magnesium oxide and silicon dioxide contained in twenty-two chondrite meteorites. The data originate in Ahrens (1965).*

Meteorite	Percent Silicon Dioxide	Percent Magnesium Oxide
1	20.77	15.65
2	22.56	16.10
3	22.71	15.81
4	22.99	16.10
5	26.39	18.73
6	27.08	19.46
7	27.32	17.89
8	27.33	18.04
9	27.57	18.45
10	27.81	19.11
11	28.69	19.77
12	29.36	21.16
13	30.25	21.19
14	31.89	23.76
15	32.88	22.06
16	33.23	23.57
17	33.28	23.98
18	33.40	23.54
19	33.52	23.74
20	33.83	23.87
21	33.95	25.27
22	34.82	23.57

computed using the bandwidth set to 0.25 as suggested by Simonoff (1996) is presented in Figure 1.9. Suppose the duration of an eruption is 4 minutes. Based on this observed data, how much confidence is there that the mean waiting time for the next eruption of the geyser is between 75 and 80 minutes? Methods for computing observed confidence levels for such problems are addressed in Chapter 5.

1.4 About This Book

The purpose of this book is to present both theory and applications of observed confidence levels in a wide variety of problems in statistical inference. In most cases, several different methodologies for computing the confidence levels will be possible. In such cases not only will the computation of the confidence levels be presented, but analytical and empirical comparisons between

Figure 1.8 *A kernel density estimate computed on the observed percentages of silica in twenty-two chondrite meteorites. The bandwidth was estimate using a two-stage plug-in procedure.*

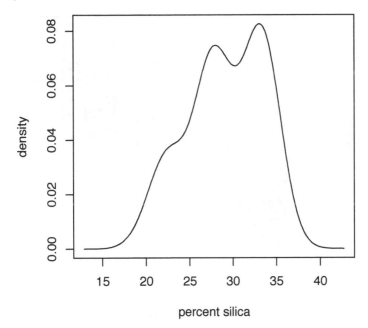

the methods will help establish those that are most reliable in a variety of situations. Practical application of the methods presented in the book include suggested methods for computation based on the R statistical computing environment.

This book will specifically focus on the methods for computing the confidence levels as outlined in Section 1.2. Though Efron and Tibshirani (1998) does provide a blueprint for this problem, there are some fundamental differences between the application of the methods given in this book, and those studied by Efron and Tibshirani (1998). These differences are highlighted in Section 7.4.

Many of the statistical inference problems presented in this book are not new, and there may be several competing methods to observed confidence levels for many of them. Multiple comparison techniques have been mentioned in previous sections. Best subset selection methods have been applied to the process capability example in the Section 1.3.2. Of course, the method of observed confidence levels bears a striking similarity to Bayesian posterior probabilities. Indeed there are connections between the two methods. See Efron and Tibshirani (1998) for more information on these connections. Some of these

Figure 1.9 *Eruption duration (minutes) versus waiting time (minutes) for the Old Faithful geyser data of Azzalini and Bowman (1990). The solid line is a local constant regression line with the bandwidth set to 0.25.*

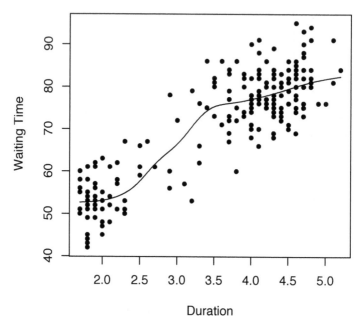

methods will be considered in Chapter 7, but an in-depth comparison of the various methods will not be attempted. The purpose of this book is to highlight the theory and applications of observed confidence levels, leaving the debate as to which method is "best" to each individual practitioner in the field.

This book is structured in terms of the increasing complexity of the problems considered. Chapter 2 discusses single parameter problems such as the bioequivalence problem studied in Section 1.3.1. There are many options for computing observed confidence levels in the single parameter case, and much of the presentation will focus on determining which of these methods is most reliable in general application.

Chapter 3 will focus on multiparameter problems, such as the process capability problem considered in Section 1.3.2. The complexity of the issues involved with these problems increases greatly, but there are still relatively simple and reliable methods that can be applied to a variety of interesting problems.

Chapter 4 considers applications within linear models and regression problems. The main thrust of this chapter concerns theory and application within linear regression models, with examples that demonstrate how the results can be extended to some particular general linear models. Bootstrap theory for the regression model is unusual with regard to the accuracy that can be achieved with confidence intervals for certain regression parameters. This chapter will assess to what extent these properties transfer to the associated observed confidence levels. This chapter also briefly discusses the issues of model selection and prediction.

Chapter 5 considers some nonparametric problems with are often associated with smoothing methods. The first problem considers nonparametric density estimation and the particular problem of determining the number of modes of an unknown density. This problem was one of the original motivating examples that lead to the development of observed confidence levels. As is demonstrated in Chapter 5, the theoretical issues related to the mode problem are not simple, and developing a reliable methodology to address this problem may best best solved by slightly changing the viewpoint and definition of a mode. The second half of this chapter addresses issues concerned with nonparametric regression techniques. As with nonparametric density estimation there are many technical issues that must be addressed. Some issues are beyond the scope of this book and are mentioned but not specifically addressed.

Chapter 6 briefly considers some additional applications of observed confidence levels. These applications are demonstrated through some simple examples, and theoretical issues are not addressed. These additional applications include problems in generalized linear models, classical nonparametric statistical inference, multivariate statistics, and survival analysis.

Chapter 7 discusses the connection that observed confidence levels have with statistical hypothesis testing, multiple comparison techniques, the method of attained confidence levels introduced by Efron and Tibshirani (1998) and Bayesian confidence levels. The purpose of this chapter is not an in-depth and exhaustive comparison. Rather, this chapter attempts to build the framework for further analysis of these methods. Some rather modest conclusions will be made in some instances. However, keeping with the general philosophy of this book, there will not be a direct attempt to argue that some methods are necessarily always better than others.

An appendix has been included that briefly reviews some aspects of asymptotic statistics. The review is not extensive, but is designed to present the basic tools used to prove many of the asymptotic results in the book.

CHAPTER 2

Single Parameter Problems

2.1 Introduction

In this chapter we consider computing observed confidence levels for the special case where there is a single parameter. The computation of observed confidence levels in the single parameter case differs greatly from the multi-parameter case in many respects, and therefore a separate development is necessary. Motivated by some basic results from measure theory, the emphasis of the study of observed confidence levels in the single parameter case centers on regions that are interval subsets of the real line. This simplifies the theoretical structure used to study the confidence levels, and allows for the development of a wide variety of potential methods for computing the levels. With many potential methods available, it becomes necessary to develop techniques for comparing competing methods. In particular, the concept of asymptotic accuracy will be the central tool used to compare competing methods for computing observed confidence levels. The theory of asymptotic expansions, particularly Edgeworth and Cornish-Fisher expansions, will play a central role in computing the measure of asymptotic accuracy. Not surprisingly, it often turns out that more accurate confidence intervals can be used to compute more accurate observed confidence levels.

Some general concepts concerning the development of observed confidence levels in the single parameter case are given in Section 2.2. In Section 2.3 a special model called the smooth function model is introduced. This model allows for an in-depth study of parameters that are differentiable functions of vector means. Several general methods for computing observed confidence levels are developed in this section. Section 2.4 considers the asymptotic properties of these methods, focusing on the concepts of consistency and asymptotic accuracy. This section also introduces the required results from the theory of asymptotic expansions. The applicability of the asymptotic results to some finite sample size situations are discussed in Section 2.5. The results of this section rely on empirical comparisons based on computer based simulations. Some practical examples are given in Section 2.6. Methods for computing observed confidence levels using the R statistical computing environment are considered in Section 2.7.

2.2 The General Case

To develop observed confidence levels for the case where there is a single parameter, the general framework from Section 1.2 is used with the restriction that $\Theta \subset \mathbb{R}$. To simplify the development in this case, it will be further assumed that Θ is a connected set, that is, it is an interval subset of $\mathbb{R}$. Most standard single parameter problems in statistical inference fall within these assumptions. These restrictions greatly simplify the types of regions that need to be considered when developing observed confidence levels for the single parameter case.

An observed confidence level is simply a function that takes a subset of Θ and maps it to a real number between 0 and 1. Formally, let α be a function and let $\mathcal{T}$ be a collection of subsets of Θ. Then an observed confidence level is a function $\alpha : \mathcal{T} \to [0,1]$. Because confidence levels are closely related to probabilities, it is reasonable to assume that α has the same axiomatic properties as a *probability measure*. That is, suppose $\mathcal{T}$ is a sigma-field of subsets of Θ. Then it will be assumed that

1. $\alpha(\Theta_i) \in [0,1]$ for all $\Theta_i \in \mathcal{T}$,

2. $\alpha(\emptyset) = 0$, and

3. If $\{\Theta_i\}_{i=1}^{\infty}$ is a countable collection of pairwise disjoint regions from $\mathcal{T}$, then

$$\alpha \left(\bigcup_{i=1}^{\infty} \Theta_i \right) = \sum_{i=1}^{\infty} \alpha(\Theta_i).$$

For most reasonable problems in statistical inference it should suffice to take $\mathcal{T}$ to be the *Borel* sigma-field on Θ, that is, the sigma-field generated by the open subsets of Θ. Given this structure, it suffices to develop observed confidence levels for interval subsets of Θ. Observed confidence levels for other regions can be obtained through operations derived from Properties 1–3, which mirror those of probability measures. See Chapter 2 of Pollard (2002) for an accessible account of probability measures and Borel sigma-fields.

To develop observed confidence levels for a general scalar parameter θ, consider a single interval region of the form $\Psi = (t_L, t_U) \in \mathcal{T}$. To compute the observed confidence level of Ψ, a confidence interval for θ based on the sample $\mathbf{X}$ is required. The general form of a confidence interval for θ based on $\mathbf{X}$ can usually be written as $C(\alpha, \boldsymbol{\omega}; \mathbf{X}) = [L(\omega_L; \mathbf{X}), U(\omega_U; \mathbf{X})]$, where ω_L and ω_U are shape parameters such that $(\omega_L, \omega_U) \in \Omega_\alpha$, for some $\Omega_\alpha \subset \mathbb{R}^2$. It can often be assumed that $L(\omega_L; \mathbf{X})$ and $U(\omega_U; \mathbf{X})$ are continuous monotonic functions of ω_L and ω_U onto Θ, respectively, conditional on the observed sample $\mathbf{X}$. See, for example, Section 9.2 of Casella and Berger (2002). If such an assumption is true, the observed confidence level of Ψ is computed by setting $\Psi = C(\alpha, \boldsymbol{\omega}; \mathbf{X})$ and solving for $\boldsymbol{\omega}$. The value of α for which $\boldsymbol{\omega} \in \Omega_\alpha$ is the observed confidence level of Ψ. For the form of the confidence interval given above, the solution is obtained by setting $\omega_L = L^{-1}(t_L; \mathbf{X})$ and $\omega_U = U^{-1}(t_U; \mathbf{X})$, conditional on $\mathbf{X}$. A unique solution will exist for both shape parameters given the assumptions on the functions L and U. Therefore, the observed confidence level of Ψ is the value of α such that $\boldsymbol{\omega} = (\omega_L, \omega_U) \in \Omega_\alpha$. Thus, the calculation of observed confidence levels in the single parameter case is equivalent to inverting the endpoints of a confidence interval for θ. Some simple examples illustrating this method is given below.

2.2.1 Normal Mean and Variance

Following the example studied at the end of Section 1.2, suppose that X_1, ...,X_n is a set of independent and identically distributed random variables from a normal distribution with mean θ and variance $\sigma^2 < \infty$. The parameter space is $\Theta = \mathbb{R}$ and an arbitrary interval region $\Psi = (t_L, t_U)$ is considered. A confidence interval for θ is given by Equation (1.5). Writing this interval in terms of the functions L and U given above yields

$$L(\omega_L; \mathbf{X}) = \hat{\theta} - t_{n-1;1-\omega_L} n^{-1/2} \hat{\sigma},$$

and

$$U(\omega_U; \mathbf{X}) = \hat{\theta} - t_{n-1;1-\omega_U} n^{-1/2} \hat{\sigma}.$$

Therefore, setting $\Psi = C(\alpha, \boldsymbol{\omega}; \mathbf{X})$ yields

$$\omega_L = L^{-1}(t_L; \mathbf{X}) = 1 - T_{n-1}\left[\frac{n^{1/2}(\hat{\theta} - t_L)}{\hat{\sigma}}\right],$$

and

$$\omega_U = U^{-1}(t_U; \mathbf{X}) = 1 - T_{n-1}\left[\frac{n^{1/2}(\hat{\theta} - t_U)}{\hat{\sigma}}\right],$$

as shown in Section 1.2, where it follows that the observed confidence level for Ψ is given by

$$\alpha(\Psi) = T_{n-1}\left[\frac{n^{1/2}(\hat{\theta} - t_L)}{\hat{\sigma}}\right] - T_{n-1}\left[\frac{n^{1/2}(\hat{\theta} - t_U)}{\hat{\sigma}}\right]. \qquad (2.1)$$

For the variance, the parameter space is $\Theta = \mathbb{R}^+$, so that the region $\Psi = (t_L, t_U)$ is assumed to follow the restriction that $0 < t_L \leq t_U < \infty$. A $100\alpha\%$ confidence interval for σ^2 is given by

$$C(\alpha, \boldsymbol{\omega}; \mathbf{X}) = \left[\frac{(n-1)\hat{\sigma}^2}{\chi^2_{n-1;1-\omega_L}}, \frac{(n-1)\hat{\sigma}^2}{\chi^2_{n-1;1-\omega_U}}\right], \qquad (2.2)$$

where $\boldsymbol{\omega} \in \Omega_\alpha$ with

$$\Omega_\alpha = \{\boldsymbol{\omega}' = (\omega_1, \omega_2) : \omega_L \in [0, 1], \omega_U \in [0, 1], \omega_U - \omega_L = \alpha\},$$

where $\chi^2_{\nu,\xi}$ is the ξth percentile of a χ^2 distribution with ν degrees of freedom. Therefore

$$L(\omega; \mathbf{X}) = \frac{(n-1)\hat{\sigma}^2}{\chi^2_{n-1;1-\omega_L}},$$

and

$$U(\omega; \mathbf{X}) = \frac{(n-1)\hat{\sigma}^2}{\chi^2_{n-1;1-\omega_U}}.$$

Setting $\Psi = C(\alpha, \omega; \mathbf{X})$ and solving for ω_L and ω_U yields

$$\omega_L = L^{-1}(t_L; \mathbf{X}) = 1 - \chi^2_{n-1}\left[\frac{(n-1)\hat{\sigma}^2}{t_L}\right],$$

and

$$\omega_U = U^{-1}(t_U; \mathbf{X}) = 1 - \chi^2_{n-1}\left[\frac{(n-1)\hat{\sigma}^2}{t_U}\right],$$

where χ^2_ν is the distribution function of a χ^2-distribution with ν degrees of freedom. This implies that the observed confidence limit for Ψ is given by

$$\alpha(\Psi) = \chi^2_{n-1}\left[\frac{(n-1)\hat{\sigma}^2}{t_L}\right] - \chi^2_{n-1}\left[\frac{(n-1)\hat{\sigma}^2}{t_U}\right]. \tag{2.3}$$

2.2.2 Correlation Coefficient

Suppose $(X_1, Y_1), \ldots, (X_n, Y_n)$ is a set of independent and identically distributed bivariate random vectors from a bivariate normal distribution with mean vector $\boldsymbol{\mu}' = (\mu_X, \mu_Y)$ and covariance matrix

$$\Sigma = \begin{bmatrix} \sigma^2_X & \sigma_{XY} \\ \sigma_{XY} & \sigma^2_Y \end{bmatrix},$$

where $V(X_i) = \sigma^2_X$, $V(Y_i) = \sigma^2_Y$ and the covariance between X and Y is σ_{XY}. Let $\theta = \sigma_{XY}/(\sigma_X \sigma_Y)$, the correlation coefficient between X and Y. The problem of constructing a reliable confidence interval for θ is usually simplified using Fisher's normalizing transformation. See Fisher (1915) and Winterbottom (1979). Using the fact that $\tanh^{-1}(\hat{\theta})$ is approximately normal with mean $\tanh^{-1}(\theta)$ and variance $(n-3)^{-1/2}$, the resulting approximate confidence interval for θ has the form

$$C(\alpha, \omega; \mathbf{X}, \mathbf{Y}) = [\tanh(\tanh^{-1}(\hat{\theta}) - z_{1-\omega_L}(n-3)^{-1/2}),$$
$$\tanh(\tanh^{-1}(\hat{\theta}) - z_{1-\omega_U}(n-3)^{-1/2})] \tag{2.4}$$

where $\hat{\theta}$ is the sample correlation coefficient given by the plug-in estimator

$$\hat{\theta} = \frac{\sum_{i=1}^{n}(X_i - \bar{X})(Y_i - \bar{Y})}{\left[\sum_{i=1}^{n}(X_i - \bar{X})^2\right]^{1/2}\left[\sum_{i=1}^{n}(Y_i - \bar{Y})^2\right]^{1/2}}, \tag{2.5}$$

and $\omega' = (\omega_L, \omega_U) \in \Omega_\alpha$ with

$$\Omega_\alpha = \{\omega' = (\omega_1, \omega_2) : \omega_L \in [0,1], \omega_U \in [0,1], \omega_U - \omega_L = \alpha\}.$$

Therefore

$$L(\omega_L; \mathbf{X}, \mathbf{Y}) = \tanh(\tanh^{-1}(\hat{\theta}) - z_{1-\omega_L}(n-3)^{-1/2})$$

and

$$U(\omega_U; \mathbf{X}, \mathbf{Y}) = \tanh(\tanh^{-1}(\hat{\theta}) - z_{1-\omega_U}(n-3)^{-1/2}).$$

Setting $\Psi = (t_L, t_U) = C(\alpha, \boldsymbol{\omega}; \mathbf{X}, \mathbf{Y})$ yields

$$\omega_L = 1 - \Phi[(n-3)^{1/2}(\tanh^{-1}(\hat{\theta}) - \tanh^{-1}(t_L))],$$

and

$$\omega_U = 1 - \Phi[(n-3)^{1/2}(\tanh^{-1}(\hat{\theta}) - \tanh^{-1}(t_U))],$$

so that the observed confidence level for Ψ is given by

$$\alpha(\Psi) = \Phi[(n-3)^{1/2}(\tanh^{-1}(\hat{\theta}) - \tanh^{-1}(t_L))] - $$
$$\Phi[(n-3)^{1/2}(\tanh^{-1}(\hat{\theta}) - \tanh^{-1}(t_U))].$$

2.2.3 Population Quantile

Suppose that $X_1, \ldots, X_n$ is a set of independent and identically distributed random variables from an absolutely continuous distribution F. Let $\xi \in (0,1)$ and define $\theta = F^{-1}(\xi)$, the ξth population quantile of F. A confidence interval for θ can be computed by inverting the classical sign test. Let $X_{(1)} \leq X_{(2)} \leq \cdots \leq X_{(n)}$ be the order statistics of the sample and let B be a binomial random variable based on n Bernoulli trials with success probability ξ. The usual point estimator of θ is $\hat{\theta} = X_{\lfloor np \rfloor + 1}$ where $\lfloor x \rfloor$ is the largest integer strictly less than x. A confidence interval for θ is given by $C(\alpha, \boldsymbol{\omega}; \mathbf{X}) = [X_{(L)}, X_{(U)}]$, where L and U are chosen so that $P(B < L) = \omega_L$ and $P(B \geq U) = 1 - \omega_U$ and $\alpha = \omega_U - \omega_L$. See Section 3.2 of Conover (1980) for examples using this method.

To derive an observed confidence level based on this confidence interval let $\Psi = (t_L, t_U)$ be an arbitrary region in the parameter space of θ. Setting $\Psi = C(\alpha, \boldsymbol{\omega}; \mathbf{X})$ indicates that L and U must be found so that $X_{(L)} \simeq t_L$ and $X_{(U)} \simeq t_U$. The corresponding observed confidence level for Ψ is then given by $\alpha(\Psi) = P(L \leq B \leq U - 1)$. Of course the observed confidence level will be approximate unless $X_{(L)} = t_L$ and $X_{(U)} = t_U$.

The normal approximation to the binomial can be used to obtain a $100\alpha\%$ approximate confidence interval for θ that is valid for large sample sizes. This interval also has the form $[X_{(L)}, X_{(U)}]$, but L and U are computed as

$$L = n\xi - z_{1-\omega_L}[n\xi(1-\xi)]^{1/2},$$

and

$$U = n\xi - z_{1-\omega_U}[n\xi(1-\xi)]^{1/2},$$

where $\alpha = \omega_U - \omega_L$. Using this interval, an approximate observed confidence level for a arbitrary region $\Psi = (t_L, t_U)$ is again obtained by finding L and U such that $X_{(L)} \simeq t_L$ and $X_{(U)} \simeq t_U$. It then follows that for these values of L and U that

$$\omega_L = 1 - \Phi \left\{ \frac{n\xi - L}{[n\xi(1 - \xi)]^{1/2}} \right\},$$

and

$$\omega_U = 1 - \Phi \left\{ \frac{n\xi - U}{[n\xi(1 - \xi)]^{1/2}} \right\},$$

so that the approximate observed confidence level is given by

$$\alpha(\Psi) = \Phi \left\{ \frac{n\xi - L}{[n\xi(1 - \xi)]^{1/2}} \right\} - \Phi \left\{ \frac{n\xi - U}{[n\xi(1 - \xi)]^{1/2}} \right\}.$$

Both of these methods rely on finding order statistics in the sample that approximate the endpoints of the region of interest $\Psi = (t_L, t_U)$. This will obviously not be possible for all observed data and regions, though picking the closest order statistic will still provide approximate observed confidence levels.

2.2.4 Poisson Rate

Suppose $X_1, \ldots, X_n$ is a set of independent and identically distributed random variables from a Poisson distribution with rate $\theta \in \Theta = \mathbb{R}^+$. Garwood (1936) suggested a $100\alpha\%$ confidence interval for θ using the form

$$C(\alpha, \boldsymbol{\omega}; \mathbf{X}) = \left[\frac{\chi^2_{2Y;\omega_L}}{2n}, \frac{\chi^2_{2(Y+1);\omega_U}}{2n} \right],$$

where

$$Y = \sum_{i=1}^{n} X_i,$$

and $\boldsymbol{\omega} = (\omega_L, \omega_U) \in \Omega_\alpha$ where

$$\Omega_\alpha = \{ \boldsymbol{\omega} = (\omega_1, \omega_2) : \omega_L \in [0, 1], \omega_U \in [0, 1], \omega_U - \omega_L = \alpha \}.$$

Therefore $L(\omega_L; \mathbf{X}) = \chi^2_{2Y;1-\omega_L}/2n$ and $U(\omega_U; \mathbf{X}) = \chi^2_{2(Y+1);1-\omega_U}/2n$. Setting $\Psi = (t_L, t_U) = C(\alpha, \boldsymbol{\omega}; \mathbf{X})$ yields $\omega_L = \chi^2_{2Y}(2nt_L)$ and $\omega_U = \chi^2_{2(Y+1)}(2nt_U)$, so that the observed confidence limit for Ψ is

$$\alpha(\Psi) = \chi^2_{2(Y+1)}(2nt_U) - \chi^2_{2Y}(2nt_L).$$

Note that for a given region and sample size, only certain levels of confidence are possible, corresponding to the different values of Y. This is related to the effect studied by Agresti and Coull (1998) and Brown, Cai and DasGupta (2003). This effect is a natural consequence of the discrete nature of the Poisson distribution. Note that the observed confidence level is still a continuous function of the endpoints of the region Ψ.

2.3 Smooth Function Model

Many standard problems in statistical inference can be studied in terms of a model for differentiable functions of vector means called the smooth function model. Hall (1992a) uses the structure afforded by this model to provide a detailed study of confidence regions, including those based on the bootstrap. This makes the smooth function model a natural starting point for the study of observed confidence levels as well. This section begins with a description of this model.

Suppose $\mathbf{X}_1, \ldots, \mathbf{X}_n$ is a set of independent and identically distributed random vectors from a d-dimensional distribution F. It is assumed that F satisfies the multivariate form of Cramér's continuity condition, which states that if ψ is the characteristic function of F, then

$$\limsup_{||\mathbf{t}|| \to \infty} |\psi(\mathbf{t})| < 1. \tag{2.6}$$

This condition is satisfied if the distribution of $\mathbf{X}_i$ is nonsingular. See page 57 of Hall (1992a) for more details on this condition. Let $\boldsymbol{\mu} = E(\mathbf{X}_i) \in \mathbb{R}^d$ and assume that $\theta = g(\boldsymbol{\mu})$ for some smooth function $g : \mathbb{R}^d \to \mathbb{R}$. Specific conditions on the smoothness of g as well as moment conditions on F will be discussed later in the chapter. At the very least it is obviously prudent to assume that the components of $\boldsymbol{\mu}$ are finite. Let $\hat{F}_n$ be the multivariate form of the empirical distribution function defined in Equation (1.3). That is

$$\hat{F}_n(\mathbf{t}) = n^{-1} \sum_{i=1}^{n} \delta(\mathbf{t}; Q(\mathbf{X}_i)),$$

for $\mathbf{t} \in \mathbb{R}^d$ where for $\mathbf{X}_i' = (X_{i1}, \ldots, X_{id})$,

$$Q(\mathbf{X}_i) = \{(t_1, \ldots, t_d)' \in \mathbb{R}^d : t_j \leq X_{ij} \text{ for } j = 1, \ldots, d\}.$$

Generalizing the result given in Equation (1.4), the plug-in estimate of $\boldsymbol{\mu}$ is given by

$$\hat{\boldsymbol{\mu}} = \int_{\mathbb{R}^d} \mathbf{t} d\hat{F}_n(\mathbf{t}) = n^{-1} \sum_{i=1}^{n} \mathbf{X}_i = \bar{\mathbf{X}},$$

so that the corresponding plug-in estimate of θ is given by $\hat{\theta} = g(\bar{\mathbf{X}})$. Let σ^2 be the asymptotic variance of $\hat{\theta}$ defined by

$$\hat{\sigma}^2 = \lim_{n \to \infty} V(n^{1/2}\hat{\theta}).$$

It is assumed that $\sigma^2 = h(\boldsymbol{\mu})$ for some smooth function $h : \mathbb{R}^d \to \mathbb{R}^+$. A plug-in estimate of σ^2 is given by $\hat{\sigma}^2 = h(\bar{\mathbf{X}})$. Parameters that can be studied using the smooth function model include means, variances, ratios of means, ratios of variances, correlations, and others. See Section 2.4 of Hall (1992a) for further details on some of these examples. There are also many problems that also fall outside of the smooth function model. These include percentiles, such as the median, and modes of continuous densities.

To develop confidence intervals for θ within this model define

$$G_n(t) = P\left[\frac{n^{1/2}(\hat{\theta} - \theta)}{\sigma} \leq t \,\middle|\, \mathbf{X}_1, \ldots, \mathbf{X}_n \sim F\right],$$

and

$$H_n(t) = P\left[\frac{n^{1/2}(\hat{\theta} - \theta)}{\hat{\sigma}} \leq t \,\middle|\, \mathbf{X}_1, \ldots, \mathbf{X}_n \sim F\right],$$

where $\mathbf{X}_1, \ldots, \mathbf{X}_n \sim F$ represents a set of n independent and identically distributed random vectors each following the distribution F. Define percentiles of the distributions as

$$g_\alpha = G_n^{-1}(\alpha) = \inf\{t \in \mathbb{R} : G_n(t) \geq \alpha\},$$

and

$$h_\alpha = H_n^{-1}(\alpha) = \inf\{t \in \mathbb{R} : H_n(t) \geq \alpha\},$$

for $\alpha \in (0, 1)$.

Exact and approximate confidence intervals for θ can be developed by considering the four theoretical critical points defined by Hall (1988). The two exact critical points are the ordinary critical point

$$\hat{\theta}_{\text{ord}}(\alpha) = \hat{\theta} - n^{-1/2}\sigma g_{1-\alpha}, \tag{2.7}$$

for the case when σ is known, and the studentized critical point

$$\hat{\theta}_{\text{stud}}(\alpha) = \hat{\theta} - n^{-1/2}\hat{\sigma} h_{1-\alpha}, \tag{2.8}$$

for the case when σ is unknown. When σ is unknown, two approximate theoretical critical points are also considered. These are the hybrid critical point

$$\hat{\theta}_{\text{hyb}}(\alpha) = \hat{\theta} - n^{-1/2}\hat{\sigma} g_{1-\alpha}, \tag{2.9}$$

where the approximation $g_{1-\alpha} \simeq h_{1-\alpha}$ is used, and the percentile critical point

$$\hat{\theta}_{\text{perc}}(\alpha) = \hat{\theta} + n^{-1/2}\hat{\sigma} g_\alpha, \tag{2.10}$$

where the approximation $g_\alpha \simeq -h_{1-\alpha}$ is used. Two additional theoretical critical points, the bias-corrected and accelerated bias-corrected critical points, require further development and will be introduced in Section 2.4. Two-sided confidence intervals can be developed using each of the four theoretical critical points defined in Equations (2.7)–(2.10). For example, if $\omega_L \in (0, 1)$ and $\omega_U \in (0, 1)$ are such that $\alpha = \omega_U - \omega_L \in (0, 1)$ then $C_{\text{ord}}(\alpha, \boldsymbol{\omega}; \mathbf{X}) = [\hat{\theta}_{\text{ord}}(\omega_L), \hat{\theta}_{\text{ord}}(\omega_U)]$ is a $100\alpha\%$ confidence interval for θ when σ is known, where for notational convenience $\mathbf{X} = \text{vec}(\mathbf{X}_1, \ldots, \mathbf{X}_n)$. Similarly

$$C_{\text{stud}}(\alpha, \boldsymbol{\omega}; \mathbf{X}) = [\hat{\theta}_{\text{stud}}(\omega_L), \hat{\theta}_{\text{stud}}(\omega_U)],$$

is a $100\alpha\%$ confidence interval for θ when σ is unknown.

The intervals $C_{\text{hyb}}(\alpha, \boldsymbol{\omega}; \mathbf{X})$ and $C_{\text{perc}}(\alpha, \boldsymbol{\omega}; \mathbf{X})$ are the corresponding approximate confidence intervals for θ based on the hybrid and percentile theoretical critical points, respectively.

For a region $\Psi = (t_L, t_U)$ the observed confidence level corresponding to each of the theoretical critical points can be computed by setting Ψ equal to the confidence interval and solving for ω_L and ω_U, as detailed in Section 2.2. For example, in the case of the ordinary theoretical critical point, setting $\Psi = C_{\mathrm{ord}}(\alpha, \boldsymbol{\omega}; \mathbf{X})$ yields the two equations

$$t_L = L(\boldsymbol{\omega}; \mathbf{X}) = \hat{\theta} - n^{-1/2} \sigma g_{1-\omega_L}, \tag{2.11}$$

and

$$t_U = U(\boldsymbol{\omega}; \mathbf{X}) = \hat{\theta} - n^{-1/2} \sigma g_{1-\omega_U}. \tag{2.12}$$

Solving for ω_L and ω_U in Equations (2.11) and (2.12) yields

$$\omega_L = L^{-1}(t_L; \mathbf{X}) = 1 - G_n \left[\frac{n^{1/2}(\hat{\theta} - t_L)}{\sigma} \right],$$

and

$$\omega_U = U^{-1}(t_U; \mathbf{X}) = 1 - G_n \left[\frac{n^{1/2}(\hat{\theta} - t_U)}{\sigma} \right],$$

so that the observed confidence level corresponding to the ordinary theoretical critical point is

$$\alpha_{\mathrm{ord}}(\Psi) = G_n \left[\frac{n^{1/2}(\hat{\theta} - t_L)}{\sigma} \right] - G_n \left[\frac{n^{1/2}(\hat{\theta} - t_U)}{\sigma} \right]. \tag{2.13}$$

Similarly, the observed confidence levels corresponding to the studentized, hybrid, and percentile theoretical critical points are

$$\alpha_{\mathrm{stud}}(\Psi) = H_n \left[\frac{n^{1/2}(\hat{\theta} - t_L)}{\hat{\sigma}} \right] - H_n \left[\frac{n^{1/2}(\hat{\theta} - t_U)}{\hat{\sigma}} \right], \tag{2.14}$$

$$\alpha_{\mathrm{hyb}}(\Psi) = G_n \left[\frac{n^{1/2}(\hat{\theta} - t_L)}{\hat{\sigma}} \right] - G_n \left[\frac{n^{1/2}(\hat{\theta} - t_U)}{\hat{\sigma}} \right], \tag{2.15}$$

and

$$\alpha_{\mathrm{perc}}(\Psi) = G_n \left[\frac{n^{1/2}(t_U - \hat{\theta})}{\hat{\sigma}} \right] - G_n \left[\frac{n^{1/2}(t_L - \hat{\theta})}{\hat{\sigma}} \right], \tag{2.16}$$

respectively.

When F is unknown the distributions G_n and H_n will also be unknown unless $n^{1/2}(\hat{\theta} - \theta)/\sigma$ or $n^{1/2}(\hat{\theta} - \theta)/\hat{\sigma}$ is a distribution free statistic over $\mathcal{F}$. See Definition 2.1.1 of Randles and Wolfe (1979). In such cases several methods can be used to estimate or approximate the observed confidence levels. The simplest approach is based on the Central Limit Theorem. Under the assumptions of the smooth function model with $\sigma^2 < \infty$, and that

$$\left. \frac{\partial}{\partial t_i} g(\mathbf{t}) \right|_{\mathbf{t} = \mu} \neq 0$$

where $\mathbf{t}' = (t_1, \ldots, t_d)$, it follows from Theorem 3.3.A of Serfling (1980) that

$$\frac{n^{1/2}(\hat{\theta} - \theta)}{\sigma} \xrightarrow{w} Z,$$

and

$$\frac{n^{1/2}(\hat{\theta} - \theta)}{\hat{\sigma}} \xrightarrow{w} Z,$$

as $n \to \infty$, where Z is a standard normal random variable. This suggests the large sample approximation $G_n(t) \simeq \Phi(t)$ and $H_n(t) \simeq \Phi(t)$ where Φ is the distribution function of a standard normal distribution. Using the normal approximation, the approximate observed confidence levels corresponding to the ordinary, studentized, hybrid and percentile theoretical critical points are

$$\hat{\alpha}_{\text{ord}}(\Psi) = \Phi\left[\frac{n^{1/2}(\hat{\theta} - t_L)}{\sigma}\right] - \Phi\left[\frac{n^{1/2}(\hat{\theta} - t_U)}{\sigma}\right], \qquad (2.17)$$

$$\begin{aligned}\hat{\alpha}_{\text{stud}}(\Psi) &= \hat{\alpha}_{\text{hyb}}(\Psi) = \hat{\alpha}_{\text{perc}}(\Psi) \\ &= \Phi\left[\frac{n^{1/2}(\hat{\theta} - t_L)}{\hat{\sigma}}\right] - \Phi\left[\frac{n^{1/2}(\hat{\theta} - t_U)}{\hat{\sigma}}\right], \qquad (2.18)\end{aligned}$$

respectively. Note that $\hat{\alpha}_{\text{perc}}$ coincides with $\hat{\alpha}_{\text{stud}}$ and $\hat{\alpha}_{\text{hyb}}$ due to the symmetry of the normal density.

When the normal approximation is not accurate enough, the bootstrap method of Efron (1979) is suggested as a method to estimate the observed confidence levels. Consider a sample $\mathbf{X}_1^* \ldots, \mathbf{X}_n^* \sim \hat{F}_n$, conditional on the original sample $\mathbf{X}_1, \ldots, \mathbf{X}_n$. Let $\hat{\theta}^* = g(\bar{\mathbf{X}}^*)$ where

$$\bar{\mathbf{X}}^* = n^{-1} \sum_{i=1}^{n} \mathbf{X}_i^*.$$

The bootstrap estimate of G_n is given by the sampling distribution of $n^{1/2}(\hat{\theta}^* - \hat{\theta})/\hat{\sigma}$ when $\mathbf{X}_1^* \ldots, \mathbf{X}_n^* \sim \hat{F}_n$, conditional on the original sample $\mathbf{X}_1, \ldots, \mathbf{X}_n$. Let P^* denote the probability measure P conditional on the sample $\mathbf{X}_1 \ldots, \mathbf{X}_n$. That is $P^*(A) = P(A|\mathbf{X}_1 \ldots, \mathbf{X}_n)$ for any event within the correct sample space. Then the bootstrap estimate of G_n is given by

$$\hat{G}_n(t) = P^*\left[\frac{n^{1/2}(\hat{\theta}^* - \hat{\theta})}{\hat{\sigma}} \leq t \,\middle|\, \mathbf{X}_1^* \ldots, \mathbf{X}_n^* \sim \hat{F}\right],$$

and the bootstrap estimate of H_n is given by

$$\hat{H}_n(t) = P^*\left[\frac{n^{1/2}(\hat{\theta}^* - \hat{\theta})}{\hat{\sigma}^*} \leq t \,\middle|\, \mathbf{X}_1^* \ldots, \mathbf{X}_n^* \sim \hat{F}\right],$$

where $\hat{\sigma}^* = h(\bar{\mathbf{X}}^*)$. Closed analytic forms for these estimates exist only in special cases. An example is given in Efron (1979). In most cases approximations to the bootstrap estimates must be computed using the familiar "resampling" algorithm of Efron (1979). See Efron and Tibshirani (1993) or Davison and Hinkley (1997) for more information on the practical computation of these estimates. Computation of these estimates using the R statistical computing environment will also be discussed in Section 2.7. Define, conditional on $\mathbf{X}_1, \ldots, \mathbf{X}_n$,

$$\hat{g}_\alpha = \hat{G}_n^{-1}(\alpha) = \inf\{t \in \mathbb{R} : \hat{G}_n(t) \geq \alpha\},$$

and

$$\hat{h}_\alpha = \hat{H}_n^{-1}(\alpha) = \inf\{t \in \mathbb{R} : \hat{H}_n(t) \geq \alpha\},$$

for $\alpha \in (0,1)$. The bootstrap estimates of the four theoretical critical points defined in Equations (2.7)–(2.10) are

$$\hat{\theta}^*_{\text{ord}}(\alpha) = \hat{\theta} - n^{-1/2}\sigma\hat{g}_{1-\alpha}, \tag{2.19}$$

$$\hat{\theta}^*_{\text{stud}}(\alpha) = \hat{\theta} - n^{-1/2}\hat{\sigma}\hat{h}_{1-\alpha}, \tag{2.20}$$

$$\hat{\theta}^*_{\text{hyb}}(\alpha) = \hat{\theta} - n^{-1/2}\hat{\sigma}\hat{g}_{1-\alpha}, \tag{2.21}$$

and

$$\hat{\theta}^*_{\text{perc}}(\alpha) = \hat{\theta} + n^{-1/2}\hat{\sigma}\hat{g}_\alpha. \tag{2.22}$$

The resulting estimated observed confidence levels based on the bootstrap ordinary, studentized, hybrid and percentile theoretical critical points are

$$\hat{\alpha}^*_{\text{ord}}(\Psi) = \hat{G}_n\left[\frac{n^{1/2}(\hat{\theta} - t_L)}{\sigma}\right] - \hat{G}_n\left[\frac{n^{1/2}(\hat{\theta} - t_U)}{\sigma}\right], \tag{2.23}$$

$$\hat{\alpha}^*_{\text{stud}}(\Psi) = \hat{H}_n\left[\frac{n^{1/2}(\hat{\theta} - t_L)}{\hat{\sigma}}\right] - \hat{H}_n\left[\frac{n^{1/2}(\hat{\theta} - t_U)}{\hat{\sigma}}\right], \tag{2.24}$$

$$\hat{\alpha}^*_{\text{hyb}}(\Psi) = \hat{G}_n\left[\frac{n^{1/2}(\hat{\theta} - t_L)}{\hat{\sigma}}\right] - \hat{G}_n\left[\frac{n^{1/2}(\hat{\theta} - t_U)}{\hat{\sigma}}\right], \tag{2.25}$$

and

$$\hat{\alpha}^*_{\text{perc}}(\Psi) = \hat{G}_n\left[\frac{n^{1/2}(t_U - \hat{\theta})}{\hat{\sigma}}\right] - \hat{G}_n\left[\frac{n^{1/2}(t_L - \hat{\theta})}{\hat{\sigma}}\right], \tag{2.26}$$

respectively.

Given the large number of methods available for computing observed confidence levels in the case of the smooth function model, the natural question as to which of the methods would be preferred in practice is obviously relevant. Some of the methods require specific assumptions about the knowledge of either H_n or G_n, while others use either the normal or the bootstrap approximation to relax these assumptions. Relevant questions then arise as to which of the approximations would be preferred, and how well the approximations work. Two routes of inquiry will provide at least partial answers to these questions. Asymptotic analytic comparisons will be developed in Section 2.4, and some limited empirical comparisons will be presented in Section 2.5. The next two subsections present examples of the applications of the smooth function model to the problem of the univariate mean and variance.

2.3.1 Univariate Mean

Let $Y_1, \ldots, Y_n$ be a set of independent and identically distributed random variables from a univariate distribution J and let $\theta = E(Y_i)$. To put this problem in the form of the smooth function model, define $\mathbf{X}_i' = (Y_i, Y_i^2)$ so that $\boldsymbol{\mu} = E(\mathbf{X}_i) = (\theta, \theta^2 + \tau^2)$ where $\tau^2 = V(Y_i)$. For a vector $\mathbf{v}' = (v_1, v_2)$ the functions g and h can be defined as $g(\mathbf{v}) = v_1$ and $h(\mathbf{v}) = v_2 - v_1^2$, so that $\hat{\theta} = \bar{Y}$ and

$$\hat{\sigma}^2 = n^{-1} \sum_{i=1}^{n} Y_i^2 - \bar{Y}^2.$$

It can be further shown that if J is a continuous distribution, then the distribution of $\mathbf{X}_i$ will satisfy the Cramér continuity condition given in Equation (2.6).

If the distribution J is taken to be a normal distribution with mean θ and variance τ^2, then the distribution G_n is a standard normal distribution and the distribution H_n is a t-distribution with $n - 1$ degrees of freedom. The ordinary confidence interval has the form

$$C_{\mathrm{ord}}(\alpha, \boldsymbol{\omega}; \mathbf{X}) = [\hat{\theta} - n^{-1/2}\sigma z_{1-\omega_L}, \hat{\theta} - n^{-1/2}\sigma z_{1-\omega_U}],$$

where z_ξ is the ξth percentile of a standard normal distribution. The studentized confidence interval is

$$C_{\mathrm{stud}}(\alpha, \boldsymbol{\omega}; \mathbf{X}) = [\hat{\theta} - n^{-1/2}\hat{\sigma} t_{n-1,1-\omega_L}, \hat{\theta} - n^{-1/2}\hat{\sigma} t_{n-1,1-\omega_U}],$$

where $\alpha = \omega_U - \omega_L$. Both of these confidence intervals have an exact confidence level of $100\alpha\%$ under the assumption that the distribution J is a normal distribution. Note that $C_{\mathrm{stud}}(\alpha, \boldsymbol{\omega}, \mathbf{X})$ matches the interval given in Equation (1.5). The approximate interval given by the hybrid critical point is

$$C_{\mathrm{hyb}}(\alpha, \boldsymbol{\omega}; \mathbf{X}) = [\hat{\theta} - n^{-1/2}\hat{\sigma} z_{1-\omega_L}, \hat{\theta} - n^{-1/2}\hat{\sigma} z_{1-\omega_U}],$$

where the percentiles from the t-distribution are approximated by the corresponding percentiles from a normal distribution. Due to the symmetry of the normal density, the interval given by the percentile critical point coincides with the hybrid interval.

If τ is known, the corresponding ordinary observed confidence level for the region $\Psi = (t_L, t_U)$ is given by

$$\alpha_{\mathrm{ord}}(\Psi) = \Phi\left[\frac{n^{1/2}(\bar{Y} - t_L)}{\tau}\right] - \Phi\left[\frac{n^{1/2}(\bar{Y} - t_U)}{\tau}\right], \qquad (2.27)$$

since $G_n = \Phi$ and $\sigma^2 = \tau^2$ in this case. The studentized observed confidence level corresponds to the observed confidence level given in Equation (2.1) from Section 2.1.1. Due to the symmetry of the normal density, the hybrid and percentile observed confidence levels for Ψ are given by

$$\alpha_{\text{hyb}}(\Psi) = \alpha_{\text{perc}}(\Psi) = \Phi\left[\frac{n^{1/2}(\bar{Y} - t_L)}{\hat{\tau}}\right] - \Phi\left[\frac{n^{1/2}(\bar{Y} - t_U)}{\hat{\tau}}\right]. \qquad (2.28)$$

If the assumption of the normality of J is not justified, the observed confidence levels given in Equations (2.27) and (2.28) can still be justified through the normal approximation. The computation of observed confidence levels based on the bootstrap will be discussed in Section 2.7.

2.3.2 Univariate Variance

Let $Y_1, \ldots, Y_n$ be a set of independent and identically distributed random variables from a univariate distribution J and let $\theta = V(Y_i)$. To put this problem in the form of the smooth function model, define $\mathbf{X}'_i = (Y_i, Y_i^2, Y_i^3, Y_i^4)$. For a vector $\mathbf{v}' = (v_1, v_2, v_3, v_4)$, define $g(\mathbf{v}) = v_2 - v_1^2$ and

$$h(\mathbf{v}) = v_4 - 4v_1v_3 + 6v_1^2v_2 - 3v_1^4 - (v_2 - v_1^2)^2.$$

See Chapter 2 of Hall (1992a) for more information on the smooth function model for the variance. The plug-in estimate of θ is

$$\hat{\theta} = g(\bar{\mathbf{X}}) = n^{-1}\sum_{i=1}^{n}(Y_i - \bar{Y})^2.$$

After some algebra, it can be shown that

$$\hat{\sigma}^2 = h(\bar{\mathbf{X}}) = n^{-1}\sum_{i=1}^{n}(Y_i - \bar{Y})^4 - \hat{\theta}^2.$$

Note that the smooth function model does not yield the interval for θ given in Equation (2.2), even when J is assumed to have a normal distribution. Indeed, if σ^2 were known, G_n is a location-scale family based on the χ^2-distribution, but the expression $n^{1/2}(\hat{\theta} - \theta)/\sigma$ is not a pivotal quantity for θ. The distribution of $n^{1/2}(\hat{\theta} - \theta)/\hat{\sigma}$ is generally intractable, even when J is a normal distribution. The normal approximations given in Equation (2.17) and (2.18) can be used here, although if J is a normal distribution the observed confidence level given in Equation (2.3) would certainly be more reliable. In the case where the distribution of J is unknown, the normal approximations, or those given by the bootstrap, can be used to compute approximate observed confidence levels for θ. Details for computing the bootstrap estimates are given in Section 2.7.

2.3.3 Correlation Coefficient

Let $(W_1, Y_1), \ldots, (W_n, Y_n)$ be a set of independent and identically distributed random vectors following a bivariate distribution J and let

$$\theta = \frac{\mu_{11} - \mu_{01}\mu_{10}}{(\mu_{20} - \mu_{10}^2)^{1/2}(\mu_{02} - \mu_{01}^2)^{1/2}}$$

where $\mu_{jk} = E(W_i^j Y_i^k)$. Hence θ is the correlation coefficient between W_i and Y_i. To set this problem within the framework of the smooth function model, define

$$\mathbf{X}_i' = (W_i, Y_i, W_iY_i, W_i^2, Y_i^2, W_i^2Y_i, W_iY_i^2, W_i^2Y_i^2, W_i^3, Y_i^3, W_i^3Y_i,$$
$$W_iY_i^3, W_i^4, Y_i^4)$$

so that

$$\boldsymbol{\mu}' = (\mu_{10}, \mu_{01}, \mu_{11}, \mu_{20}, \mu_{02}, \mu_{21}, \mu_{12}, \mu_{22}, \mu_{30}, \mu_{03}, \mu_{31}, \mu_{13}, \mu_{40}, \mu_{04}).$$

For a vector $\mathbf{v}' = (v_1, \ldots, v_{15})$ define

$$g(\mathbf{v}) = \frac{v_3 - v_1v_2}{(v_4 - v_1^2)^{1/2}(v_5 - v_2^2)^{1/2}},$$

Therefore $\theta = g(\boldsymbol{\mu})$, where g is a smooth function of $\boldsymbol{\mu}$. The plug-in estimate of θ is given in Equation (2.5). The asymptotic variance of $\hat{\theta}$ is given in Section 27.8 of Cramér (1946) as

$$\sigma^2 = \frac{\theta^2}{4}\left(\frac{\tau_{40}}{\tau_{20}^2} + \frac{\tau_{04}}{\tau_{02}^2} + \frac{2\tau_{22}}{\tau_{11}^2} + \frac{4\tau_{22}}{\tau_{11}^2} - \frac{4\tau_{31}}{\tau_{11}\tau_{20}} - \frac{4\tau_{13}}{\tau_{11}\tau_{02}}\right),$$

where $\tau_{jk} = E[(W_i - \mu_{10})^j(Y_i - \mu_{01})^k]$. Note that

$$\tau_{jk} = \sum_{i=0}^{j}\sum_{l=0}^{k}\binom{j}{i}\binom{k}{l}(-1)^{j-i}(-1)^{k-l}\mu_{il}\mu_{10}^{j-i}\mu_{01}^{k-l}.$$

It then follows that σ^2 is a smooth function of $\boldsymbol{\mu}$ as well. The plug-in estimate for σ^2 is given by

$$\hat{\sigma}^2 = \frac{\hat{\theta}^2}{4}\left(\frac{\hat{\tau}_{40}}{\hat{\tau}_{20}^2} + \frac{\hat{\tau}_{04}}{\hat{\tau}_{02}^2} + \frac{2\hat{\tau}_{22}}{\hat{\tau}_{11}^2} + \frac{4\hat{\tau}_{22}}{\hat{\tau}_{11}^2} - \frac{4\hat{\tau}_{31}}{\hat{\tau}_{11}\hat{\tau}_{20}} - \frac{4\hat{\tau}_{13}}{\hat{\tau}_{11}\hat{\tau}_{02}}\right),$$

where

$$\hat{\tau}_{jk} = n^{-1}\sum_{i=1}^{n}(W_i - \bar{W})^j(Y_i - \bar{Y})^k.$$

The confidence interval given in Equation (2.4) is derived from similar considerations as these. See Winterbottom (1979), who shows that the transformation of Fisher (1915) aids in the convergence of the standardized distribution to normality when J is a bivariate normal distribution. In general the standardized and studentized distributions do not have a closed form, and the bootstrap methodology is helpful in constructing approximate nonparametric confidence intervals for θ. Efron (1979) points out that Fisher's transformation is still useful as the bootstrap is more effective when the statistic of interest is closer to being pivotal.

2.4 Asymptotic Comparisons

As seen in Sections 2.2 and 2.3 there may be several methods available for computing an observed confidence level for any given parameter. Indeed, as was pointed out in Section 2.2, any function that maps regions to the unit interval such that the three properties of a probability measure are satisfied is technically a method for computing an observed confidence level. In this book, the focus will be on observed confidence levels that are derived from confidence intervals that at least guarantee their coverage level asymptotically. Even under this restriction there may be many methods to choose from, and techniques for comparing the methods become paramount in importance.

This motivates the question as to what properties we would wish the observed confidence levels to possess. Certainly the issue of consistency would be relevant in that an observed confidence level computed on a region $\Theta_0 = (t_{0L}, t_{0U})$ that contains θ should converge to 1 as the sample size becomes large. Correspondingly, an observed confidence level computed on a region $\Theta_1 = (t_{1L}, t_{1U})$ that does not contain θ should converge to 0 as the sample size becomes large. The issue of consistency is relatively simple to decide within the smooth function model studied in Section 2.3. The normal approximation provides the simplest case and is a good starting point. Consider the ordinary observed confidence level given in Equation (2.17). Note that

$$\Phi\left[\frac{n^{1/2}(\hat{\theta} - t_{0L})}{\sigma}\right] = \Phi\left[\frac{n^{1/2}(\theta - t_{0L})}{\sigma} + \frac{n^{1/2}(\hat{\theta} - \theta)}{\sigma}\right]$$

where $n^{1/2}(\theta - t_{0L})/\sigma \to \infty$ and $n^{1/2}(\hat{\theta} - \theta)/\sigma \xrightarrow{w} Z$ as $n \to \infty$, where Z is a standard normal random variable. It is clear that the second sequence is bounded in probability, so that the first sequence dominates. It follows that

$$\Phi\left[\frac{n^{1/2}(\hat{\theta} - t_{0L})}{\sigma}\right] \xrightarrow{p} 1$$

as $n \to \infty$. Similarly, it can be shown that

$$\Phi\left[\frac{n^{1/2}(\hat{\theta} - t_{0U})}{\sigma}\right] \xrightarrow{p} 0$$

as $n \to \infty$, so that it follows that $\hat{\alpha}_{\text{ord}}(\Theta_0) \xrightarrow{p} 1$ as $n \to \infty$ when $\theta \in \Theta_0$. A similar argument, using the fact that $\hat{\sigma}/\sigma \xrightarrow{p} 1$ as $n \to \infty$ can be used to show that $\hat{\alpha}_{\text{stud}}(\Theta_0) \xrightarrow{p} 1$ as $n \to \infty$, as well. Arguments to show that $\alpha_{\text{ord}}(\Theta_0)$ $\alpha_{\text{stud}}(\Theta_0)$ $\alpha_{\text{hyb}}(\Theta_0)$ and $\alpha_{\text{perc}}(\Theta_0)$ are also consistent follow in a similar manner, though one must assume that $G_n \overset{w}{\to} G$ and $H_n \overset{w}{\to} H$ for some limiting distributions G and H. Typically, both H and G are the standard normal distribution. A result similar to the one given on page 124 of Lehmann (1999) is used to complete the proof. Demonstrating the consistency of the bootstrap estimates is somewhat more involved, and will not be presented in detail here. Under various sets of regularity conditions it can often be shown, for example, that $\rho(\hat{H}_n, H_n) \overset{p}{\to} 0$ as $n \to \infty$ where ρ is a metric on the space of distribution functions. See Section 3.2 of Shao and Tu (1995) for a summary of these results. The triangle inequality, the fact that $\rho(H_n, H) \overset{p}{\to} 0$ as $n \to \infty$ along with an argument similar to the ones outlined above, can be used to prove consistency in this case.

Beyond consistency, it is desirable for an observed confidence level to provide an accurate representation of the level of confidence there is that $\theta \in \Psi$, given the observed sample $\mathbf{X} = \text{vec}(\mathbf{X}_1, \ldots, \mathbf{X}_n)$. Considering the definition of an observed confidence level, it is clear that if Ψ corresponds to a $100\alpha\%$ confidence interval for θ, conditional on $\mathbf{X}$, the observed confidence level for Ψ should be α. When σ is known the interval $C_{\text{ord}}(\alpha, \boldsymbol{\omega}; \mathbf{X})$ will be used as the standard for a confidence interval for θ. Hence, a measure $\tilde{\alpha}$ of an observed confidence level is *accurate* if $\tilde{\alpha}[C_{\text{ord}}(\alpha, \boldsymbol{\omega}; \mathbf{X})] = \alpha$. For the case when σ is unknown the interval $C_{\text{stud}}(\alpha, \boldsymbol{\omega}; \mathbf{X})$ will be used as the standard for a confidence interval for θ, and an arbitrary measure $\tilde{\alpha}$ is defined to be accurate if $\tilde{\alpha}[C_{\text{stud}}(\alpha, \boldsymbol{\omega}; \mathbf{X})] = \alpha$. Using this definition, it is clear that α_{ord} and α_{stud} are accurate when σ is known and unknown, respectively. When σ is known and $\tilde{\alpha}[C_{\text{ord}}(\alpha, \boldsymbol{\omega}; \mathbf{X})] \neq \alpha$ or when σ is unknown and $\tilde{\alpha}[C_{\text{stud}}(\alpha, \boldsymbol{\omega}; \mathbf{X})] \neq \alpha$ one can analyze how close close $\tilde{\alpha}[C_{\text{ord}}(\alpha, \boldsymbol{\omega}; \mathbf{X})]$ or $\tilde{\alpha}[C_{\text{stud}}(\alpha, \boldsymbol{\omega}; \mathbf{X})]$ is to α using asymptotic expansion theory. In particular, if σ is known then a measure of confidence $\tilde{\alpha}$ is said to be k^{th}-*order accurate* if $\tilde{\alpha}[C_{\text{ord}}(\alpha, \boldsymbol{\omega}; \mathbf{X})] = \alpha + O(n^{-k/2})$. To simplify the presentation, all limits in the sample size n will be taken as $n \to \infty$. Similarly, if σ is unknown a measure $\tilde{\alpha}$ is said to be k^{th}-order accurate if $\tilde{\alpha}[C_{\text{stud}}(\alpha, \boldsymbol{\omega}; \mathbf{X})] = \alpha + O(n^{-k/2})$. Within the smooth function model, this analysis is based on the Edgeworth expansion theory of Bhattacharya and Ghosh (1978) and the related theory for the bootstrap outlined in Hall (1988, 1992a). We will assume the framework of the smooth function model along with the assumptions that for a specified positive integer v, g has $v + 2$ continuous derivatives in a neighborhood of $\boldsymbol{\mu}$ and that $E(\|\mathbf{X}\|^{v+2}) < \infty$. See Chapter 2 of Hall (1992a) for more information on these conditions. Under these assumptions, the *Edgeworth expansions* for G_n and H_n are given by

$$G_n(t) = \Phi(t) + \sum_{i=1}^{v} n^{-i/2} p_i(t) \phi(t) + O(n^{-(v+1)/2}), \qquad (2.29)$$

and

$$H_n(t) = \Phi(t) + \sum_{i=1}^{v} n^{-i/2} q_i(t) \phi(t) + O(n^{-(v+1)/2}), \qquad (2.30)$$

where $\phi(t) = \Phi'(t)$. The functions p_i and q_i are polynomials of degree $3i - 1$ whose coefficients are functions of the moments of F. The polynomials are odd for even i and are even for odd i. The specific form of these polynomials, which can be quite complicated, will typically not be important to develop the results in this section. Some examples of these polynomials are given at the end of this section.

Inversion of the Edgeworth expansions given in Equations (2.29) and (2.30) result in the *Cornish-Fisher expansions* for g_α and h_α, which are given by

$$g_\alpha = z_\alpha + \sum_{i=1}^{v} n^{-i/2} p_{i1}(z_\alpha) + O(n^{-(v+1)/2}), \qquad (2.31)$$

and

$$h_\alpha = z_\alpha + \sum_{i=1}^{v} n^{-i/2} q_{i1}(z_\alpha) + O(n^{-(v+1)/2}). \qquad (2.32)$$

The functions p_{i1} and q_{i1} are polynomials of degree $i + 1$ whose coefficients are functions of the moments of F, and are related to the polynomials used in the expansions given in Equations (2.29) and (2.30). In particular, $p_{11}(x) = -p_1(x)$ and $q_{11}(x) = -q_1(x)$. As with the Edgeworth expansions, the polynomials are odd for even i and are even for odd i. See Barndorff-Nielsen and Cox (1989), Bhattacharya and Ghosh (1978), Hall (1988, 1992a), and Withers (1983) for more specific information on these expansions.

The observed confidence levels based on the normal approximations given in Equations (2.17) and (2.18) are a good starting point for the analysis of asymptotic accuracy. Suppose σ is known. If $\alpha = \omega_U - \omega_L$ then

$$\hat{\alpha}_{\text{ord}}[C_{\text{ord}}(\alpha, \boldsymbol{\omega}; \mathbf{X})] = \Phi\left\{ \frac{n^{1/2}[\hat{\theta} - \hat{\theta}_{\text{ord}}(\omega_L)]}{\sigma} \right\} -$$

$$\Phi\left\{ \frac{n^{1/2}[\hat{\theta} - \hat{\theta}_{\text{ord}}(\omega_U)]}{\sigma} \right\}$$

$$= \Phi(g_{1-\omega_L}) - \Phi(g_{1-\omega_U}). \qquad (2.33)$$

If $G_n = \Phi$, then $\hat{\alpha}_{\text{ord}}[C_{\text{ord}}(\alpha, \boldsymbol{\omega}; \mathbf{X})] = \alpha$, and the method is accurate. When $G_n \neq \Phi$ the Cornish-Fisher expansion given in Equation (2.31), along with a simple Taylor expansion for Φ yields

$$\Phi(g_{1-\omega}) = 1 - \omega - n^{-1/2} p_1(z_\omega) \phi(z_\omega) + O(n^{-1}),$$

for an arbitrary value of $\omega \in (0, 1)$. It is then clear that

$$\hat{\alpha}_{\text{ord}}[C_{\text{ord}}(\alpha, \boldsymbol{\omega}; \mathbf{X})] = \alpha + n^{-1/2} \Delta(\omega_L, \omega_U) + O(n^{-1}), \qquad (2.34)$$

where

$$\Delta(\omega_L, \omega_U) = p_1(z_{\omega_U}) \phi(z_{\omega_U}) - p_1(z_{\omega_L}) \phi(z_{\omega_L}).$$

One can observe that $\hat{\alpha}_{\text{ord}}$ is first-order accurate, unless the first-order term in Equation (2.34) is functionally zero. If it happens that $\omega_L = \omega_U$ or $\omega_L = $

$1 - \omega_U$, then it follows that $p_1(z_{\omega_L})\phi(z_{\omega_L}) = p_1(z_{\omega_U})\phi(z_{\omega_U})$ since p_1 is an even function and the first-order term vanishes. The first case corresponds to a degenerate interval with confidence measure equal to 0. The second case corresponds to the situation where $\hat{\theta}$ corresponds to the midpoint of the interval (t_L, t_U). Otherwise, the term is typically nonzero.

When σ is unknown and $\alpha = \omega_U - \omega_L$ we have that

$$
\begin{aligned}
\hat{\alpha}_{\text{stud}}[C_{\text{stud}}(\alpha, \boldsymbol{\omega}; \mathbf{X})] &= \Phi\left\{ \frac{n^{1/2}[\hat{\theta} - \hat{\theta}_{\text{stud}}(\omega_L)]}{\hat{\sigma}} \right\} - \\
&\quad \Phi\left\{ \frac{n^{1/2}[\hat{\theta} - \hat{\theta}_{\text{stud}}(\omega_U)]}{\hat{\sigma}} \right\} \\
&= \Phi(h_{1-\omega_L}) - \Phi(h_{1-\omega_U}).
\end{aligned}
$$

A similar argument to the one given above yields

$$
\hat{\alpha}_{\text{stud}}[C_{\text{stud}}(\alpha, \boldsymbol{\omega}; \mathbf{X})] = \alpha + n^{-1/2}\Lambda(\omega_L, \omega_U) + O(n^{-1}),
$$

where

$$
\Lambda(\omega_L, \omega_U) = q_1(z_{\omega_U})\phi(z_{\omega_U}) - q_1(z_{\omega_L})\phi(z_{\omega_L}).
$$

Therefore, all of the methods based on the normal approximation are first-order accurate. If the distributions G_n or H_n are known, then one need not use the normal approximation, as the measures defined in Equations (2.13) - (2.16) can be used. When σ is known, we have that $\alpha_{\text{ord}}[C_{\text{ord}}(\alpha, \boldsymbol{\omega}; \mathbf{X})] = \alpha$, and when σ is unknown it follows that $\alpha_{\text{stud}}[C_{\text{stud}}(\alpha, \boldsymbol{\omega}; \mathbf{X})] = \alpha$. To observe the effect that the approximations used in computing α_{hyb} and α_{perc} have on the accuracy of the measures, we derive expansions similar to those derived above. For α_{hyb} we have from Equation (2.15) that

$$
\begin{aligned}
\alpha_{\text{hyb}}[C_{\text{stud}}(\alpha, \boldsymbol{\omega}; \mathbf{X})] &= G_n\left\{ \frac{n^{1/2}[\hat{\theta} - \hat{\theta}_{\text{stud}}(\omega_L)]}{\hat{\sigma}} \right\} - \\
&\quad G_n\left\{ \frac{n^{1/2}[\hat{\theta} - \hat{\theta}_{\text{stud}}(\omega_U)]}{\hat{\sigma}} \right\} \\
&= G_n(h_{1-\omega_L}) - G_n(h_{1-\omega_U}).
\end{aligned}
$$

Using arguments similar to those in Section 4.5 of Hall (1988), the expansions in Equations (2.29) and (2.32) yield

$$
G_n(h_{1-\omega}) = 1 - \omega + n^{-1/2}\phi(z_\omega)[p_1(z_\omega) - q_1(z_\omega)] + O(n^{-1}),
$$

for an arbitrary $\omega \in (0, 1)$. It then follows that

$$
\alpha_{\text{hyb}}[C_{\text{stud}}(\alpha, \boldsymbol{\omega}; \mathbf{X})] = \alpha + n^{-1/2}\Lambda(\omega_L, \omega_U) - n^{-1/2}\Delta(\omega_L, \omega_U) + O(n^{-1}),
$$
$$
(2.35)
$$

which shows that α_{hyb} is first-order accurate. Note that Equation (1.3) of Hall (1988) implies that $p_1(t) - q_1(t) \neq 0$ except for special cases of F and θ. Therefore, the first-order term is typically nonzero. Similar arguments can be used to show that

$$
\alpha_{\mathrm{perc}}[C_{\mathrm{stud}}(\alpha,\boldsymbol{\omega};\mathbf{X})] = G_n\left\{\frac{n^{1/2}[\hat{\theta}_{\mathrm{stud}}(\omega_U)-\hat{\theta}]}{\hat{\sigma}}\right\} -
$$

$$
G_n\left\{\frac{n^{1/2}[\hat{\theta}_{\mathrm{stud}}(\omega_L)-\hat{\theta}]}{\hat{\sigma}}\right\}
$$

$$
= \alpha + n^{-1/2}\Lambda(\omega_L,\omega_U) +
$$

$$
n^{-1/2}\Delta(\omega_L,\omega_U) + O(n^{-1}) \qquad (2.36)
$$

which is also first-order accurate. It is clear that when G_n and H_n are known, α_{stud} and α_{ord} give the most accurate measures of confidence. When G_n and H_n are unknown, one can use the normal approximations given in Equations (2.17) and (2.18), which are all first-order accurate. Alternatively, one can also use the bootstrap methods given in Equations (2.23)–(2.26).

Hall (1988) shows that under some additional conditions on F and θ that Edgeworth expansions also exist for the bootstrap estimates $\hat{G}_n$ and $\hat{H}_n$. These expansions are given by

$$
\hat{G}_n(t) = \Phi(t) + \sum_{i=1}^{v} n^{-i/2}\hat{p}_i(t)\phi(t) + O_p(n^{-(v+1)/2}), \qquad (2.37)
$$

and

$$
\hat{H}_n(t) = \Phi(t) + \sum_{i=1}^{v} n^{-i/2}\hat{q}_i(t)\phi(t) + O_p(n^{-(v+1)/2}). \qquad (2.38)
$$

The functions $\hat{p}_i$ and $\hat{q}_i$ are the same as the functions p_i and q_i, except that the moments of F have been replaced by the moments of $\hat{F}_n$, which correspond to the sample moments. In the usual case, the consistency of the sample moments insures that $\hat{p}_i(t) = p_i(t) + O_p(n^{-1/2})$ and $\hat{q}_i(t) = q_i(t) + O_p(n^{-1/2})$. See Theorem 2.2.3.B of Serfling (1980). Note that this implies, in particular, that

$$
\hat{G}_n(t) = \Phi(t) + n^{-1/2}p_1(t)\phi(t) + O_p(n^{-1}) = G_n(t) + O_p(n^{-1}),
$$

with a similar relationship for $\hat{H}_n(t)$ and $H_n(t)$. Therefore, the expansions for the bootstrap estimates of $G_n(t)$ and $H_n(t)$ match the expansions for $G_n(t)$ and $H_n(t)$ to first-order. For the case when σ is known these results yield

$$
\hat{\alpha}^*_{\mathrm{ord}}[C_{\mathrm{ord}}(\alpha,\boldsymbol{\omega};\mathbf{X})] = \hat{G}_n\left\{\frac{n^{1/2}[\hat{\theta}-\hat{\theta}_{\mathrm{ord}}(\omega_L)]}{\sigma}\right\} -
$$

$$
\hat{G}_n\left\{\frac{n^{1/2}[\hat{\theta}-\hat{\theta}_{\mathrm{ord}}(\omega_U)]}{\sigma}\right\}
$$

$$
= \hat{G}_n(g_{1-\omega_L}) - \hat{G}_n(g_{1-\omega_U})
$$

$$
= \alpha + O_p(n^{-1}).
$$

This indicates that $\hat{\alpha}^*_{\text{ord}}$ is second-order accurate in probability. Therefore, when σ is known, the bootstrap method will asymptotically provide a more accurate measure of confidence than the measure based on the normal approximation. When σ is unknown we have that

$$
\begin{aligned}
\hat{\alpha}^*_{\text{stud}}[C_{\text{stud}}(\alpha, \boldsymbol{\omega}; \mathbf{X})] &= \hat{H}_n\left\{ \frac{n^{1/2}[\hat{\theta} - \hat{\theta}_{\text{stud}}(\omega_L)]}{\hat{\sigma}} \right\} - \\
&\qquad \hat{H}_n\left\{ \frac{n^{1/2}[\hat{\theta} - \hat{\theta}_{\text{stud}}(\omega_U)]}{\hat{\sigma}} \right\} \\
&= \hat{H}_n(h_{1-\omega_L}) - \hat{H}_n(h_{1-\omega_U}) \\
&= \alpha + O_p(n^{-1}).
\end{aligned}
$$

Therefore $\hat{\alpha}^*_{\text{stud}}$ is also second-order accurate in probability, and outperforms all of the methods based on the normal approximation. It can further be shown that since the expansions for G_n and $\hat{G}_n$ match to order n^{-1}, the expansions for $\hat{\alpha}^*_{\text{hyb}}[C_{\text{stud}}(\alpha, \boldsymbol{\omega}; \mathbf{X})]$ and $\hat{\alpha}^*_{\text{perc}}[C_{\text{stud}}(\alpha, \boldsymbol{\omega}; \mathbf{X})]$ match those given in Equations (2.35) and (2.36), so that these methods are first-order accurate in probability. The results here mirror those found for bootstrap confidence intervals by Hall (1988). When G_n and H_n are unknown, the bootstrap ordinary and studentized methods provide better asymptotic approximations than the normal approximation.

It is well known that confidence intervals based on the hybrid and percentile critical points are less accurate than confidence intervals based on the studentized critical point. A first attempt to improve the accuracy of confidence intervals based on the percentile critical point was given in Efron (1981a). The resulting method is the bias-corrected method, which uses the critical point

$$
\hat{\theta}_{\text{bc}}(\alpha) = \hat{\theta}_{\text{perc}}[\beta(\alpha)] = \hat{\theta} + n^{-1/2}\hat{\sigma}g_{\beta(\alpha)},
$$

where $\beta(\alpha) = \Phi(z_\alpha + \delta_{\text{bc}})$ and $\delta_{\text{bc}} = 2\Phi^{-1}[G_n(0)]$ is the bias-correction. Setting $\Psi = C_{\text{bc}}(\alpha, \boldsymbol{\omega}; \mathbf{X}) = [\hat{\theta}_{\text{bc}}(\omega_L), \hat{\theta}_{\text{bc}}(\omega_U)]$ yields

$$
\begin{aligned}
\alpha_{\text{bc}}(\Psi) &= \Phi\left\{ \Phi^{-1}\left[G_n\left(\frac{n^{1/2}(t_U - \hat{\theta})}{\hat{\sigma}} \right) \right] - \delta_{\text{bc}} \right\} - \\
&\qquad \Phi\left\{ \Phi^{-1}\left[G_n\left(\frac{n^{1/2}(t_L - \hat{\theta})}{\hat{\sigma}} \right) \right] - \delta_{\text{bc}} \right\},
\end{aligned}
$$

for the case when G_n is known. If G_n is unknown, then the bootstrap estimate, given by

$$
\begin{aligned}
\hat{\alpha}^*_{\text{bc}}(\Psi) &= \Phi\left\{ \Phi^{-1}\left[\hat{G}_n\left(\frac{n^{1/2}(t_U - \hat{\theta})}{\hat{\sigma}} \right) \right] - \hat{\delta}_{\text{bc}} \right\} - \\
&\qquad \Phi\left\{ \Phi^{-1}\left[\hat{G}_n\left(\frac{n^{1/2}(t_L - \hat{\theta})}{\hat{\sigma}} \right) \right] - \hat{\delta}_{\text{bc}} \right\},
\end{aligned}
$$

can be used, where $\hat{\delta}_{\mathrm{bc}} = 2\Phi^{-1}[\hat{G}_n(0)]$ is the bootstrap estimate of the bias-correction. To assess the accuracy of α_{bc} we note that

$$\begin{aligned}
\alpha_{\mathrm{bc}}[C_{\mathrm{stud}}(\alpha, \boldsymbol{\omega}; \mathbf{X})] &= \Phi\{\Phi^{-1}[G_n(-h_{1-\omega_U})] - \delta_{\mathrm{bc}}\} - \\
&\quad \Phi\{\Phi^{-1}[G_n(-h_{1-\omega_L})] - \delta_{\mathrm{bc}}\}.
\end{aligned}$$

Suppose $\omega \in (0,1)$ is arbitrary, then

$$\Phi\{\Phi^{-1}[G_n(-h_{1-\omega})] - \delta_{\mathrm{bc}}\} = \Phi\{z_\omega + n^{-1/2}[q_1(z_\omega) + p_1(z_\omega)] - \delta_{\mathrm{bc}} + O(n^{-1})\}.$$

Hall (1988) shows that $\delta_{\mathrm{bc}} = 2n^{-1/2}p_1(0) + O(n^{-1})$ so that

$$\Phi\{\Phi^{-1}[G_n(-h_{1-\omega})] - \delta_{\mathrm{bc}}\} = \omega + n^{-1/2}[p_1(z_\omega) + q_1(z_\omega) - 2p_1(0)]\phi(z_\omega) + O(n^{-1}).$$

It then follows that

$$\begin{aligned}
\alpha_{\mathrm{bc}}[C_{\mathrm{stud}}(\alpha, \boldsymbol{\omega}; \mathbf{X})] &= \alpha + n^{-1/2}\Delta(\omega_L, \omega_U) + n^{-1/2}\Lambda(\omega_L, \omega_U) \\
&\quad + 2n^{-1/2}p_1(0)[\phi(z_{\omega_U}) - \phi(z_{\omega_L})] \\
&\quad + O(n^{-1}),
\end{aligned} \tag{2.39}$$

so that α_{bc} is first-order accurate. Note that the bias-correction term does not eliminate the first-order term of the expansion in Equation (2.39). The effect of the bias-correction term is to eliminate the constant terms from the polynomials p_1 and q_1. See Hall (1988, Section 2.3) for further details. Similar arguments can be used to show that the bootstrap estimate $\hat{\alpha}_{\mathrm{bc}}^*$ has the same property.

To address deficiencies of the bias-corrected method, partially in response to Schenker (1985), Efron (1987) introduced the accelerated bias-corrected method, which uses the critical point

$$\hat{\theta}_{\mathrm{abc}}(\alpha) = \hat{\theta}_{\mathrm{perc}}[\psi(\alpha)] = \hat{\theta} + n^{-1/2}\hat{\sigma}g_{\psi(\alpha)},$$

where

$$\psi(\alpha) = \Phi\left[\frac{1}{2}\delta_{\mathrm{bc}} + \frac{\frac{1}{2}\delta_{\mathrm{bc}} + z_\alpha}{1 - a(\frac{1}{2}\delta_{\mathrm{bc}} + z_\alpha)}\right].$$

The term a is called the acceleration constant and is discussed in detail in Efron (1987) and Hall (1988, Section 2.4). In terms of the current notation, Hall (1988) shows that $-az_\alpha^2 = n^{-1/2}[2p_1(0) - p_1(z_\alpha) - q_1(z_\alpha)]$. Setting $C_{\mathrm{abc}}(\alpha, \boldsymbol{\omega}; \mathbf{X}) = [\hat{\theta}_{\mathrm{abc}}(\omega_L), \hat{\theta}_{\mathrm{abc}}(\omega_U)] = \Psi$ yields

$$\begin{aligned}
\alpha_{\mathrm{abc}}(\Psi) &= \Phi\left\{\frac{\Phi^{-1}\left[G_n\left(\frac{n^{1/2}(t_U - \theta)}{\hat{\sigma}}\right)\right] - \frac{1}{2}\delta_{\mathrm{bc}}}{1 + a\left[\Phi^{-1}\left(G_n\left(\frac{n^{1/2}(t_U - \theta)}{\hat{\sigma}}\right)\right) - \frac{1}{2}\delta_{\mathrm{bc}}\right]} - \frac{1}{2}\delta_{\mathrm{bc}}\right\} \\
&\quad - \Phi\left\{\frac{\Phi^{-1}\left[G_n\left(\frac{n^{1/2}(t_L - \theta)}{\hat{\sigma}}\right)\right] - \frac{1}{2}\delta_{\mathrm{bc}}}{1 + a\left[\Phi^{-1}\left(G_n\left(\frac{n^{1/2}(t_L - \theta)}{\hat{\sigma}}\right)\right) - \frac{1}{2}\delta_{\mathrm{bc}}\right]} - \frac{1}{2}\delta_{\mathrm{bc}}\right\}
\end{aligned} \tag{2.40}$$

If G_n is unknown, then the bootstrap estimate of α_{abc}, denoted $\hat{\alpha}_{\mathrm{abc}}^*$, is computed by replacing G_n with $\hat{G}_n$ and δ_{bc} with $\hat{\delta}_{\mathrm{bc}}$ in Equation (2.40). The acceleration constant a can be estimated using one of the many methods described by Efron (1987), Frangos and Schucany (1990) or Hall (1988). To show that α_{abc} is second-order accurate, note that the development of Equation (2.36) implies that for an arbitrary $\omega \in (0, 1)$,

$$G_n\left[\frac{n^{1/2}(\hat{\theta}_{\text{stud}}(\omega) - \hat{\theta})}{\hat{\sigma}}\right] = \omega + n^{-1/2}[q_1(z_\omega) + p_1(z_\omega)]\phi(z_\omega) + O(n^{-1}).$$

Using an argument based on Taylor expansions, it can be shown that

$$\Phi^{-1}(\omega + \lambda) = z_\omega + \frac{\lambda}{\phi(z_\omega)} + O(\lambda^2)$$

as $\lambda \to 0$. Therefore, it follows that

$$\Phi^{-1}\left\{G_n\left[\frac{n^{1/2}(\hat{\theta}_{\text{stud}}(\omega) - \hat{\theta})}{\hat{\sigma}}\right]\right\} = z_\omega + n^{-1/2}[q_1(z_\omega) + p_1(z_\omega)] + O(n^{-1}).$$

Hall (1988) shows that $\delta_{\text{bc}} = 2n^{-1/2}p_1(0) + O(n^{-1})$ so that

$$\Phi^{-1}\left\{G_n\left[\frac{n^{1/2}(\hat{\theta}_{\text{stud}}(\omega) - \hat{\theta})}{\hat{\sigma}}\right]\right\} - \tfrac{1}{2}\delta_{\text{bc}} = z_\omega + n^{-1/2}[q_1(z_\omega) + p_1(z_\omega)]$$
$$- p_1(0)] + O(n^{-1}). \quad (2.41)$$

Hall (1988) further shows that $a = n^{-1/2}[p_1(z_\omega) + q_1(z_\omega) - 2p_1(0)]/z_\omega^2$ so that

$$a\left(\Phi^{-1}\left\{G_n\left[\frac{n^{1/2}(\hat{\theta}_{\text{stud}}(\omega) - \hat{\theta})}{\hat{\sigma}}\right]\right\} - \tfrac{1}{2}\delta_{\text{bc}}\right) =$$
$$n^{-1/2}[p_1(z_\omega) + q_1(z_\omega) - 2p_1(0)]/z_\omega + O(n^{-1}).$$

A simple Taylor expansion can then be used to show that

$$\left[1 + a\left(\Phi^{-1}\left\{G_n\left[\frac{n^{1/2}(\hat{\theta}_{\text{stud}}(\omega) - \hat{\theta})}{\hat{\sigma}}\right]\right\} - \tfrac{1}{2}\delta_{\text{bc}}\right)\right]^{-1} =$$
$$1 - n^{-1/2}[p_1(z_\omega) + q_1(z_\omega) - 2p_1(0)]/z_\omega + O(n^{-1}). \quad (2.42)$$

Combining Equations (2.41) and (2.42) the yields

$$\frac{\Phi^{-1}\left\{G_n\left[\frac{n^{1/2}(\hat{\theta}_{\text{stud}}(\omega) - \hat{\theta})}{\hat{\sigma}}\right]\right\} - \tfrac{1}{2}\delta_{\text{bc}}}{1 + a[\Phi^{-1}\left\{G_n\left[\frac{n^{1/2}(\hat{\theta}_{\text{stud}}(\omega) - \hat{\theta})}{\hat{\sigma}}\right]\right\} - \tfrac{1}{2}\delta_{\text{bc}}]} - \tfrac{1}{2}\delta_{\text{bc}} = z_\omega + O(n^{-1}).$$

Therefore,

$$\Phi\left(\frac{\Phi^{-1}\left\{G_n\left[\frac{n^{1/2}(\hat{\theta}_{\text{stud}}(\omega) - \hat{\theta})}{\hat{\sigma}}\right]\right\} - \tfrac{1}{2}\delta_{\text{bc}}}{1 + a[\Phi^{-1}\left\{G_n\left[\frac{n^{1/2}(\hat{\theta}_{\text{stud}}(\omega) - \hat{\theta})}{\hat{\sigma}}\right]\right\} - \tfrac{1}{2}\delta_{\text{bc}}]} - \tfrac{1}{2}\delta_{\text{bc}}\right) = \omega + O(n^{-1}).$$

It then follows that the accelerated bias-corrected method is second-order accurate. The result follows for the bootstrap estimate as well, because the asymptotic expansion for G_n and $\hat{G}_n$ match to order n^{-1}, and both a and δ_{bc} can be estimated with error of order $O_p(n^{-1})$.

2.4.1 Univariate Mean

Observed confidence levels for the univariate mean based on the ordinary, studentized, hybrid and percentile confidence intervals were explored in Section 2.3.1. Using the assumptions and notation of that section, the development of the measures of asymptotic accuracy imply that α_{ord} is accurate when τ is known and α_{stud} is accurate when τ is unknown. The measures α_{hyb} and α_{perc} are first-order accurate. Therefore, under the assumption of normality, where G_n and H_n have a closed analytical form, there is a clear preference for α_{ord} and α_{stud}. If F is not normal then accurate methods for computing observed confidence levels are only available under the assumption that the distributions G_n or H_n can be derived explicitly, or can be approximated very accurately through simulation. Suppose F and τ are known, then G_n can be derived using a simple location-scale transformation on F. See Section 3.5 of Casella and Berger (2002). Note however that except in the case where $n^{1/2}(\bar{Y} - \theta)/\tau$ is a pivotal quantity for θ, the distribution G_n will be a function of θ and the calculation of α_{ord} will not be possible. See Section 9.2.2 of Casella and Berger (2002). If F is known but τ is unknown, then a closed analytical form for H_n only exists in certain cases, and again the result will only be useful if $n^{1/2}(\bar{Y} - \theta)/\hat{\tau}$ is a pivotal quantity for θ that does not depend on τ. Outside these special cases it is not possible to use these methods to compute accurate observed confidence levels for θ, and approximate methods based on the normal approximation or the bootstrap are required. Because τ is typically unknown this section will concentrate on this case. Results for the case where τ is known can be derived in a similar manner.

It has been established that the observed confidence level based on the normal approximation, $\hat{\alpha}_{\mathrm{stud}}$ as well as the bootstrap measures $\hat{\alpha}^*_{\mathrm{hyb}}$ and $\hat{\alpha}^*_{\mathrm{perc}}$ are first-order accurate while the studentized bootstrap measure $\hat{\alpha}^*_{\mathrm{stud}}$ is second-order accurate. The price of the increased accuracy of the studentized bootstrap method is in the calculation of the measure. Bootstrap calculations are typically more complex than those required for the normal approximation. Further, the studentized bootstrap approach is even more complex, often increasing the computation time by an order of magnitude or more. The normal approximation does not, in principle, require a computer to perform the required calculations. Bootstrap calculations usually require computer-based simulations. It is useful to know when the normal approximation will suffice. The first-order error terms all involve either the function $\Delta(\omega_L, \omega_U)$ or $\Lambda(\omega_L, \omega_U)$, which are in turn based on the quadratic functions $p_1(t)$ and $q_1(t)$. Let $\gamma = E\{(Y_i - \theta)^3\}/\sigma^3$ be the standardized skewness of F. Hall (1988) then shows that $p_1(t) = -\frac{1}{6}\gamma(t^2 - 1)$ and $q_1(t) = \frac{1}{6}\gamma(2t^2 + 1)$, so that

$$\Delta(\omega_L, \omega_U) = -\frac{1}{6}\gamma[z_{\omega_U}^2 \phi(z_{\omega_U}) - z_{\omega_L}^2 \phi(z_{\omega_L}) - \phi(z_{\omega_U}) + \phi(z_{\omega_L})],$$

and

$$\Lambda(\omega_L, \omega_U) = \frac{1}{6}\gamma[2z_{\omega_U}^2 \phi(z_{\omega_U}) - 2z_{\omega_L}^2 \phi(z_{\omega_L}) + \phi(z_{\omega_U}) - \phi(z_{\omega_L})].$$

The importance of these results is that one can observe that the major contributor to error in the observed confidence levels for the mean is related to the skewness of the parent population F. In particular, if F is symmetric then all of the methods $\hat{\alpha}_{\mathrm{stud}}$, $\hat{\alpha}_{\mathrm{stud}}^*$, $\hat{\alpha}_{\mathrm{perc}}^*$, $\hat{\alpha}_{\mathrm{hyb}}^*$ are second-order accurate. Of course, these measures of accuracy are based on the asymptotic viewpoint. That is, one can expect little difference in the accuracy of these methods when the sample size is large. For smaller sample sizes the bootstrap methods are still usually more accurate than the normal approximation. Examples of these types of results are given in Section 2.6.

Of course, methods outside the smooth function model can be used to compute observed confidence limits for the mean. In particular, the assumption that F is symmetric implies that the mean and median of the distribution F coincide. In this case an exact nonparametric confidence region for the median, such as the one given in Section 3.6 of Hollander and Wolfe (1999), can be used to derive accurate observed confidence levels for θ. See Section 6.1 for further applications of these methods.

2.4.2 Univariate Variance

Many of the issues concerning the computation of observed confidence levels for a univariate variance mirror those of the mean studied in Section 2.4.1. In particular, it is possible to compute accurate observed confidence levels only in special cases, such as when F is known to be normal, in which Equation (2.3) provides an exact method. Outside these special cases approximate methods must be employed. As with the case of the univariate mean, $\hat{\alpha}_{\mathrm{stud}}$, $\hat{\alpha}_{\mathrm{hyb}}^*$ and $\hat{\alpha}_{\mathrm{perc}}^*$ are all first-order accurate and $\hat{\alpha}_{\mathrm{stud}}^*$ is second-order accurate. Using the notation of Section 2.3.2, Hall (1992a, Section 2.6) shows that

$$p_1(t) = \tau^{-1} + (\gamma^2 - \lambda/6)(t^2 - 1)/\tau^3,$$

where $\tau^2 = E(Y_i - \psi)^4/\theta^2 - 1$, $\lambda = E[(Y_i - \psi)^2/\theta - 1]^3$, $\gamma = E(Y_i - \psi)^3/\theta^{3/2}$ and $\psi_i = E(Y_i)$. It follows that substantial simplifications do not occur in the form of $\Delta(\omega_L, \omega_U)$ except for special cases of ω_L and ω_U or F. In particular note that the first-order error in this case is a function of the skewness and kurtosis of F. The polynomial $q_1(t)$ is usually quite different from $p_1(t)$. Indeed, Hall (1992a, Section 2.6) shows that $q_1(t)$ is a function of the first six standardized moments of F. It is therefore not a simple matter to predict whether the first-order error term will be negligible based on an observed set of data. Therefore, in the case of the variance functional, second-order accurate methods are usually preferable unless specific information about F is known.

2.4.3 Population Quantile

Assume the framework outlined in Section 2.2.3, where $\theta = F^{-1}(\xi)$, the ξth quantile of F. The sample quantile $\hat{\theta} = X_{\lfloor np \rfloor + 1}$ does not fall within the smooth function model described in Section 2.3, and as such the asymptotic accuracy results of this section may not hold for observed confidence levels for population quantiles. Hall and Martin (1989) provide theoretical studies of confidence intervals for this problem. The findings are particularly interesting as they exhibit the distinct differences in asymptotic behavior that can be observed within and outside the smooth function model. In this case the confidence interval based on the inversion of the sign test is equivalent to the bootstrap percentile method confidence interval for θ which has coverage accuracy of order $n^{-1/2}$. For further results see Falk and Kaufman (1991). While this result matches the asymptotic behavior of the bootstrap percentile method exhibited in this section, Hall and Martin (1989) show that the coverage accuracy cannot be improved by adjusting the nominal rate of the confidence interval. This implies that applying the accelerated bias correction to the interval will not improve the asymptotic accuracy of the interval. Hall and Martin (1989) also conclude that the studentized interval is not a practical alternative in this case because there is no simple estimate of the standard error of $\hat{\theta}$, and the bootstrap estimate of the standard error of $\hat{\theta}$ converges at the slow rate of $O(n^{-1/4})$. See Hall and Martin (1987) and Maritz and Jarrett (1978). This leaves no hope of finding an observed confidence level based on this confidence interval for the population quantile.

There has been some improvement in this problem that arises from using the smoothed bootstrap. The smoothed bootstrap replaces empirical distribution function as the estimate of the population with a continuous estimate, usually based on a kernel density estimate. These methods will not be discussed here, but more details can be found in Falk and Reiss (1989), Hall, DiCiccio, and Romano (1989), and Ho and Lee (2005a,b). See also Cheung and Lee (2005) and Shoemaker and Pathak (2001) for some alternate approaches to implementing the bootstrap to the population quantile problem.

2.5 Empirical Comparisons

To investigate how the asymptotic results of Section 2.4 translate to finite sample behavior, a small empirical study was performed using the mean functional studied in Sections 2.3.1 and 2.4.1. Samples of size $n = 5$, 10, 20 and 50 were generated from six normal mixtures. The normal mixtures used are a subset of those studied in Marron and Wand (1992). In particular, the standard normal, skewed unimodal, strongly skewed, kurtotic, outlier and skewed bimodal distributions were used. A plot of the densities of these normal mixtures is given in Figure 2.1. One can observe from the plot that these normal mixtures allows for the study of the effect of several types of skewness and kurtosis in the parent population. For more information of these densities see Marron and Wand (1992). For each sample, the observed confidence level for the mean to be within the region $\Psi = (-\frac{1}{4}, \frac{1}{2})$ was computed. This region is indicated in Figure 2.1, along with the mean of each of the densities. Note that this allows the study of the effect when the regions covers the actual parameter values, as well as cases when it does not. The measures computed were α_{stud}, $\hat{\alpha}_{\text{stud}}$, $\hat{\alpha}^*_{\text{stud}}$, $\hat{\alpha}^*_{\text{hyb}}$, $\hat{\alpha}^*_{\text{perc}}$, $\hat{\alpha}^*_{\text{bc}}$, and $\hat{\alpha}^*_{\text{abc}}$. For the case when τ is unknown, α_{stud} is the only accurate method as defined in Section 2.4. Therefore α_{stud} will provide the benchmark for the evaluating the performance of the methods based on the normal approximation and bootstrap estimates. The calculation of α_{stud} requires that H_n be known. For the case of the standard normal population, H_n is a t-distribution with $n-1$ degrees of freedom. For the remaining distributions the distribution H_n was approximated empirically using full knowledge of the population. For the bootstrap estimates, G_n and H_n were estimated using the usual bootstrap estimates that do not rely on the knowledge of the population. For each sample size and population, 1000 samples were generated. For each generated sample, the observed confidence level for $\theta \in \Psi$ using each of the methods described above was computed. These levels were averaged over the 1000 samples.

The results of the study are given in Table 2.1. For the standard normal distribution, one can observe that $\hat{\alpha}^*_{\text{stud}}$ has an average observed confidence level closest to the average for α_{stud}. All of the remaining methods tend to overstate the confidence level, though the difference becomes quite small as n approaches 50. The fact that there are no major differences between the remaining methods may largely be due to the fact that since the distribution is symmetric, all of the methods are second-order accurate. The fact that $\hat{\alpha}^*_{\text{stud}}$ performs slightly better in this case indicates that perhaps the bootstrap studentized method is better able to model the finite sample behavior of the distribution H_n than the remaining methods. Note further that the observed confidence levels increase as the sample size increases, which indicates that the method is consistent since the mean is within the region. For the skewed unimodal distribution we begin to see some differences between the methods. In this case only $\hat{\alpha}^*_{\text{stud}}$ and $\hat{\alpha}^*_{\text{abc}}$ are second order accurate due to the fact that the distribution is not symmetric. One can indeed observe that the $\hat{\alpha}^*_{\text{stud}}$ and $\hat{\alpha}^*_{\text{abc}}$ measures are the most accurate, with $\hat{\alpha}^*_{\text{stud}}$ having a slight edge in most cases, except for the case when $n = 5$. This general behavior is carried over to the strongly skewed distribution with the exception that the overall confidence levels are much smaller, due to the fact that the mean in this case is much farther away from the region than for the skewed distribution. Note that in both of these cases, the observed confidence levels decrease with the sample size. This general behavior is observed for the kurtotic uni-modal, outlier, and skewed bimodal distributions as well, except for the two largest sample sizes for the kurtotic unimodal distribution, where $\hat{\alpha}^*_{abc}$ has the smallest error. Therefore, it appears that the studentized and accelerated bias-correction methods are the preferred measures for observed confidence levels for the mean, with a slight edge given to the studentized method in most situations.

2.6 Examples

2.6.1 A Problem from Biostatistics

Consider the bioequivalence experiment described in Section 1.3.1. Recall that A_i, N_i and P_i are the blood hormone levels in patient i after wearing the approved, new and placebo patches, respectively. Define a random vector $\mathbf{X}_i \in \mathbb{R}^9$ as

$$\mathbf{X}'_i = (A_i, N_i, P_i, A_i^2, N_i^2, P_i^2, A_iN_i, A_iP_i, N_iP_i),$$

with mean vector $\boldsymbol{\mu} = E(\mathbf{X}_i)$. Note that the components of $\boldsymbol{\mu}$ can be used to compute the means and variances of A_i, N_i, P_i, as well as covariances between the observations. The bioequivalence parameter θ can be defined as a smooth function of $\boldsymbol{\mu}$ as $\theta = g(\boldsymbol{\mu}) = (\mu_N - \mu_A)/(\mu_A - \mu_P)$, where $E(A_i) = \mu_A$, $E(N_i) = \mu_N$, $E(P_i) = \mu_P$, and it is assumed that $\mu_A \neq \mu_P$, which implies that the approved patch has some effect. The plug-in estimator of θ is the same as is given in Section 1.3.1. Using the asymptotic variance of a ratio, as given on page 245 of Casella and Berger (2002), one can show that the asymptotic variance of $\hat{\theta}$ is

$$\lim_{n \to \infty} V(n^{1/2}\hat{\theta}) = h(\boldsymbol{\mu}) = \left(\frac{\mu_N - \mu_A}{\mu_A - \mu_P}\right)^2 \left[\frac{\sigma_N^2 + \sigma_A^2 - 2\gamma_{N,A}}{(\mu_N - \mu_A)^2}\right.$$
$$\left. + \frac{\sigma_P^2 + \sigma_A^2 - 2\gamma_{P,A}}{(\mu_A - \mu_P)^2} - 2\frac{\gamma_{N,A} + \gamma_{A,P} - \gamma_{N,P} - \sigma_A^2}{(\mu_N - \mu_A)(\mu_A - \mu_P)}\right],$$

where $\sigma_A^2 = V(A_i)$, $\sigma_N^2 = V(N_i)$, $\sigma_P^2 = V(P_i)$, $\gamma_{N,A} = C(N_i, A_i)$, $\gamma_{A,P} = C(A_i, P_i)$ and $\gamma_{N,P} = C(N_i, P_i)$, which is also a smooth function of $\boldsymbol{\mu}$. There-fore the problem falls within the smooth function model described in Sec-tion 2.3 as long as the joint distribution of (A_i, N_i, P_i) is continuous and $E(A_i^k N_i^k P_i^k) < \infty$ for a sufficiently large value of k. This implies that the asymptotic comparisons of Section 2.4 should be valid for this problem.

Recall that the FDA requirement for bioequivalence is that a 90% confi-dence interval for θ lie in the range $[-0.2, 0.2]$. Therefore, consider com-puting observed confidence levels for θ on the regions $\Theta_0 = (-\infty, -0.2)$, $\Theta_1 = [-0.2, 0.2]$, and $\Theta_2 = (0.2, \infty)$. Bioequivalence would then be accepted if the observed confidence level of Θ_1 exceeds 0.90. Note that this condition is

Figure 2.1 *The densities of the normal mixtures from Marron and Wand (1992) used in the empirical studies of Section 2.5: (a) Standard Normal, (b) Skewed Unimodal, (c) Strongly Skewed, (d) Kurtotic Unimodal, (e) Outlier, and (f) Skewed Bimodal. The dashed lines indicate the boundaries of the region Ψ used in the simulation. The dotted lines indicated the location of the mean of the distribution.*

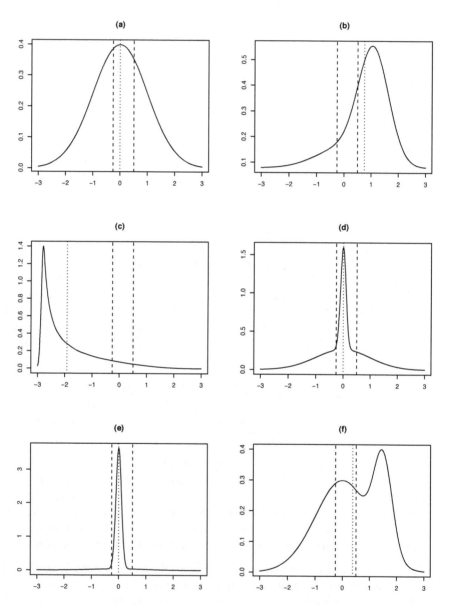

Table 2.1 *Average observed confidence levels in percent from 1000 simulated samples from the indicated distributions and sample sizes. The measure whose average level is closest to the average level of α_{stud} is italicized.*

Distribution	n	α_{stud}	$\hat{\alpha}_{\text{stud}}$	$\hat{\alpha}^*_{\text{stud}}$	$\hat{\alpha}^*_{\text{hyb}}$	$\hat{\alpha}^*_{\text{perc}}$	$\hat{\alpha}^*_{\text{bc}}$	$\hat{\alpha}^*_{\text{abc}}$
Standard	5	41.4	44.2	*40.3*	45.7	45.7	45.4	45.2
Normal	10	56.7	58.3	*56.3*	59.2	59.3	59.1	59.0
	20	70.9	71.8	*70.8*	72.2	72.3	72.3	72.0
	50	88.6	88.8	*88.5*	89.0	88.9	88.9	88.9
Skewed	5	25.8	25.2	23.1	*25.6*	24.8	25.3	25.2
Unimodal	10	25.5	23.7	*24.3*	23.7	23.1	24.1	24.2
	20	18.7	16.9	*18.4*	16.8	16.5	17.3	17.7
	50	7.5	6.6	*7.5*	6.4	6.6	6.9	7.3
Strongly	5	3.4	1.7	*3.1*	1.3	1.5	1.8	2.0
Skewed	10	1.6	0.2	*1.4*	0.1	0.2	0.3	0.4
	20	0.2	0.0	*0.2*	0.0	0.0	0.0	0.0
	50	0.0	0.0	0.0	0.0	0.0	0.0	0.0
Kurtotic	5	53.0	56.2	*50.7*	57.0	58.3	57.0	56.6
Unimodal	10	66.7	68.5	*65.3*	69.1	69.8	68.7	68.6
	20	79.6	80.4	78.8	80.7	81.0	80.5	*80.2*
	50	93.0	93.2	92.8	93.3	93.3	93.1	*93.1*
Outlier	5	26.1	*27.2*	22.7	29.0	27.2	28.4	28.2
	10	38.3	39.2	*38.0*	40.5	39.3	40.3	40.2
	20	51.8	52.4	*51.9*	53.3	52.5	52.9	53.0
	50	72.8	73.1	*72.8*	73.4	73.2	73.3	73.3
Skewed	5	24.9	*26.4*	22.9	28.5	26.8	27.9	27.8
Bimodal	10	36.3	37.2	*36.1*	38.5	37.3	38.2	37.9
	20	49.8	50.4	*49.8*	51.2	50.4	50.9	51.0
	50	71.3	71.7	*71.4*	71.8	71.6	71.7	71.8

not equivalent to the condition on the confidence intervals. The reason for this is that the confidence intervals used to evaluate bioequivalence are forced to have equal tails, whereas the confidence interval used to compute the observed confidence level will generally not have equal tails. It could be argued, however, that the condition for bioequivalence based on the observed confidence levels is more intuitive. Table 2.2 gives the observed confidence levels for each region using the $\hat{\alpha}_{\text{stud}}$, $\hat{\alpha}^*_{\text{stud}}$, $\hat{\alpha}^*_{\text{hyb}}$, $\hat{\alpha}^*_{\text{perc}}$, $\hat{\alpha}^*_{\text{bc}}$, and $\hat{\alpha}^*_{\text{abc}}$ measures. The details

Table 2.2 *Observed confidence levels for the bioequivalence data used in the example. The observed confidence levels are reported as a percentage.*

Measure	Θ_0	Θ_1	Θ_2
$\hat{\alpha}_{\text{stud}}$	9.00	90.76	0.24
$\hat{\alpha}^*_{\text{stud}}$	7.02	86.11	6.87
$\hat{\alpha}^*_{\text{hyb}}$	11.33	88.78	0.00
$\hat{\alpha}^*_{\text{perc}}$	5.73	92.99	1.28
$\hat{\alpha}^*_{\text{bc}}$	5.11	93.41	1.48
$\hat{\alpha}^*_{\text{abc}}$	4.59	92.97	2.42

of how these measures were computed can be found in Section 2.7. The bias-correction and acceleration constants were computed using Equations (14.14) and (14.15) of Efron and Tibshirani (1993). The estimated values of these parameters are $\hat{\delta}_{\text{bc}} = 0.056$ and $\hat{a} = 0.028$. One can observe from Table 2.2 that all of the measures, except the studentized measure, indicate bioequivalence. In particular, one can observe that the correction provided by the bias-corrected and the accelerated bias-corrected methods is very slight. The studentized method has both good theoretical properties, and is typically the most reliable method studied in the empirical study. This method indicates that the patches are not bioequivalent. This is consistent with the conclusions of Efron and Tibshirani (1993). Note that the BC_a interval in Efron and Tibshirani (1993) does not indicate bioequivalence, but that the accelerated bias-corrected observed confidence level does. This is due to the nonequivalence of the two conditions arising from the fact that the BC_a interval in Efron and Tibshirani (1993) is forced to have equal tail probabilities.

2.6.2 A Problem from Ecology

Consider the problem of estimating the extinction probability for the Whooping Crane, as described in Section 1.3.4. Transforming the data as $\log(x_i/x_{i-1})$ for $i = 1, \ldots, n$ creates an sequence of independent and identically distributed random variables under the assumption that all of the time increments $t_i - t_{i-1}$ are constant for $i = 1, \ldots, n$. The extinction parameter is a smooth function of the parameters β and τ, which can be written as smooth functions of a vector mean based on the increments of the process. However, note that the extinction probability is not differentiable at $\beta = 0$, and therefore the application of the smooth function model for this problem may be suspect. Figure 2.2 presents a plot of θ as a function of β for fixed values of τ and d to demonstrate this point. It can easily be seen that if β is less than zero, the extinction probability will be asymptotically degenerate. In the current application an

Figure 2.2 *Plot the of the extinction probability θ as a function of β for fixed values of τ and d.*

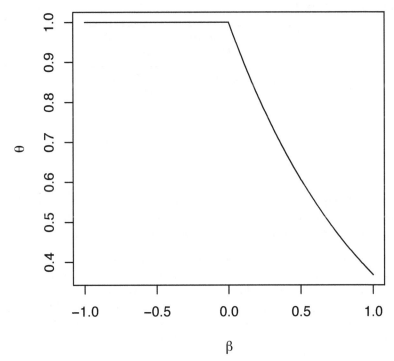

initial assumption will be made that β is far enough away from 0 to not cause any substantial difficulties.

In the application of the bootstrap to solve this problem note that it is vitally important not to resample from the population counts, but rather to sample from the increments in the population counts or from the transformed counts given by $\log(x_i/x_{i-1})$. A resampled realization can be created by starting with the observed initial population count of 18 and incrementally adding these differences to create a new realization. It is also important to account for certain problems that may occur when resampling these differences. For example, when reconstructing a realization of the process it may turn out that the resampled differences creates a process realization whose count dips below zero, or whose final count is below the threshold v. In either case the estimated extinction probability is obviously 1.

Using these methods, the bootstrap estimates of the standard error of the extinction parameter are 0.2610 for $v = 10$ and 0.3463 for $v = 100$. Observed confidence levels computed using the methods outlined in this chapter are presented in Table 2.3. The bias-corrections were estimated to be $\hat{\delta}_{bc} = 0.1319$

Table 2.3 *Observed confidence levels, in percent, that the probability of crossing thresholds of $v = 10$ and $v = 100$ for the Whooping Crane population is less than 0.10 (Θ_1) and 0.05 (Θ_2).*

	$v = 10$		$v = 100$	
	Θ_1	Θ_2	Θ_1	Θ_2
$\hat{\alpha}_{\text{stud}}$	64.9	57.6	53.5	47.7
$\hat{\alpha}^*_{\text{stud}}$	99.9	99.9	54.8	41.8
$\hat{\alpha}^*_{\text{hyb}}$	100.0	100.0	61.8	47.8
$\hat{\alpha}^*_{\text{perc}}$	88.5	86.4	56.7	48.2
$\hat{\alpha}^*_{\text{bc}}$	85.7	83.4	57.5	49.0

for $v = 10$ and $\hat{\delta}_{\text{bc}} = -0.0211$ for $v = 100$. The estimates of the acceleration constant is virtually zero for both $v = 10$ and $v = 100$, which means that virtually no further correction will be made over just using the bias correction.

One can observe from the table that when the threshold is set at 10, the generally more reliable methods indicate a great deal of confidence that the probability that the threshold will be passed is less than both 0.05 and 0.10. There is some degree of disagreement between the methods in this study, with the normal approximation providing far less confidence than the remaining measures. This is particularly surprising when one notes that neither the bias-corrected or accelerated bias-corrected methods show much correction over the percentile method. This difference could be due to the variability of $\hat{\beta}$ being large enough to cause problems when $\hat{\beta}$ approaches 0. When the threshold is 100 the methods show more agreement and of course less confidence that the probability of crossing the threshold is below the two values.

2.6.3 A Problem from Astronomy

The chondrite meteorite data of Ahrens (1965) provides an interesting example of the use of observed confidence levels for correlation coefficients. Of particular interest to geochemists who study such meteorites is their classification according to the chemical makeup of the meteorites. Ahrens (1965) was interested in the association between the iron, silicon and magnesium content of the meteorites by observing the percentage of silicon dioxide (SiO_2) and magnesium oxide (MgO) for each of the twenty-two sample meteorites. The data are given in Table 1.5 and a scatterplot of the percentage of each of the compounds in the samples is given in Figure 2.3. The observed correlation between between the percentage of SiO_2 and MgO is 0.970262. Correlations are often categorized by their strength. For example, consider the regions for the parameter space of the correlation coefficient $\Theta = [-1, 1]$ given by

Figure 2.3 *The percentage of silicon dioxide (SiO2) and magnesium oxide (MgO) for each of the twenty-two sample meteorites considered in Ahrens (1965).*

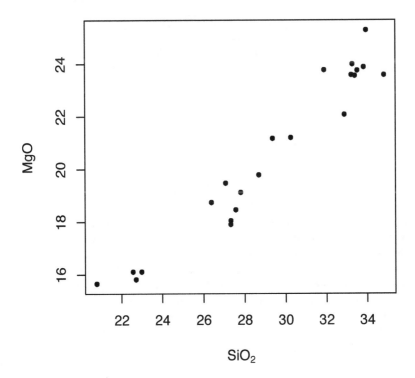

$\Theta_0 = [1.00, 0.90)$, $\Theta_1 = [0.90, 0.95)$, $\Theta_2 = [0.95, 0.99)$, $\Theta_3 = [0.99, 1.00]$ The region Θ_1 can be interpreted as indicating a strong association, Θ_2 as indicating a very strong association, and Θ_3 as indicating an extremely strong association. See Devore (2006) for an example of a broader range of regions. Therefore, the observed correlation suggests a very strong positive assocation.

The correlation coefficient falls within the smooth function model under the assumption that the joint distribution of the percentages is absolutely continuous. See Section 2.3.3. Therefore, the methods discussed in this chapter can be used to estimate observed confidence levels for each of the regions listed above. The estimated observed confidence levels for these regions are given in Table 2.4. It is clear from these estimates that there is strong confidence that the association is very strong, though it appears that many of the methods may be overstating the observed confidence when compared to the more reliable studentized observed confidence level.

Table 2.4 *Observed confidence levels for the correlation between the percentage of silicon dioxide (SiO_2) and magnesium oxide (MgO) for each of the twenty-two sample meteorites considered by Ahrens (1965). The observed confidence levels are reported as a percentage. All bootstrap methods are based on 10,000 resamples.*

Measure	Θ_1	Θ_2	Θ_3
$\hat{\alpha}_{\text{stud}}$	1.22	97.36	1.37
$\hat{\alpha}^*_{\text{stud}}$	11.80	73.99	10.68
$\hat{\alpha}^*_{\text{hyb}}$	0.18	97.42	2.14
$\hat{\alpha}^*_{\text{perc}}$	2.19	97.45	0.36
$\hat{\alpha}^*_{\text{bc}}$	4.82	95.08	0.09
$\hat{\alpha}^*_{\text{abc}}$	2.12	97.56	0.31

2.6.4 A Problem from Anthropometric Studies

Consider the problem of classifying the mean height of players in the National Basketball Association (NBA) for the 2005–2006 season with respect to the heights of typical American males of approximately the same age. The National Health and Nutrition Examination Survey (NHANES), sponsored by the Centers of Disease Control and Prevention of the United States, collects anthropometric and health related data on study participants. Based on the data from 1988-1994, the mean height of males aged 20–29 is estimated to be 176.1cm with a standard deviation of 10.9cm. The data can be obtained from the NHANES website (http://www.cdc.gov/nchs/nhanes.htm). A random sample of the heights of ten players in the NBA from the 2005–2006 season was taken using data from the NBA website (http://www.nba.com). The data are given in Table 2.5. There are ties in the data due to the fact that the heights were measured to the nearest inch, and then converted to centimeters. Of interest is to determine whether the mean height of players in the NBA during the 2005–2006 season fall within the following regions: $\Theta_0 = (-\infty, 143.4]$, $\Theta_1 = (143.4, 154.3]$, $\Theta_2 = (154.3, 165.2]$, $\Theta_3 = (165.2, 187.0)$, $\Theta_4 = [187.0, 197.9)$, $\Theta_5 = [197.9, 208.8)$ and $\Theta_6 = [208.8, \infty)$. The region Θ_3 includes all values of the mean which would be within one standard deviation of the of the mean height for the NHANES data. The regions Θ_2 and Θ_4 include all values for the mean that are between one and two standard deviations below, and above the the mean height from the NHANES data, respectively. The regions Θ_1 and Θ_5 include all values for the mean that are between two and three standard deviations below, and above the the mean height from the NHANES data, respectively. Finally, the regions Θ_0 and Θ_6 include all values for the mean that are more than three standard deviations below, and above the the mean height from the NHANES data, respectively. The average of the heights in Table 2.5 is 199.39 with a standard deviation of 10.93, which indicates a possible preference for the region Θ_5. The measures

Table 2.5 *The heights of the ten NBA players from the 2005–2006 season sampled for the example.*

Player	Height (cm)
Allen Iverson	182.88
Milt Palacio	190.50
DeSagana Diop	213.36
Theron Smith	203.20
Moochie Norris	185.42
Devin Brown	195.58
Tim Duncan	210.82
Sean Marks	208.28
Vitaly Potapenko	208.28
Kerry Kittles	195.58

Table 2.6 *Observed confidence levels for the NBA player height data used in the example. The observed confidence levels are reported as a percentage.*

Measure	Θ_0	Θ_1	Θ_2	Θ_3	Θ_4	Θ_5	Θ_6
$\hat{\alpha}_{\text{stud}}$	0.00	0.00	0.00	0.01	33.30	66.36	0.33
$\hat{\alpha}_{\text{stud}}^*$	0.00	0.00	0.00	0.60	34.60	64.80	0.00
$\hat{\alpha}_{\text{hyb}}^*$	0.00	0.00	0.00	0.00	36.90	62.80	0.30
$\hat{\alpha}_{\text{perc}}^*$	0.00	0.00	0.00	0.00	32.10	66.20	1.70
$\hat{\alpha}_{\text{bc}}^*$	0.00	0.00	0.00	0.00	36.68	62.82	0.50
$\hat{\alpha}_{\text{abc}}^*$	0.00	0.00	0.00	0.00	37.20	62.80	0.00

$\hat{\alpha}_{\text{stud}}$, $\hat{\alpha}_{\text{stud}}^*$, $\hat{\alpha}_{\text{hyb}}^*$, $\hat{\alpha}_{\text{perc}}^*$, $\hat{\alpha}_{\text{bc}}^*$, and $\hat{\alpha}_{\text{abc}}^*$ were computed for each of the regions, and are reported in Table 2.6. One can easily observe that there is little or no confidence that the mean height of players in the NBA during the 2005–2006 season is in any of the regions Θ_0–Θ_3 or Θ_6. Further, there is approximately twice, depending on which measure is used, the confidence that the mean height of players in the NBA during the 2005–2006 season is in the region Θ_5 than there is in the region Θ_4. According to the data from the NBA website, the actual mean height of *all* players in the NBA for the 2005–2006 season is 201.26cm, which is in region Θ_5. It is clear that all of the measures used for this example indicate this possibility rather well.

2.7 Computation Using R

The computations required for the examples in Sections 2.6.1 and 2.6.4 were implemented using the R statistical computing environment. The R package is freely available online at http://www.r-project.org. Distributions for Mac OS X, Unix, Linux and Windows are available. The R statistical software is especially useful for bootstrap computations because of the ability to write user-defined functions. The standard distribution of R also supports a vast array of mathematical functions, including singular value decomposition, which is required to compute studentized observed confidence levels in the multiparameter case. The software also includes a library with functions designed for performing bootstrap calculations. This section will assume that the reader has a basic working knowledge of R. An introduction to the R environment is given by Dalgaard (2002). An introduction to using the bootstrap library is given in Chapter 11 of Davison and Hinkley (1997). This section will demonstrate how to compute the observed confidence levels for the example studied in Section 2.6.1. The computations for the remaining examples in Section 2.6 can be implemented in a similar fashion.

Before implementing the calculations in R, a basic understanding of bootstrap computation is required. The bootstrap estimate of G_n is defined in Section 2.3 as

$$\hat{G}_n(t) = P^* \left[\frac{n^{1/2}(\hat{\theta}^* - \hat{\theta})}{\hat{\sigma}} \le t \,\middle|\, \mathbf{X}_1^*, \dots, \mathbf{X}_n^* \sim \hat{F}_n \right].$$

This distribution rarely has a closed analytic form and is usually approximated using a computer-based simulation, as follows. Simulate b samples of size n from the distribution $\hat{F}_n$, conditional on $\mathbf{X}_1, \dots, \mathbf{X}_n$. Because the empirical distribution places a discrete mass of n^{-1} at each of the observed sample points $\mathbf{X}_1, \dots, \mathbf{X}_n$, generating each of these samples can be achieved by randomly selecting, with replacement, n times from the original sample $\mathbf{X}_1, \dots, \mathbf{X}_n$. Each of these generated samples is called a *resample*. See Efron and Tibshirani (1993) and Davison and Hinkley (1997) for more information on such simulations. For each resample, compute $Z_i^* = n^{1/2}(\hat{\theta}_i^* - \hat{\theta})/\hat{\sigma}$, where $\hat{\theta}_i^*$ is the estimator $\hat{\theta}$ computed on the ith resample. The distribution $\hat{G}_n(t)$ can then be approximated using the empirical distribution of $Z_1^*, \dots, Z_b^*$. That is

$$\hat{G}_n(t) \simeq b^{-1} \sum_{i=1}^{b} \delta(t, [Z_i^*, \infty)). \tag{2.43}$$

It follows from Equation (2.25) that $\hat{\alpha}_{\text{hyb}}^*$ can be approximated with

$$
\begin{aligned}
\hat{\alpha}_{\text{hyb}}^*(\Psi) \quad \simeq \quad & b^{-1} \sum_{i=1}^{b} \delta \left[\frac{n^{1/2}(\hat{\theta} - t_L)}{\hat{\sigma}}; [Z_i^*, \infty) \right] \\
- \quad & b^{-1} \sum_{i=1}^{b} \delta \left[\frac{n^{1/2}(\hat{\theta} - t_U)}{\hat{\sigma}}; [Z_i^*, \infty) \right].
\end{aligned}
$$

Note that

$$\frac{n^{1/2}(\hat{\theta} - t)}{\hat{\sigma}} \geq Z_i^* = \frac{n^{1/2}(\hat{\theta}_i^* - \hat{\theta})}{\hat{\sigma}},$$

if and only if $\hat{\theta}_i^* \leq 2\hat{\theta} - t$. Therefore, $\hat{\alpha}_{\text{hyb}}^*$ can be equivalently computed using the simpler approximation

$$\begin{aligned}
\hat{\alpha}_{\text{hyb}}^*(\Psi) &\approx b^{-1} \sum_{i=1}^{b} \delta(2\hat{\theta} - t_L; [\hat{\theta}_i^*, \infty)) \\
&- b^{-1} \sum_{i=1}^{b} \delta(2\hat{\theta} - t_U; [\hat{\theta}_i^*, \infty)).
\end{aligned}$$

The measure $\hat{\alpha}_{\text{perc}}^*$ can be approximated in a similar way. From Equations (2.26) and (2.43), $\hat{\alpha}_{\text{perc}}^*$ can be approximated with

$$\begin{aligned}
\hat{\alpha}_{\text{perc}}^*(\Psi) &\simeq b^{-1} \sum_{i=1}^{b} \delta\left[\frac{n^{1/2}(t_U - \hat{\theta})}{\hat{\sigma}}; [Z_i^*, \infty)\right] \\
&- b^{-1} \sum_{i=1}^{b} \delta\left[\frac{n^{1/2}(t_L - \hat{\theta})}{\hat{\sigma}}; [Z_i^*, \infty)\right].
\end{aligned}$$

Note that

$$\frac{n^{1/2}(t - \hat{\theta})}{\hat{\sigma}} \geq Z_i^* = \frac{n^{1/2}(\hat{\theta}_i^* - \hat{\theta})}{\hat{\sigma}},$$

if and only if $\hat{\theta}_i^* \leq t$. Therefore, $\hat{\alpha}_{\text{perc}}^*$ can be equivalently computed using the approximation

$$\begin{aligned}
\hat{\alpha}_{\text{perc}}^*(\Psi) &\simeq b^{-1} \sum_{i=1}^{b} \delta(t_U; [\hat{\theta}_i^*, \infty)) \\
&- b^{-1} \sum_{i=1}^{b} \delta(t_L; [\hat{\theta}_i^*, \infty)).
\end{aligned}$$

The choice of b in these calculations is arbitrary, though larger values of b will provide a better approximation. There are no specific guidelines for choosing b for computing observed confidence levels, though the choice of b for computing confidence intervals is well documented. Typically b should be at least 1000, though modern desktop computers can easily handle $b = 10000$ or $b = 20000$ in a reasonable amount of time, depending on the complexity of the statistic $\hat{\theta}$. See Booth and Sarkar (1998) and Hall (1986) for further details.

In the R statistical computing environment the `boot` function, which is a member of the `boot` library, provides an easy way to obtain the values of $\delta(2\hat{\theta} - t; [\hat{\theta}_i^*, \infty))$ and $\delta(t; [\hat{\theta}_i^*, \infty))$. In order to implement the `boot` function from the bootstrap library in R, a function must be written that reads in the original data and a vector that contains indices corresponding to the observations to be included in the current resample. The function should return the value of the statistic for the current resample. The boot function in R is very flexible and allows for multidimensional data and statistics. To construct the function required by the boot function for the bioequivalence experiment in Section 2.6.1, it will be assumed that an 8×3 matrix called bioeq contains the data from Table 25.1 of Efron and Tibshirani (1993). The rows correspond to the different patients, and the columns of bioeq correspond to the observations from the placebo, approved and new patches, respectively. Given this structure, the function

```
bioboothyb <- function(data,i,tval) {
    d <- data[i,]
    that <- sum(data[,3]-data[,2])/sum(data[,2]-data[,1])
    thats <- sum(d[,3]-d[,2])/sum(d[,2]-d[,1])
    if(thats<=(2.0*that-tval)) delta <- 1
        else delta <- 0
    delta       }
```

will return $\delta(2\hat{\theta}-\text{tval}; [\hat{\theta}_i^*, \infty))$. Assume that the values of t_L and t_U are stored in the R objects tL and tU, respectively. Then the values of $\delta(2\hat{\theta} - t_L; [\hat{\theta}_i^*, \infty))$ and $\delta(2\hat{\theta} - t_U; [\hat{\theta}_i^*, \infty))$ for $i = 1, \ldots, b$ can be stored in the R objects dLhyb and dUhyb using the commands

```
dLhyb <- boot(data=bioeq,statistic=bioboothyb,R=1000,tval=tL)$t
dUhyb <- boot(data=bioeq,statistic=bioboothyb,R=1000,tval=tU)$t
```

The specification R=1000 indicates that $b = 1000$ resamples should be used in the bootstrap calculations. The value of tval is not used by the bootstrap function itself, but is passed directly to the bioboothyb function. The values of bioboothyb for each resample are extracted from the boot object using the $t designation. The values of $\hat{\alpha}_{\text{hyb}}^*(\Theta_i)$ are then approximated with the command mean(dUhyb) - mean(dLhyb). The measure $\hat{\alpha}_{\text{perc}}^*$ can be computed using the function

```
biobootperc <- function(data,i,tval) {
    d <- data[i,]
    that <- sum(data[,3]-data[,2])/sum(data[,2]-data[,1])
    thats <- sum(d[,3]-d[,2])/sum(d[,2]-d[,1])
    if(thats<=tval) delta <- 1
        else delta <- 0
    delta       }
```

in conjunction with commands similar to those used above for computing the hybrid measure.

The computation of $\hat{\alpha}_{\text{stud}}^*(\Theta_i)$ is somewhat more complicated as the plug-in estimator of the asymptotic standard error must be computed for each resample. That is, for each resample the value $T_i^* = n^{1/2}(\hat{\theta}_i^* - \hat{\theta})/\hat{\sigma}_i^*$ must be computed where $\hat{\sigma}_i^*$ is the plug-in estimate of σ computed using the ith resample. If the analytic form of the plug-in estimator is known, then it can be implemented directly in the R function that is used in the boot function. In many cases, however, the estimate is either not known, or has a very complicated form. In these cases the bootstrap standard error estimate can be used. Note that this is an example of the iterated, or nested, bootstrap described by Hall and Martin (1988). For each bootstrap resample, a new set of resamples is taken from the generated resample to estimate the standard error. In R this is a simple process because functional recursion is allowed. The bootstrap estimate of the standard error of $\hat{\theta}$ is given by the standard deviation of the bootstrap distribution of $\hat{\theta}$. That is,

$$\hat{\mathrm{SE}}(\hat{\theta}) = \left\{ \int_{\mathbb{R}} \left[t - \int_{\mathbb{R}} t d\hat{V}_n(t) \right]^2 d\hat{V}_n(t) \right\}^{1/2},$$

where $\hat{V}_n(t)$ is the bootstrap estimate of the distribution of $\hat{\theta}$ which can be approximated using the empirical distribution function of $\hat{\theta}_1^*, \ldots, \hat{\theta}_b^*$. Therefore the standard error of $\hat{\theta}$ can be approximated with

$$\hat{\mathrm{SE}}(\hat{\theta}) \simeq \left\{ (b-1)^{-1} \sum_{i=1}^b \left[\hat{\theta}_i^* - b^{-1} \sum_{j=1}^b \hat{\theta}_j^* \right]^2 \right\}^{1/2}.$$

Therefore, to estimate a standard error using the R boot function, a function must be written that returns the value of the statistic being studied calculated on the resample. For the bioequivalence problem this function is

```
biobootse <- function(data,i) {
    d <- data[,i]
    thats <- sum(d[,3]-d[,2])/sum(d[,2]-d[,1])
    thats    }
```

The bootstrap estimate of the standard error based on the original sample can then be computed using the boot function and the following command

```
biosehat <- sd(boot(data=bioeq,statistic=biobootse,R=100)$t)
```

To develop the studentized measure, note that the bootstrap estimate of the distribution H_n can be approximated with

$$\hat{H}_n(t) \simeq b^{-1} \sum_{i=1}^b \delta(t; [T_i^*, \infty))$$

where $T_i^* = n^{1/2}(\hat{\theta}_i^* - \hat{\theta})/\hat{\sigma}_i^*$ and $\hat{\theta}_i^*$ and $\hat{\sigma}_i^*$ are computed using the ith resample. It follows from Equation (2.24) that

$$\hat{\alpha}_{\text{stud}}^*(\Psi) \simeq b^{-1}\sum_{i=1}^{b}\delta\left[\frac{n^{1/2}(\hat{\theta}-t_L)}{\hat{\sigma}};[T_i^*,\infty)\right]$$
$$- b^{-1}\sum_{i=1}^{b}\delta\left[\frac{n^{1/2}(\hat{\theta}-t_U)}{\hat{\sigma}};[T_i^*,\infty)\right].$$

Note that

$$\frac{n^{1/2}(\hat{\theta}-t)}{\hat{\sigma}} \geq T_i^* = \frac{n^{1/2}(\hat{\theta}_i^*-\hat{\theta})}{\hat{\sigma}_i^*},$$

if and only if $t \leq \hat{\sigma}(\hat{\theta}-\hat{\theta}_i^*)/\hat{\sigma}_i^* + \hat{\theta}$. Therefore, $\hat{\alpha}_{\text{stud}}^*(\Psi)$ can be approximated with

$$\hat{\alpha}_{\text{stud}}^*(\Psi) \simeq b^{-1}\sum_{i=1}^{b}\delta[\hat{\sigma}(\hat{\theta}_i^*-\hat{\theta})/\hat{\sigma}_i^* + \hat{\theta};[t_L,\infty)]$$
$$- b^{-1}\sum_{i=1}^{b}\delta[\hat{\sigma}(\hat{\theta}_i^*-\hat{\theta})/\hat{\sigma}_i^* + \hat{\theta};[t_U,\infty)].$$

An R function that returns $\delta[\hat{\sigma}(\hat{\theta}_i^*-\hat{\theta})/\hat{\sigma}_i^* + \hat{\theta};[t,\infty)]$ is given by

```
biobootstud <- function(data,i,tval,sehat) {
    d <- data[i,]
    that <- sum(data[,3]-data[,2])/sum(data[,2]-data[,1])
    thats <- sum(d[,3]-d[,2])/sum(d[,2]-d[,1])
    sehats <- sd(boot(data=d,statistic=biobootse,R=100)$t)
    if(tval <= sehat*(that-thats)/sehats + that) delta <- 1
        else delta <- 0
    delta     }
```

The values of $\delta[\hat{\sigma}(\hat{\theta}_i^*-\hat{\theta})/\hat{\sigma}_i^* + \hat{\theta};[t_L,\infty)]$ and $\delta[\hat{\sigma}(\hat{\theta}_i^*-\hat{\theta})/\hat{\sigma}_i^* + \hat{\theta};[t_U,\infty)]$ can then be stored in the R objects dLstud and dUstud using the commands

```
dLstud <- boot(data=bioeq,statistic=biobootstud,
        R=1000,tval=tL,sehat=biosehat)&t
dUstud <- boot(data=bioeq,statistic=biobootstud,
        R=1000,tval=tU,sehat=biosehat)&t
```

and the value of $\hat{\alpha}_{\text{stud}}^*(\Theta_i)$ is then approximated with the command

```
mean(dLstud) - mean(dUstud)
```

Note that biosehat was computed earlier in the section.

The computation of the accelerated and bias-corrected measure $\hat{\alpha}_{\text{abc}}^*$ is computed using similar ideas, with the main exception that the bias-correction term δ_{bc} and the acceleration constant a must be estimated from the data. Recall that $\delta_{\text{bc}} = 2\Phi^{-1}[G_n(0)]$ so that a bootstrap estimate of the bias-correction can be computed as $\hat{\delta}_{bc} = 2\Phi^{-1}[\hat{G}_n(0)]$, where $\hat{G}_n$ is a bootstrap estimate of G_n. The approximation of $\hat{G}_n(t)$ given in Equation (2.43) yields

$$\hat{G}_n(0) \simeq b^{-1}\sum_{i=1}^{b}\delta(0;[Z_i^*,\infty))$$

$$= b^{-1}\sum_{i=1}^{b}\delta(\hat{\theta};[\hat{\theta}_i^*,\infty)).$$

A function suitable to compute $\hat{G}_n(0)$ for use in the `boot` function is given by

```
biobootbc <- function(data,i) {
    d <- data[i,]
    that <- sum(data[,3]-data[,2])/
        sum(data[,2]-data[,1])
    thats <- sum(d[,3]-d[,2])/sum(d[,2]-d[,1])
    if(that >= thats) delta <- 1
        else delta <- 0
    delta       }
```

where the bootstrap estimate of the bias-correction can then be approximated using the command

```
bc <- 2.0*qnorm(mean(boot(data=bioeq,statistic=biobootbc,
            R=100)$t))
```

The most common method for computing an estimate of the acceleration constant is to use the jackknife method given by Equation (14.15) of Efron and Tibshirani (1993). This estimate is given by

$$\hat{a} = \frac{\sum_{i=1}^{n}(\hat{\theta}_{(\cdot)} - \hat{\theta}_{(i)})^3}{6[\sum_{i=1}^{n}(\hat{\theta}_{(\cdot)} - \hat{\theta}_{(i)})^2]^{3/2}},$$

where $\hat{\theta}_{(i)}$ is the statistic $\hat{\theta}$ computed on the original sample with the ith observation missing and

$$\hat{\theta}_{(\cdot)} = n^{-1}\sum_{i=1}^{n}\hat{\theta}_{(i)}.$$

An R function to compute $\hat{a}$ for the bioequivalence example is given by

```
bioaccel <- function(data) {
    n <- nrow(data)
    thetai <- matrix(0,n,1)
    for(i in 1:n) {
        thetai[i] <- sum(data[-i,3]-data[-i,2])/
                sum(data[-i,2]-data[-i,1])       }
        thetadot <- mean(thetai)
        accel <- sum((thetadot - thetai)^3)/
```

```
                    6*sum((thetadot-thetai)^2)^(1.5)
        accel       }
```

The remaining calculations to compute $\hat{a}^*_{\text{abc}}$ can the be completed using the functions described above.

2.8 Exercises

2.8.1 Theoretical Exercises

1. Suppose $X_1, \ldots, X_n$ is a random sample from an exponential location family of densities of the form

$$f(x) = e^{-(x-\theta)} \delta(x; [\theta, \infty)),$$

   where $\theta \in \Theta = \mathbb{R}$.

   (a) Let $X_{(1)}$ be the first order-statistic of the sample $X_1, \ldots, X_n$. That is $X_{(1)} = \min\{X_1, \ldots, X_n\}$. Prove that

   $$C(\alpha, \omega; \mathbf{X}) = [X_{(1)} + n^{-1}\log(1 - \omega_L), X_{(1)} + n^{-1}\log(1 - \omega_U)]$$

   is a $100\alpha\%$ confidence interval for θ when $\omega_U - \omega_L = \alpha$ where $\omega_L \in [0, 1]$ and $\omega_U \in [0, 1]$. *Hint:* Use the fact that the density of $X_{(1)}$ is

   $$f(x_{(1)}) = ne^{-n(x_{(1)} - \theta)} \delta(x_{(1)}; [\theta, \infty))$$

   (b) Use the confidence interval given in Part (a) to derive an observed confidence level for an arbitrary region $\Psi = (t_L, t_U) \subset \mathbb{R}$ where $t_L < t_U$.

2. Suppose $X_1, \ldots, X_n$ is a random sample from a uniform density of the form

$$f(x) = \theta^{-1}\delta(x; (0, \theta)),$$

   where $\theta \in \Theta = (0, \infty)$.

   (a) Find a $100\alpha\%$ confidence interval for θ when $\omega_U - \omega_L = \alpha$ where $\omega_L \in [0, 1]$ and $\omega_U \in [0, 1]$.

   (b) Use the confidence interval given in Part (a) to derive an observed confidence level for an arbitrary region $\Psi = (t_L, t_U) \subset \mathbb{R}$ where $0 < t_L < t_U$.

3. Consider a random sample $\mathbf{W}_1, \ldots, \mathbf{W}_n$ from a continuous bivariate distribution F where $\mathbf{W}_i = (W_{i1}, W_{i2})$. Let $\boldsymbol{\eta} = (\eta_1, \eta_2)' = E(\mathbf{W}_i)$ be the bivariate mean of F and define $\theta = \eta_1/\eta_2$. This problem will demonstrate that this parameter falls within the smooth function model.

   (a) Define a new sequence of random vectors as

   $$\mathbf{X}'_i = (W_{i1}, W_{i2}, W_{i1}^2, W_{i2}^2, W_{i1}W_{i2})$$

   for $i = 1, \ldots, n$. Compute the mean vector of $\mathbf{X}_i$ using the notation $\mu_{kl} = E(W_{i1}^k W_{i2}^l)$.

(b) Construct a function g such that $\theta = g(\mu)$ where $\mu = E(\mathbf{X}_i)$. What is the corresponding plug-in estimate of θ?

(c) Compute the asymptotic variance of $n^{1/2}\hat{\theta}$ and show that this variance can be written at a smooth function of μ. What is the plug-in estimate of σ^2?

4. Using the Edgeworth expansion for $G_n(t)$ and the Cornish-Fisher expansion of h_ξ, prove that

$$G_n(h_{1-w}) = 1 - w + n^{-1/2}\phi(z_w)[p_1(z_w) - q_1(z_w)] + O(n^{-1}).$$

5. Using a Taylor expansion, prove that

$$\Phi^{-1}(w + \lambda) = z_w + \frac{\lambda}{\phi(z_w)} + O(\lambda^2)$$

as $\lambda \to 0$.

2.8.2 Applied Exercises

1. The table below gives the midterm and final exam scores for ten randomly selected students in a large basic statistics class. Assume that the assumptions of the smooth function model hold for this population. The observed correlation between the midterm and final exam scores is 0.7875. Compute observed confidence levels that the correlation is within the the regions $\Theta_0 = [-1, 0]$, $\Theta_1 = (0, 0.50)$, $\Theta_2 = [0.50, 0.80)$, $\Theta_3 = [0.80, 0.90)$, $\Theta_4 = [0.90, 0.95)$, $\Theta_5 = [0.95, 0.99)$ and $\Theta_6 = [0.99, 1.00]$. Use the percentile, hybrid, and studentized methods to compute the observed confidence levels. What region has the largest level of confidence? Are there substantial differences between the methods used to compute the observed confidence levels?

Student	1	2	3	4	5	6	7	8	9	10
Midterm	77	70	94	80	94	65	69	76	69	71
Final	78	82	91	91	94	76	79	71	76	78

2. Two precision measuring instruments are assumed to both give unbiased estimates of the length of a small object. Researchers at a lab are interested in comparing the variability of the two instruments. Measurements of seven objects are taken with each instrument. Each object is thought to have a length of 1.27mm. The measurements from each instrument are given in the table below. Define the parameter of interest for this problem to be the ratio of standard deviations of Instrument 1 to Instrument 2 (labeled I1 and I2 in the table). Calculate observed confidence levels using the percentile, hybrid, and studentized methods that the ratio is in the regions $\Theta_1 = (0, \frac{1}{2})$, $\Theta_2 = [\frac{1}{2}, \frac{3}{4})$, $\Theta_3 = [\frac{3}{4}, \frac{9}{10})$, $\Theta_4 = [\frac{9}{10}, 1)$, $\Theta_5 = [1, \frac{10}{9})$, $\Theta_6 = [\frac{10}{9}, \frac{4}{3})$, $\Theta_7 = [\frac{4}{3}, 2)$ and $\Theta_8 = [2, \infty)$.

Object	1	2	3	4	5	6	7
I1	1.2776	1.2745	1.2616	1.2474	1.2678	1.2548	1.2788
I2	1.2726	1.2687	1.2648	1.2677	1.2703	1.2713	1.2739

3. A tire company inspects twenty-five warranty records for a particular brand of tire. The company will replace the tires in question if they fail under normal operating conditions after less than 50,000 miles. The company also offers an extended warranty that will replace the tires if they fail under normal operating conditions after less than 60,000 miles. The mileage at failure of each tire that each replaced under one of these warranties was obtained from the warranty data. The data are given in the table below (in thousands of miles). Let θ be the mean number of miles at failure (in thousands of miles) for tires replaced under these warranty plans. Compute observed confidence levels using the percentile, hybrid, studentized, and accelerated and bias-corrected methods that θ is within each of the regions $\Theta_1 = (0, 40)$, $\Theta_1 = (40, 45)$, $\Theta_1 = (45, 50)$, $\Theta_1 = (50, 60)$.

58.18	20.29	44.33	45.15	52.08
51.07	48.41	56.26	56.46	57.90
59.83	34.79	59.83	50.56	49.63
54.85	58.88	56.36	38.92	53.00
52.69	56.11	56.02	58.88	54.69

4. The purity of a chemical compound used in manufacturing is an important quality characteristic that is monitored carefully by a manufacturing company. Suppose the manufacturing company has an acceptance sampling plan with a supplier for a large shipment of the compound that is written in terms of observed confidence levels. For any shipment, n samples of the compound from the shipment are analyzed by chemists from the company. Using the observed purity levels for the n samples, observed confidence levels for computed that the mean purity level is with the following regions: $\Theta_1 = (0, 0.95)$, $\Theta_2 = [0.95, 0.98)$, $\Theta_3 = [0.98, 0.99)$ and $\Theta_4 = [0.99, 1.00]$. The shipment will be accepted if the observed confidence level of Θ_4 is above 80% or if the sum of the observed confidence levels for Θ_3 and Θ_4 exceed 95%. Suppose the data in the table below correspond to $n = 16$ observed purity levels for a shipment of the compound. Compute observed confidence levels for each of the regions using the percentile, hybrid, studentized, and accelerated and bias-corrected methods. Would the shipment be accepted according to each of the methods for computing the observed confidence levels? Comment on any differences.

0.9943	0.9910	0.9920	0.9917	0.9999	0.9999	0.9922	0.9768
0.9917	0.9773	0.9999	0.9974	0.9995	0.9641	0.9977	0.9990

5. Fifteen samples of water are taken from a lake that is thought to be contaminated by a local industry. The pH of the water is determined for each sample. The data are given in the table below. The range of the pH scale has been divided into three regions: $\Theta_a = (0, 6.8)$, which corresponds to acidic readings, $\Theta_n = [6.8, 7.2]$, which corresponds to neutral readings, and $\Theta_b = (7.2, 14)$, which corresponds to basic readings. Compute observed confidence levels that the mean reading from the lake is in each of these regions using the normal, percentile, hybrid, studentized and accelerated and bias-corrected methods. If the lake water should be neutral, does it appear that that the lake has been contaminated to cause it to be acidic?

7.71	7.38	7.87	6.54	6.67
6.84	7.88	7.90	7.81	6.23
6.87	6.81	6.88	6.79	7.62

CHAPTER 3

Multiple Parameter Problems

3.1 Introduction

This chapter considers computing observed confidence levels for the more general case of vector parameters. While the single parameter case necessarily focuses on interval regions of the parameter space, this chapter will consider more general subsets of $\mathbb{R}^p$. As in the case of scalar parameters, there are several potential methods that exist for computing the levels, and the measure of asymptotic accuracy developed in Chapter 2 will play a central role in the comparison of these methods. These comparisons once again depend heavily on asymptotic expansion theory, in this case developed from multivariate Edgeworth expansions. As in the previous chapter, observed confidence levels based on studentized confidence regions will be shown to have superior asymptotic accuracy.

Following the development of the general case for scalar parameters considered in Section 2.2, it suffices to consider observed confidence levels in the parameter vector case as a probability measure on a sigma-field $\mathfrak{I}$ of subsets of Θ. For most problems it is more than sufficient to let $\mathfrak{I}$ be the Borel sets in $\mathbb{R}^p$. Even within this class of subsets the typical regions considered tend to have a relatively simple structure. In Section 3.3, for example, regions that are a finite union of convex sets will be considered exclusively.

The general problem of computing observed confidence levels in the vector parameter case is much more complicated than in the scalar parameter case as a typical region Ψ and a confidence region associated with $\boldsymbol{\theta}$ can have many complex shapes, even when restricted to a fairly simple class. Hence the problem of solving $\mathbf{C}(\alpha, \boldsymbol{\omega}; \mathbf{X}) = \Psi$ is difficult to solve in general. Fortunately, as in the case of scalar parameters, standard confidence regions can be used which simplify matters significantly. In many cases, particularly when the bootstrap is being used, the computation of the observed confidence levels is relatively simple, and may not require too much specific knowledge about the shape of the region of interest. These issues are discussed in Sections 3.2 and 3.6.

Several general methods for computing confidence regions are presented in Section 3.2. Each of these methods are used to develop corresponding observed confidence levels. The development of the methods are based on a smooth function model, which in this case is a simple generalization of the model used in Chapter 2. The asymptotic accuracy of the proposed methods are studied in Section 3.3. Due to the complexity of the multiparameter case, adjustments to the methods such as those given by bias-correction and accelerated bias-correction are quite complicated and will not be addressed. Empirical comparisons between the methods are considered in Section 3.4, and some examples are given in Section 3.5. The calculation of observed confidence levels in the multiparameter case using the R statistical computing environment is addressed in Section 3.6.

3.2 Smooth Function Model

The smooth function model considered in this chapter is essentially the same as in Section 2.3 with the exception that $\boldsymbol{\theta}$ can be a p-dimensional vector. In particular, let $\boldsymbol{\mu} = E(\mathbf{X}_i) \in \mathbb{R}^d$ and assume that $\boldsymbol{\theta} = g(\boldsymbol{\mu})$ for some smooth function $g : \mathbb{R}^d \to \mathbb{R}^p$. Recall that the plug-in estimate of $\boldsymbol{\mu}$ is $\bar{\mathbf{X}}$, so that the corresponding plug-in estimate of $\boldsymbol{\theta}$ is given by $\hat{\boldsymbol{\theta}} = g(\bar{\mathbf{X}})$. Let $\boldsymbol{\Sigma}$ be the asymptotic covariance matrix of $\hat{\boldsymbol{\theta}}$ defined by

$$
\begin{aligned}
\boldsymbol{\Sigma} &= \lim_{n \to \infty} V(n^{1/2} \hat{\boldsymbol{\theta}}) \\
&= \lim_{n \to \infty} E\{n^{1/2} [\hat{\boldsymbol{\theta}} - E(\hat{\boldsymbol{\theta}})][\hat{\boldsymbol{\theta}} - E(\hat{\boldsymbol{\theta}})]'\}.
\end{aligned}
$$

We assume that $\boldsymbol{\Sigma} = h(\boldsymbol{\mu})$ for some smooth function $h : \mathbb{R}^d \to \mathbb{R}^p \times \mathbb{R}^p$. A plug-in estimate of $\boldsymbol{\Sigma}$ is given by $\hat{\boldsymbol{\Sigma}} = h(\bar{\mathbf{X}})$. It will be assumed that $\boldsymbol{\Sigma}$ is positive definite and that $\hat{\boldsymbol{\Sigma}}$ is positive definite with probability one. Define

$$
G_n(\mathbf{t}) = P[n^{1/2} \boldsymbol{\Sigma}^{-1/2}(\hat{\boldsymbol{\theta}} - \boldsymbol{\theta}) \le \mathbf{t} | \mathbf{X}_1, \dots \mathbf{X}_n \sim F], \tag{3.1}
$$

and

$$
H_n(\mathbf{t}) = P[n^{1/2} \hat{\boldsymbol{\Sigma}}^{-1/2}(\hat{\boldsymbol{\theta}} - \boldsymbol{\theta}) \le \mathbf{t} | \mathbf{X}_1, \dots \mathbf{X}_n \sim F], \tag{3.2}
$$

where $\mathbf{t} \in \mathbb{R}^p$. Assume that G_n and H_n are absolutely continuous and let $g_n(\mathbf{t})$ and $h_n(\mathbf{t})$ be the p-dimensional densities corresponding to $G_n(\mathbf{t})$ and $H_n(\mathbf{t})$, respectively. Define $\mathcal{G}_\alpha$ to be any region of $\mathbb{R}^p$ that satisfies

$$
P[n^{1/2} \boldsymbol{\Sigma}^{-1/2}(\hat{\boldsymbol{\theta}} - \boldsymbol{\theta}) \in \mathcal{G}_\alpha | \mathbf{X}_1, \dots \mathbf{X}_n \sim F] = \alpha,
$$

and similarly define $\mathcal{H}_\alpha$ to be a region of $\mathbb{R}^p$ that satisfies

$$
P[n^{1/2} \hat{\boldsymbol{\Sigma}}^{-1/2}(\hat{\boldsymbol{\theta}} - \boldsymbol{\theta}) \in \mathcal{H}_\alpha | \mathbf{X}_1, \dots, \mathbf{X}_n \sim F] = \alpha.
$$

If $\boldsymbol{\Sigma}$ is known, then a $100\alpha\%$ confidence region for $\boldsymbol{\theta}$ that is a multivariate analog of a confidence interval based on the ordinary critical point in Equation (2.7) is given by

$$
C_{\text{ord}}(\alpha; \mathbf{X}) = \{\hat{\boldsymbol{\theta}} - n^{-1/2} \boldsymbol{\Sigma}^{1/2} \mathbf{t} : \mathbf{t} \in \mathcal{G}_\alpha\}, \tag{3.3}
$$

where $\mathbf{X} = \text{vec}(\mathbf{X}_1, \dots, \mathbf{X}_n)$. The shape of the confidence region is determined by the shape of $\mathcal{G}_\alpha$. Therefore, the shape vector $\boldsymbol{\omega}$ will not be used in this chapter. When $\boldsymbol{\Sigma}$ is unknown, a $100\alpha\%$ confidence region for $\boldsymbol{\theta}$ that is a multivariate analog of a confidence interval based on the studentized critical point given in Equation (2.8) is given by

$$\mathbf{C}_{\text{stud}}(\alpha; \mathbf{X}) = \{\hat{\boldsymbol{\theta}} - n^{-1/2}\hat{\boldsymbol{\Sigma}}^{1/2}\mathbf{t} : \mathbf{t} \in \mathcal{H}_\alpha\}.$$

Note that if the regions $\mathcal{G}_\alpha$ and $\mathcal{H}_\alpha$ are known, then the regions $\mathbf{C}_{\text{ord}}(\alpha; \mathbf{X})$ and $\mathbf{C}_{\text{stud}}(\alpha; \mathbf{X})$ are exact $100\alpha\%$ confidence regions for $\boldsymbol{\theta}$. As in the scalar parameter case, it is not uncommon for the studentized region $\mathcal{H}_\alpha$ to be unknown, but for the ordinary region $\mathcal{G}_\alpha$ to be known. In such a case one might use the approximation $\mathcal{G}_\alpha \simeq \mathcal{H}_\alpha$ for the case when $\boldsymbol{\Sigma}$ is unknown. This results in a multivariate analog of a confidence interval based on the hybrid approximate critical point given in Equation (2.9) given by

$$\mathbf{C}_{\text{hyb}}(\alpha; \mathbf{X}) = \{\hat{\boldsymbol{\theta}} - n^{-1/2}\hat{\boldsymbol{\Sigma}}^{1/2}\mathbf{t} : \mathbf{t} \in \mathcal{G}_\alpha\}.$$

This region is exact only if

$$P[n^{1/2}\hat{\boldsymbol{\Sigma}}^{-1/2}(\hat{\boldsymbol{\theta}} - \boldsymbol{\theta}) \in \mathcal{G}_\alpha | \mathbf{X}_1, \ldots, \mathbf{X}_n \sim F] = \alpha.$$

An analog of percentile intervals will be discussed later in Section 3.3. If both the regions $\mathcal{G}_\alpha$ and $\mathcal{H}_\alpha$ are unknown, then approximate regions can be constructed using the normal approximation. Consider the assumptions of the smooth function model along with the additional assumption that

$$\left. \frac{\partial g_i(\mathbf{t})}{\partial t_j} \right|_{\mathbf{t}=\mu} \neq 0$$

for $i = 1, \ldots, p$ and $j = 1, \ldots, p$, where $g(\mathbf{t}) = [g_1(\mathbf{t}), \ldots, g_p(\mathbf{t})]'$ and $\mathbf{t} = (t_1, \ldots, t_d)' \in \mathbb{R}^d$. Under these assumptions Theorem 3.3.A of Serfling (1980) implies that

$$n^{1/2}\boldsymbol{\Sigma}^{-1/2}(\hat{\boldsymbol{\theta}} - \boldsymbol{\theta}) \xrightarrow{w} N_p(\mathbf{0}, \mathbf{I}), \qquad (3.4)$$

and

$$n^{1/2}\hat{\boldsymbol{\Sigma}}^{-1/2}(\hat{\boldsymbol{\theta}} - \boldsymbol{\theta}) \xrightarrow{w} N_p(\mathbf{0}, \mathbf{I}), \qquad (3.5)$$

as $n \to \infty$, where $N_p(\mathbf{0}, \mathbf{I})$ represents a p-variate normal distribution with mean vector at the origin and identity covariance matrix. Define $\mathcal{N}_\alpha$ to be any region of $\mathbb{R}^p$ that satisfies $P(\mathbf{Z} \in \mathcal{N}_\alpha) = \alpha$, where $\mathbf{Z} \sim N_p(\mathbf{0}, \mathbf{I})$. The weak convergence exhibited in Equations (3.4) and (3.5) suggests that for large n the confidence regions $\mathbf{C}_{\text{ord}}(\alpha; \mathbf{X})$ and $\mathbf{C}_{\text{stud}}(\alpha; \mathbf{X})$ can be approximated with

$$\hat{\mathbf{C}}_{\text{ord}}(\alpha; \mathbf{X}) = \{\hat{\boldsymbol{\theta}} - n^{-1/2}\boldsymbol{\Sigma}^{1/2}\mathbf{t} : \mathbf{t} \in \mathcal{N}_\alpha\},$$

and

$$\hat{\mathbf{C}}_{\text{stud}}(\alpha; \mathbf{X}) = \{\hat{\boldsymbol{\theta}} - n^{-1/2}\hat{\boldsymbol{\Sigma}}^{1/2}\mathbf{t} : \mathbf{t} \in \mathcal{N}_\alpha\},$$

for the cases when $\boldsymbol{\Sigma}$ is known, and unknown, respectively.

To compute an observed confidence level for an arbitrary region $\Psi \subset \Theta$ based on the confidence region $\mathbf{C}_{\text{ord}}(\alpha; \mathbf{X})$ for the case when $\boldsymbol{\Sigma}$ is known the equation $\Psi = \{\hat{\boldsymbol{\theta}} - n^{-1/2}\boldsymbol{\Sigma}^{1/2}\mathbf{t} : \mathbf{t} \in \mathcal{G}_\alpha\}$ is solved for α. Equivalently, we find a value of α such that $n^{1/2}\boldsymbol{\Sigma}^{-1/2}(\hat{\boldsymbol{\theta}} - \boldsymbol{\Psi}) = \mathcal{G}_\alpha$ where $n^{1/2}\boldsymbol{\Sigma}^{-1/2}(\hat{\boldsymbol{\theta}} - \boldsymbol{\Psi})$ denotes the linear transformation of the region $\boldsymbol{\Psi}$, that is

$$n^{1/2}\boldsymbol{\Sigma}^{-1/2}(\hat{\boldsymbol{\theta}} - \boldsymbol{\Psi}) = \{n^{1/2}\boldsymbol{\Sigma}^{-1/2}(\hat{\boldsymbol{\theta}} - \mathbf{t}) : \mathbf{t} \in \boldsymbol{\Psi}\}.$$

Therefore, if $g_n(\mathbf{t})$ is known, then an observed confidence level for $\boldsymbol{\Psi}$ is given by

$$\alpha_{\text{ord}}(\boldsymbol{\Psi}) = \int_{n^{1/2}\boldsymbol{\Sigma}^{-1/2}(\hat{\boldsymbol{\theta}} - \boldsymbol{\Psi})} g_n(\mathbf{t})d\mathbf{t}. \tag{3.6}$$

If $g_n(\mathbf{t})$ is unknown, then the normal approximation can be used to compute

$$\hat{\alpha}_{\text{ord}}(\boldsymbol{\Psi}) = \int_{n^{1/2}\boldsymbol{\Sigma}^{-1/2}(\hat{\boldsymbol{\theta}} - \boldsymbol{\Psi})} \phi_p(\mathbf{t})d\mathbf{t},$$

where $\phi_p(\mathbf{t})$ is the p-variate standard normal density. Similarly, if $\boldsymbol{\Sigma}$ is unknown, then an observed confidence level for $\boldsymbol{\Psi}$ can be computed as

$$\alpha_{\text{stud}}(\boldsymbol{\Psi}) = \int_{n^{1/2}\hat{\boldsymbol{\Sigma}}^{-1/2}(\hat{\boldsymbol{\theta}} - \boldsymbol{\Psi})} h_n(\mathbf{t})d\mathbf{t},$$

when $h_n(\mathbf{t})$ is known. If $h_n(\mathbf{t})$ is unknown but $g_n(\mathbf{t})$ is known, then the observed confidence level for $\boldsymbol{\Psi}$ can be computed using the hybrid region as

$$\alpha_{\text{hyb}}(\boldsymbol{\Psi}) = \int_{n^{1/2}\hat{\boldsymbol{\Sigma}}^{-1/2}(\hat{\boldsymbol{\theta}} - \boldsymbol{\Psi})} g_n(\mathbf{t})d\mathbf{t}.$$

Finally, if both $g_n(\mathbf{t})$ and $h_n(\mathbf{t})$ are unknown, then the normal approximation can be used, which yields

$$\hat{\alpha}_{\text{stud}}(\boldsymbol{\Psi}) = \int_{n^{1/2}\hat{\boldsymbol{\Sigma}}^{-1/2}(\hat{\boldsymbol{\theta}} - \boldsymbol{\Psi})} \phi_p(\mathbf{t})d\mathbf{t}.$$

If $g_n(\mathbf{t})$ and $h_n(\mathbf{t})$ are unknown and the normal approximation is not considered accurate enough, the bootstrap method of Efron (1979) is again suggested as a method to estimate observed confidence levels. This development closely follows that given in Section 2.3. Consider a sample $\mathbf{X}_1^*, \ldots, \mathbf{X}_n^* \sim \hat{F}_n$, conditional on the original sample $\mathbf{X}_1, \ldots, \mathbf{X}_n$. Let $\hat{\boldsymbol{\theta}}^* = g(\bar{\mathbf{X}}^*)$ where

$$\bar{\mathbf{X}}^* = n^{-1} \sum_{i=1}^{n} \mathbf{X}_i^*.$$

The bootstrap estimate of G_n is given by the sampling distribution of $n^{1/2}\hat{\boldsymbol{\Sigma}}^{-1/2}(\hat{\boldsymbol{\theta}}^* - \hat{\boldsymbol{\theta}})$ when $\mathbf{X}_1^*, \ldots, \mathbf{X}_n^* \sim \hat{F}$, conditional of $\mathbf{X}_1, \ldots, \mathbf{X}_n$. Hence

$$\hat{G}_n(\mathbf{t}) = P^*[n^{1/2}\hat{\boldsymbol{\Sigma}}^{-1/2}(\hat{\boldsymbol{\theta}}^* - \hat{\boldsymbol{\theta}}) \le \mathbf{t}|\mathbf{X}_1^*, \ldots, \mathbf{X}_n^* \sim \hat{F}_n],$$

where once again $P^*(A) = P(A|\mathbf{X}_1, \ldots, \mathbf{X}_n)$. Similarly, the bootstrap estimate of H_n is given by

$$\hat{H}_n(\mathbf{t}) = P^*[n^{1/2}\hat{\boldsymbol{\Sigma}}^{*-1/2}(\hat{\boldsymbol{\theta}}^* - \hat{\boldsymbol{\theta}}) \le \mathbf{t}|\mathbf{X}_1^*, \ldots, \mathbf{X}_n^* \sim \hat{F}_n],$$

where $\hat{\boldsymbol{\Sigma}}^* = h(\bar{\mathbf{X}}^*)$. In most cases these estimates will be discrete, and $\hat{g}_n(\mathbf{t})$

and $\hat{h}_n(\mathbf{t})$ will be taken to be the corresponding mass functions of $\hat{G}_n(\mathbf{t})$ and $\hat{H}_n(\mathbf{t})$, respectively. The corresponding bootstrap estimates of $\alpha_{\text{ord}}(\Psi)$, $\alpha_{\text{stud}}(\Psi)$ and $\alpha_{\text{hyb}}(\Psi)$ are

$$\hat{\alpha}^*_{\text{ord}}(\Psi) = \int_{n^{1/2}\boldsymbol{\Sigma}^{-1/2}(\hat{\boldsymbol{\theta}}-\Psi)} \hat{g}_n(\mathbf{t})dt, \qquad (3.7)$$

$$\hat{\alpha}^*_{\text{stud}}(\Psi) = \int_{n^{1/2}\hat{\boldsymbol{\Sigma}}^{-1/2}(\hat{\boldsymbol{\theta}}-\Psi)} \hat{h}_n(\mathbf{t})dt, \qquad (3.8)$$

and

$$\hat{\alpha}^*_{\text{hyb}}(\Psi) = \int_{n^{1/2}\hat{\boldsymbol{\Sigma}}^{-1/2}(\hat{\boldsymbol{\theta}}-\Psi)} \hat{g}_n(\mathbf{t})dt, \qquad (3.9)$$

respectively. The integrals in Equations (3.7)–(3.9) are taken to be Lebesgue-Stieltjes integrals so that the cases when the distributions $\hat{g}_n(\mathbf{t})$ and $\hat{h}_n(\mathbf{t})$ are discrete and continuous can be covered simultaneously. The asymptotic behavior of these estimates is studied in Section 3.3. The estimates $\hat{g}_n(\mathbf{t})$ and $\hat{h}_n(\mathbf{t})$ rarely exist is a closed analytic form, so that the practical computation of these estimates is typically handled using computer-based simulations. These computational issues will be discussed in Section 3.6.

Efron and Tibshirani (1998) provide an additional measure of confidence that is based on the bootstrap percentile method confidence region first introduced by Efron (1979). Let $V_n(\mathbf{t}) = P(\hat{\boldsymbol{\theta}} \leq \mathbf{t}|\mathbf{X}_1,\ldots,\mathbf{X}_n \sim F)$ with density (or mass function) $v_n(\mathbf{t})$. The bootstrap estimate of $V_n(\mathbf{t})$ is $\hat{V}_n(\mathbf{t}) = P^*(\hat{\boldsymbol{\theta}}^* \leq \mathbf{t}|\mathbf{X}_1^*,\ldots,\mathbf{X}_n^* \sim \hat{F}_n)$ with density (or mass function) $\hat{v}_n(\mathbf{t})$. A bootstrap percentile method confidence region for $\boldsymbol{\theta}$ is then given by any region $\mathcal{V}_\alpha$ such that

$$\int_{\mathcal{V}_\alpha} \hat{v}_n(\mathbf{t})dt = \alpha, \qquad (3.10)$$

conditional on $\mathbf{X}_1,\ldots,\mathbf{X}_n$. Setting $\mathcal{V}_\alpha = \Psi$ implies that the observed confidence level for Ψ based on the bootstrap percentile method confidence interval is

$$\hat{\alpha}^*_{\text{perc}}(\Psi) = \int_\Psi \hat{v}_n(\mathbf{t})dt. \qquad (3.11)$$

In application, this method can be taken to be an analog of the percentile method studied in Sections 2.3 and 2.4. The strength of this method lies in its computational simplicity when using resampling algorithms to approximate $\hat{\alpha}^*_{\text{perc}}(\Psi)$. However, as Hall (1987) and Efron and Tibshirani (1998) note, the measure is not as reliable as some other methods. The issue of reliability for all of the methods presented in this section, will be considered in Section 3.3.

3.2.1 Multivariate Mean

Let $\mathbf{Y}_1,\ldots,\mathbf{Y}_n$ be a set of independent and identically distributed random vectors from a d-dimensional distribution F and let $\boldsymbol{\theta} = E(\mathbf{Y}_i)$. To put this problem in the form of the smooth function model, define $\mathbf{X}_i = \text{vec}(\mathbf{Y}_i, \mathbf{Y}_i \mathbf{Y}_i')$, so that $\boldsymbol{\mu} = E(\mathbf{X}_i) = \text{vec}(\boldsymbol{\theta}, \boldsymbol{\Xi})$ where $\boldsymbol{\Xi} = E(\mathbf{Y}_i \mathbf{Y}_i')$. Let $g : \mathbb{R}^{d(d+1)} \to \mathbb{R}^d$ be a function defined as $g(\text{vec}(\mathbf{v}_1, \mathbf{v}_2)) = \mathbf{v}_1$ where $\mathbf{v}_1 \in \mathbb{R}^d$ and $\mathbf{v}_2 \in \mathbb{R}^{d^2}$. Then $g(\text{vec}(\boldsymbol{\theta}, \boldsymbol{\Xi})) = \boldsymbol{\theta}$, the mean vector of $\mathbf{Y}_i$. The corresponding plug-in estimate of $\boldsymbol{\theta}$ is then given by $\hat{\boldsymbol{\theta}} = \bar{\mathbf{Y}}$. Similarly let $h : \mathbb{R}^{d(d+1)} \to \mathbb{R}^d \times \mathbb{R}^d$ be a function defined as $h(\text{vec}(\boldsymbol{\theta}, \boldsymbol{\Xi})) = \boldsymbol{\Xi} - \boldsymbol{\theta}\boldsymbol{\theta}' = \boldsymbol{\Sigma}$, the asymptotic covariance matrix of $\hat{\boldsymbol{\theta}}$. The corresponding plug-in estimate of $\boldsymbol{\Sigma}$ is the usual sample covariance matrix given by

$$\hat{\boldsymbol{\Sigma}} = n^{-1} \sum_{i=1}^{n} (\mathbf{X}_i - \bar{\mathbf{X}})(\mathbf{X}_i - \bar{\mathbf{X}})'.$$

As with the univariate case studied in Section 2.3.1, if F is absolutely continuous then the distribution of $\mathbf{X}_i$ will satisfy the Cramèr continuity condition given in Equation (2.6).

If F is a d-dimensional normal distribution with mean vector $\boldsymbol{\theta}$ and covariance matrix $\boldsymbol{\Sigma}$, then the distribution G_n is an n-dimensional normal distribution with mean vector $\mathbf{0}$ and identity covariance matrix. In the case where the covariance matrix $\boldsymbol{\Sigma}$ is not known, the distribution H_n is rarely considered directly since the distribution of $(\bar{\mathbf{X}} - \boldsymbol{\mu})'\hat{\boldsymbol{\Sigma}}^{-1}(\bar{\mathbf{X}} - \boldsymbol{\mu})$ is related to the univariate F-distribution, and can be used to derive confidence regions for $\boldsymbol{\mu}$. Contours of equal density for the distribution $G_n(\mathbf{t})$ are ellipsoidal regions in $\mathbb{R}^p$, and hence confidence regions for $\boldsymbol{\theta}$ in this case are usually taken to be ellipsoidal regions as well. This allows for the development of confidence regions that generalize the use of equal tail probabilities in the single parameter case. However, as in the one parameter case, this restriction is not necessary for the construction of confidence sets, and confidence regions of any reasonable shape are possible. While this is generally not an important practical factor in the construction of confidence sets in the multiparameter case, it is crucial to the computation of observed confidence levels. For simplicity consider the case where $\boldsymbol{\Sigma}$ is known. From Equation (3.3) it follows that the linear transformation $\hat{\boldsymbol{\theta}} - n^{-1/2}\boldsymbol{\Sigma}^{1/2}\mathbf{N}_\alpha$ is a confidence region for $\boldsymbol{\theta}$ under the assumption that F is a $\mathbf{N}_p(\boldsymbol{\theta}, \boldsymbol{\Sigma})$ distribution. Therefore, for a region $\boldsymbol{\Psi}$, the ordinary observed confidence level is given by Equation (3.6) as

$$\alpha_{\text{ord}}(\boldsymbol{\Psi}) = \int_{n^{1/2}\boldsymbol{\Sigma}^{-1/2}(\hat{\boldsymbol{\theta}} - \boldsymbol{\Psi})} \phi_p(\mathbf{t})dt.$$

A specific example will allow for easier visualization of this situation. Consider the case where $p = 2$, $n = 4$, $\hat{\boldsymbol{\mu}}' = (1, 2)$, and

$$\boldsymbol{\Sigma} = \begin{bmatrix} 1 & \frac{1}{2} \\ \frac{1}{2} & 1 \end{bmatrix},$$

where it is of interest to ascertain the amount of confidence there is that $0 < \mu_1 < \mu_2 < \infty$. That is $\boldsymbol{\Psi} = \{\boldsymbol{\mu}' = (\mu_1, \mu_2) : 0 < \mu_1 < \mu_2 < \infty\}$. This region is plotted in Figure 3.1. The observed confidence level of the region is computed by integrating the standard bivariate normal density over the

Figure 3.1 *The multivariate mean example. The region* Ψ *is indicated by the grey shaded area and the observed mean is indicated by the* $+$.

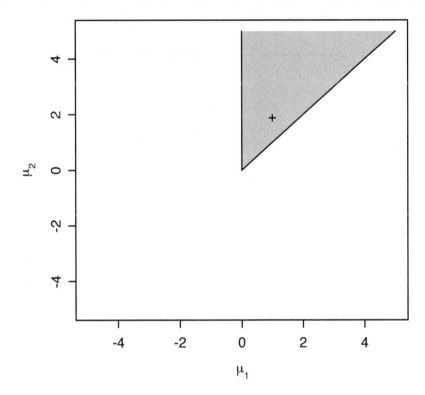

linear transformation $2\Sigma^{-1/2}[(1,2)' - \Psi]$. This transformed region is plotted in Figure 3.2. The observed confidence level is then computed by integrating the shaded region in Figure 3.2 over the density $\phi_p(\mathbf{t})$. This yields an observed confidence level equal to approximately 95.51%. Note that the region Ψ does not have to be linear. For example, consider the region $\Psi = \{\boldsymbol{\mu}' = (\mu_1, \mu_2) : \mu_1^2 + \mu_2^2 \leq 9\}$. This region is plotted in Figure 3.3, and the linear transformation $2\Sigma^{-1/2}[(1,2)' - \Psi]$ is plotted in Figure 3.4. The observed confidence level is then computed by integrating the density $\phi_p(\mathbf{t})$ over the region in Figure 3.4. This results in an observed confidence level equal to approximately 89.33%.

3.3 Asymptotic Accuracy

This section will use the same measure of asymptotic accuracy that is developed for the scalar parameter case in Section 2.4, with the exception that the confidence sets discussed in Section 3.2 will be used in place of the

Figure 3.2 *The multivariate mean example. The linear transformation of the region* $\boldsymbol{\Psi}$ *given by* $2\boldsymbol{\Sigma}^{-1/2}[(1,2)' - \boldsymbol{\Psi}]$ *is indicated by the grey shaded area and the corresponding linear transformation of the observed mean is indicated by the* $+$. *For comparison, the dotted line contains a standard bivariate normal random vector with a probability of 0.99.*

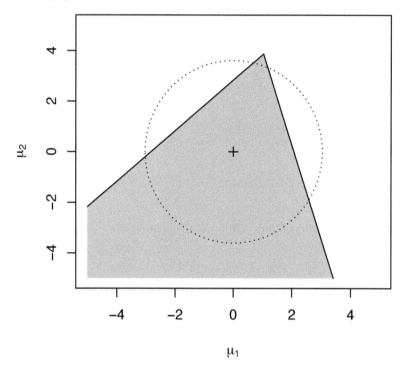

standard confidence intervals used in Chapter 2. Therefore, when the covariance matrix $\boldsymbol{\Sigma}$ is known, the confidence region $\mathbf{C}_{\mathrm{ord}}(\alpha; \mathbf{X})$ will be used as the standard confidence region for $\boldsymbol{\theta}$, and a confidence measure $\tilde{\alpha}$ is accurate if $\tilde{\alpha}[\mathbf{C}_{\mathrm{ord}}(\alpha; \mathbf{X})] = \alpha$. If $\tilde{\alpha}[\mathbf{C}_{\mathrm{ord}}(\alpha; \mathbf{X})] \neq \alpha$, but $\tilde{\alpha}[\mathbf{C}_{\mathrm{ord}}(\alpha; \mathbf{X})] = \alpha + O(n^{-k/2})$, for a positive integer k, as $n \to \infty$, then the measure is k^{th}-order accurate. For the case when $\boldsymbol{\Sigma}$ is unknown, the confidence region $\mathbf{C}_{\mathrm{stud}}(\alpha; \mathbf{X})$ will be used as the standard confidence region for $\boldsymbol{\theta}$ and a confidence measure $\tilde{\alpha}$ will be accurate if $\tilde{\alpha}[\mathbf{C}_{\mathrm{stud}}(\alpha; \mathbf{X})] = \alpha$, with a similar definition for k^{th}-order accuracy as given above. From these definitions it is clear that α_{ord} and α_{stud} are accurate.

As with the case of scalar parameters the asymptotic analysis of kth-order accurate models is usually based on the Edgeworth expansion theory of Bhattacharya and Ghosh (1978) and the related theory for the bootstrap outlined in Hall (1988, 1992a). In particular, the arguments in this section will closely follow the multivariate Edgeworth expansion theory used in Hall (1992a, Chap-

Figure 3.3 *The multivariate mean example. The nonlinear region* Ψ *is indicated by the grey shaded area and the observed mean is indicated by the* +.

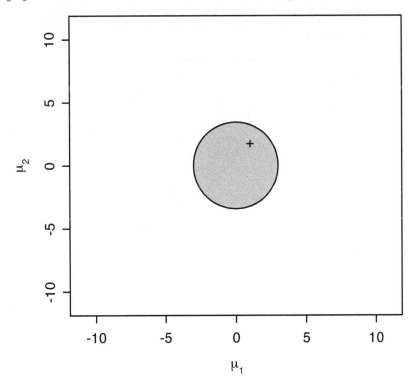

ter 4). Further information on multivariate Edgeworth expansions can also be found in Barndorff-Nielsen and Cox (1989). Under suitable conditions on F and θ and the smooth function model described in Section 3.2, if $\mathcal{S} \subset \mathbb{R}^p$ is a finite union of convex sets, then

$$
\begin{aligned}
P[n^{1/2}\boldsymbol{\Sigma}^{-1/2}(\hat{\boldsymbol{\theta}} - \boldsymbol{\theta}) \in \mathcal{S}] &= \int_{\mathcal{S}} g_n(\mathbf{t})d\mathbf{t} \\
&= \int_{\mathcal{S}} \left[1 + \sum_{i=1}^{k} n^{-i/2} r_i(\mathbf{t}) \right] \phi_p(\mathbf{t})d\mathbf{t} \\
&\quad + o(n^{-k/2}),
\end{aligned}
\tag{3.12}
$$

Figure 3.4 *The multivariate mean example. The linear transformation of the nonlinear region* Ψ *given by* $2\boldsymbol{\Sigma}^{-1/2}[(1,2)' - \Psi]$ *is indicated by the grey shaded area and the corresponding linear transformation of the observed mean is indicated by the* +. *For comparison, the dotted line contains a standard bivariate normal random vector with a probability of 0.99.*

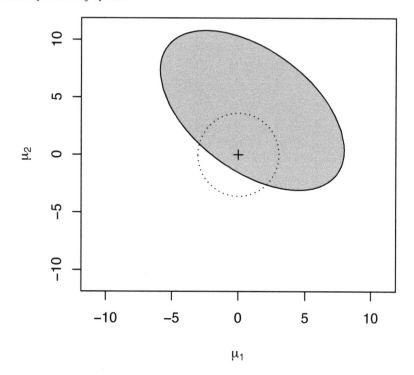

and

$$P[n^{1/2}\hat{\boldsymbol{\Sigma}}^{-1/2}(\hat{\boldsymbol{\theta}} - \boldsymbol{\theta}) \in \mathcal{S}] = \int_{\mathcal{S}} h_n(\mathbf{t}) d\mathbf{t}$$

$$= \int_{\mathcal{S}} \left[1 + \sum_{i=1}^{k} n^{-i/2} s_i(\mathbf{t})\right] \phi_p(\mathbf{t}) d\mathbf{t}$$

$$+ o(n^{-k/2}). \tag{3.13}$$

The asymptotic expansions given in Equations (3.12) and (3.13) are known as multivariate Edgeworth expansions. The functions r_i and s_i are polynomials in the components of $\mathbf{t}$ of degree $3i$ with coefficients that depend on the moments of F. The functions are even (odd) when i is even (odd). To simplify the notation, let

$$\Delta_i(\mathcal{S}) = \int_{\mathcal{S}} r_i(\mathbf{t})\phi_p(\mathbf{t}) d\mathbf{t},$$

and

$$\Lambda_i(\mathcal{S}) = \int_{\mathcal{S}} s_i(\mathbf{t})\phi_p(\mathbf{t})dt,$$

for $i \in \{0, 1, 2, \ldots\}$, where $r_0(\mathbf{t}) = s_0(\mathbf{t}) = 1$ for all $\mathbf{t} \in \mathbb{R}^p$. From Equation (3.12) it is clear that if $n^{1/2}\boldsymbol{\Sigma}^{-1/2}[\hat{\boldsymbol{\theta}} - \mathbf{C}_{\mathrm{ord}}(\alpha; \mathbf{X})]$ is a finite union of convex sets, then

$$\Delta_0\{n^{1/2}\boldsymbol{\Sigma}^{-1/2}[\hat{\boldsymbol{\theta}} - \mathbf{C}_{\mathrm{ord}}(\alpha; \mathbf{X})]\}+$$
$$n^{-1/2}\Delta_1\{n^{1/2}\boldsymbol{\Sigma}^{-1/2}[\hat{\boldsymbol{\theta}} - \mathbf{C}_{\mathrm{ord}}(\alpha; \mathbf{X})]\} = \alpha + o(n^{-1/2}). \quad (3.14)$$

Since $\Delta_0\{n^{1/2}\boldsymbol{\Sigma}^{-1/2}[\hat{\boldsymbol{\theta}} - \mathbf{C}_{\mathrm{ord}}(\alpha; \mathbf{X})]\} = \hat{\alpha}_{\mathrm{ord}}[\mathbf{C}_{\mathrm{ord}}(\alpha; \mathbf{X})]$ it follows that

$$\hat{\alpha}_{\mathrm{ord}}[\mathbf{C}_{\mathrm{ord}}(\alpha; \mathbf{X})] = \alpha + O(n^{-1/2}),$$

and therefore $\hat{\alpha}_{\mathrm{ord}}$ is first-order accurate for the case when $\boldsymbol{\Sigma}$ is known. A similar argument shows that $\hat{\alpha}_{\mathrm{stud}}$ is first-order accurate for the case when $\boldsymbol{\Sigma}$ is unknown. To analyze α_{hyb}, note that

$$\alpha_{\mathrm{hyb}}[\mathbf{C}_{\mathrm{stud}}(\alpha; \mathbf{X})] = \int_{\mathcal{H}_\alpha} g_n(\mathbf{t})dt$$
$$= \Delta_0(\mathcal{H}_\alpha) + n^{-1/2}\Delta_1(\mathcal{H}_\alpha) + O(n^{-1}). \quad (3.15)$$

Equation (3.13) implies that

$$\Delta_0(\mathcal{H}_\alpha) = \Lambda_0(\mathcal{H}_\alpha) = \alpha - n^{-1/2}\Lambda_1(\mathcal{H}_\alpha) + O(n^{-1}). \quad (3.16)$$

Substituting the result from Equation (3.16) into Equation (3.15) yields

$$\alpha_{\mathrm{hyb}}[\mathbf{C}_{\mathrm{stud}}(\alpha; \mathbf{X})] = \alpha + n^{-1/2}[\Delta_1(\mathcal{H}_\alpha) - \Lambda_1(\mathcal{H}_\alpha)] + O(n^{-1}), \quad (3.17)$$

which implies that α_{hyb} is first-order accurate. In some special cases it is possible that the first-order term is 0. For example, since r_1 and s_1 are odd functions, it follows that $\Delta_1(\mathcal{H}_\alpha) = \Lambda_1(\mathcal{H}_\alpha) = 0$ whenever $\mathcal{H}_\alpha$ is a sphere centered at the origin. See Equation (4.6) of Hall (1992a). In general, the first order term is nonzero.

To analyze the observed confidence levels based on bootstrap confidence regions, Edgeworth expansions are required for the integrals of the bootstrap estimates of $g_n(\mathbf{t})$ and $h_n(\mathbf{t})$. As with univariate Edgeworth expansions, Hall (1992a) argues that under sufficient conditions of $\boldsymbol{\theta}$ and F, it follows that if $\mathcal{S} \subset \mathbb{R}^p$ is a finite union of convex sets, then

$$P^*[n^{1/2}\hat{\boldsymbol{\Sigma}}^{-1/2}(\hat{\boldsymbol{\theta}}^* - \hat{\boldsymbol{\theta}}) \in \mathcal{S}|\mathbf{X}_1^*, \ldots, \mathbf{X}_n^* \sim \hat{F}_n] =$$
$$\int_{\mathcal{S}} \hat{g}_n(\mathbf{t})dt = \int_{\mathcal{S}} \left[1 + \sum_{i=1}^{k} n^{-i/2}\hat{r}_i(\mathbf{t})\right]\phi_p(\mathbf{t})dt + o_p(n^{-k/2}), \quad (3.18)$$

and

$$P^*[n^{1/2}\hat{\Sigma}^{*-1/2}(\hat{\theta}^* - \hat{\theta}) \in \mathcal{S}|\mathbf{X}_1^*, \ldots, \mathbf{X}_n^* \sim \hat{F}_n] =$$

$$\int_{\mathcal{S}} \hat{h}_n(\mathbf{t})d\mathbf{t} = \int_{\mathcal{S}} \left[1 + \sum_{i=1}^{k} n^{-i/2}\hat{s}_i(\mathbf{t})\right] \phi_p(\mathbf{t})d\mathbf{t} + o_p(n^{-k/2}), \quad (3.19)$$

The functions $\hat{r}_i$ and $\hat{s}_i$ are the same polynomial functions as r_i and s_i, respectively, with the population moments replaced by the corresponding sample moments. Theorems 2.2.B and 3.3.A of Serfling (1980) imply that $\hat{r}_i(\mathbf{t}) = r_i(\mathbf{t}) + O_p(n^{-1/2})$ and $\hat{s}_i(\mathbf{t}) = s_i(\mathbf{t}) + O_p(n^{-1/2})$ for all $\mathbf{t} \in \mathbb{R}^p$. Therefore, it follows from Equations (3.18) and (3.19) that

$$P^*[n^{1/2}\hat{\Sigma}^{-1/2}(\hat{\theta}^* - \hat{\theta}) \in \mathcal{S}|\mathbf{X}_1^*, \ldots, \mathbf{X}_n^* \sim \hat{F}_n] =$$

$$\Delta_0(\mathcal{S}) + n^{-1/2}\Delta_1(\mathcal{S}) + O_p(n^{-1}), \quad (3.20)$$

and

$$P^*[n^{1/2}\hat{\Sigma}^{*-1/2}(\hat{\theta}^* - \hat{\theta}) \in \mathcal{S}|\mathbf{X}_1^*, \ldots, \mathbf{X}_n^* \sim \hat{F}_n] =$$

$$\Lambda_0(\mathcal{S}) + n^{-1/2}\Lambda_1(\mathcal{S}) + O_p(n^{-1}). \quad (3.21)$$

Suppose Σ is known. It follows from Equation (3.20) that

$$\hat{\alpha}_{\text{ord}}^*[\mathbf{C}_{\text{ord}}(\alpha; \mathbf{X})] = \int_{\mathcal{G}_\alpha} \hat{g}_n(\mathbf{t})d\mathbf{t} = \Delta_0(\mathcal{G}_\alpha) + n^{-1/2}\Delta_1(\mathcal{G}_\alpha) + O_p(n^{-1}).$$

Equation (3.13) implies that $\Delta_0(\mathcal{G}_\alpha) = \alpha - n^{-1/2}\Delta_1(\mathcal{G}_\alpha) + O(n^{-1})$. It follows that $\hat{\alpha}_{\text{ord}}^*[\mathbf{C}_{\text{ord}}(\alpha; \mathbf{X})] = \alpha + O_p(n^{-1})$, and therefore $\hat{\alpha}_{\text{ord}}^*$ is second-order accurate in probability. A similar argument can be used to establish that $\hat{\alpha}_{\text{stud}}^*$ is second-order accurate in probability when Σ is unknown, using the expansion in Equation (3.21). For the hybrid bootstrap estimate, we have from Equation (3.20) that

$$\hat{\alpha}_{\text{hyb}}^*[\mathbf{C}_{\text{stud}}(\alpha; \mathbf{X})] = \int_{\mathcal{H}_\alpha} \hat{g}_n(\mathbf{t})d\mathbf{t}$$

$$= \Delta_0(\mathcal{H}_\alpha) + n^{-1/2}\Delta_1(\mathcal{H}_\alpha) + O_p(n^{-1}),$$

whose first two terms match those in the expansion in Equation (3.15). It follows that $\hat{\alpha}_{\text{hyb}}^*$ is first-order accurate in probability, matching the accuracy of α_{hyb}. To analyze the observed confidence level based on the percentile method bootstrap confidence region, note that

$$\hat{\alpha}_{\text{perc}}^*(\Psi) = P^*(\hat{\theta}^* \in \Psi|\mathbf{X}_1^*, \ldots, \mathbf{X}_n^* \sim \hat{F}_n)$$

$$= P^*[n^{1/2}\hat{\Sigma}^{-1/2}(\hat{\theta}^* - \hat{\theta}) \in n^{1/2}\hat{\Sigma}^{-1/2}(\Psi - \hat{\theta})|\mathbf{X}_1^*, \ldots, \mathbf{X}_n^* \sim \hat{F}_n]$$

$$= \int_{n^{1/2}\hat{\Sigma}^{-1/2}(\Psi - \hat{\theta})} \hat{g}_n(\mathbf{t})d\mathbf{t}.$$

Therefore, Equation (3.20) implies that

$$\hat{\alpha}^*_{\text{perc}}[\mathbf{C}_{\text{stud}}(\alpha; \mathbf{X})] = \int_{-\mathcal{H}_\alpha} \hat{g}_n(\mathbf{t})d\mathbf{t}$$

$$= \Delta_0(-\mathcal{H}_\alpha) + n^{1/2}\Delta_1(-\mathcal{H}_\alpha) + O_p(n^{-1}).$$

Now

$$\Delta_0(-\mathcal{H}_\alpha) = \int_{-\mathcal{H}_\alpha} \phi_p(\mathbf{t})d\mathbf{t} = \int_{\mathcal{H}_\alpha} \phi_p(\mathbf{t})d\mathbf{t} = \Delta_0(\mathcal{H}_\alpha),$$

since $\phi_p(\mathbf{t})$ is a spherically symmetric density. Because $r_1(\mathbf{t})$ is an odd function and $\phi_p(\mathbf{t})$ is an even function, it follows that

$$\Delta_1(-\mathcal{H}_\alpha) = \int_{-\mathcal{H}_\alpha} r_1(\mathbf{t})\phi_p(\mathbf{t})d\mathbf{t} = -\int_{\mathcal{H}_\alpha} r_1(\mathbf{t})\phi_p(\mathbf{t})d\mathbf{t} = -\Delta_1(\mathcal{H}_\alpha)$$

Combining these results with Equation (3.20) yields

$$\hat{\alpha}^*_{\text{perc}}[\mathbf{C}(\alpha; \mathbf{X})] = \alpha - n^{-1/2}[\Delta_1(\mathcal{H}_\alpha) + \Lambda_1(\mathcal{H}_\alpha)] + O_p(n^{-1}), \qquad (3.22)$$

which implies that $\hat{\alpha}^*_{\text{perc}}$ is first-order accurate in probability. From these results it is clear that from an asymptotic viewpoint, when g_n and h_n are unknown, the bootstrap estimates $\hat{\alpha}^*_{\text{ord}}$ and $\hat{\alpha}^*_{\text{stud}}$ are preferred when Σ is known, and unknown, respectively.

3.4 Empirical Comparisons

The asymptotic comparisons of Section 3.3 reveal a clear preference for the bootstrap estimates of the studentized method for computing observed confidence levels when g_n, h_n, and Σ are unknown. This section will investigate the finite sample accuracy of these methods empirically using computer-based simulations. The two parameter case will be the focus of the investigation. In particular, we will consider $\boldsymbol{\theta}$ to be a bivariate mean vector and the region of interest to be the triangular region

$$\Psi = \{\boldsymbol{\theta}' = (\theta_1, \theta_2) : 0 \leq \theta_1 \leq \theta_2 < \infty\}.$$

Four choices for the parent distribution F will be considered. The first is a bivariate normal density with mean vector $\boldsymbol{\theta}' = (0.5, 1.0)$ and identity covariance matrix. The second distribution is also a bivariate normal density with mean vector $\boldsymbol{\theta}' = (0.5, 1.0)$ and unit variances, but the covariance between the random variables is 0.5. The third and fourth distributions considered are the skewed and kurtotic bivariate normal mixtures from Wand and Jones (1993). Contour plots of these densities, along with the region Ψ, are presented in Figure 3.5.

The empirical study consisted of generating 100 samples of size 10, 25 and 50 from each of the four densities and computing the observed confidence levels α_{stud}, $\hat{\alpha}_{\text{stud}}$, $\hat{\alpha}^*_{\text{stud}}$, $\hat{\alpha}^*_{\text{hyb}}$, and $\hat{\alpha}^*_{\text{perc}}$ on each sample. The average observed

Figure 3.5 *Contour plots of the four bivariate densities used in the empirical study: (a) Independent Bivariate Normal, (b) Dependent Bivariate Normal, (c) Skewed Bivariate, and (d) Kurtotic Bivariate. The dashed lines indicate the boundary of the region* Θ_0 *used in the study. The point marked with* + *indicates the location of the mean vector* $\boldsymbol{\theta}' = (0.5, 1.0)$.

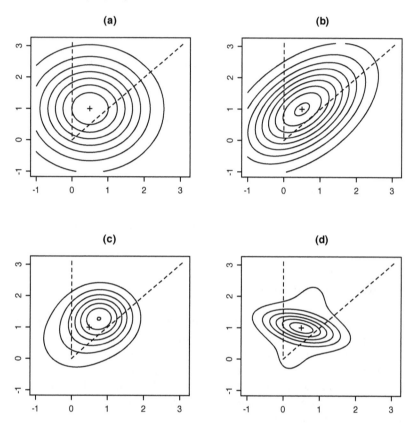

confidence level over the 100 samples for each method was then computed. To compute the confidence level α_{stud}, the density $h_n(\mathbf{t})$ was approximated using simulation when the multivariate normal theory was not applicable. The bootstrap estimates all used $b = 1000$ resamples. Section 3.6 contains information on computing observed confidence levels for multivariate regions using R.

The results of the study are given in Table 3.1. The normal and bootstrap methods are compared to α_{stud}, which was shown to be accurate in Section 3.3. The bootstrap studentized method $\hat{\alpha}^*_{\text{stud}}$ has an average observed confidence level closest to the average level for the correct method in all of the cases studied. The $\hat{\alpha}_{\text{stud}}$, $\hat{\alpha}^*_{\text{hyb}}$, and $\hat{\alpha}^*_{\text{perc}}$ methods tended to overstate the confidence

Table 3.1 *Average observed confidence levels (in percent) from 100 simulated samples for the indicated distributions and sample sizes. The measure whose average level is closest to the average level of α_{stud} is italicized.*

Distribution	n	α_{stud}	$\hat{\alpha}_{\text{stud}}$	$\hat{\alpha}^*_{\text{stud}}$	$\hat{\alpha}^*_{\text{hyb}}$	$\hat{\alpha}^*_{\text{perc}}$
Independent Normal	10	64.12	67.85	*62.88*	68.98	68.86
	25	84.50	85.64	*84.45*	86.13	86.11
	50	94.69	94.99	*94.69*	95.24	95.24
Dependent Normal	10	71.33	74.40	*69.87*	75.53	75.54
	25	90.42	91.42	*90.23*	91.82	91.70
	50	98.48	98.73	*98.41*	98.75	98.73
Skewed	10	74.15	80.25	*72.79*	80.93	81.17
	25	92.06	93.54	*91.85*	93.75	93.66
	50	98.24	98.47	*98.10*	98.53	98.42
Kurtotic	10	62.88	65.78	*61.44*	66.92	66.28
	25	82.53	83.54	*82.18*	83.73	83.56
	50	95.38	95.58	*95.28*	95.63	95.63

level on average, while $\hat{\alpha}^*_{\text{stud}}$ understated the confidence level, on average. It appears that the bootstrap studentized method does generally provide the most accurate measure of observed confidence. However, all of the methods provided a reasonable measure of observed confidence, particularly when the sample size is moderately large. As a final note, it can be observed from Table 3.1 that all of the methods appear to be consistent, as all of the average observed confidence levels are converging to one as the sample size increases.

3.5 Examples

3.5.1 A Problem from Biostatistics

Consider the profile analysis problem presented in Section 1.3.3. For the first part of this example, represent the set of four measurements from the ith male subject in the study is represented as $\mathbf{X}_i$, a random vector in $\mathbb{R}^4$. Assuming that $\mathbf{X}_1, \ldots, \mathbf{X}_{16}$ is a set of independent and identically distributed random variables from a distribution F with mean vector $\boldsymbol{\theta} = E(\mathbf{X}_i)$, then the problem easily falls within the smooth function model of Section 3.2 using the development from the multivariate mean example in Section 3.2.1. Without setting a specific parametric model for the trend in the mean vector, the focus of this first investigation will be on the orderings of the components of $\boldsymbol{\theta}$.

Table 3.2 *Observed confidence levels (in percent) for the trend in the mean vector for the dental data of Potthoff and Roy (1964) by gender. Only regions with non-zero observed confidence levels are included in the table.*

Region	Ordering	Male Subjects			Female Subjects		
		$\hat{\alpha}^*_{stud}$	$\hat{\alpha}^*_{hyb}$	$\hat{\alpha}^*_{perc}$	$\hat{\alpha}^*_{stud}$	$\hat{\alpha}^*_{hyb}$	$\hat{\alpha}^*_{perc}$
Θ_1	$\theta_1 \leq \theta_2 \leq \theta_3 \leq \theta_4$	78.6	92.8	94.9	86.7	99.3	99.9
Θ_2	$\theta_1 \leq \theta_2 \leq \theta_4 \leq \theta_3$	2.9	0.0	0.0	4.2	0.0	0.0
Θ_3	$\theta_1 \leq \theta_3 \leq \theta_4 \leq \theta_2$	3.7	0.5	0.0	1.8	0.4	0.0
Θ_4	$\theta_1 \leq \theta_3 \leq \theta_4 \leq \theta_2$	0.2	0.0	0.0	0.1	0.0	0.0
Θ_5	$\theta_1 \leq \theta_4 \leq \theta_2 \leq \theta_3$	0.1	0.0	0.0	0.3	0.0	0.0
Θ_7	$\theta_2 \leq \theta_1 \leq \theta_3 \leq \theta_4$	9.8	6.7	4.9	3.9	0.3	0.1
Θ_8	$\theta_2 \leq \theta_1 \leq \theta_4 \leq \theta_3$	3.7	0.0	0.2	0.9	0.0	0.0
Θ_9	$\theta_2 \leq \theta_3 \leq \theta_1 \leq \theta_4$	0.2	0.0	0.0	1.1	0.0	0.0
Θ_{11}	$\theta_2 \leq \theta_4 \leq \theta_1 \leq \theta_3$	0.6	0.0	0.0	0.2	0.0	0.0
Θ_{12}	$\theta_2 \leq \theta_4 \leq \theta_3 \leq \theta_1$	0.1	0.0	0.0	0.0	0.0	0.0
Θ_{14}	$\theta_3 \leq \theta_1 \leq \theta_4 \leq \theta_2$	0.1	0.0	0.0	0.1	0.0	0.0
Θ_{15}	$\theta_3 \leq \theta_2 \leq \theta_1 \leq \theta_4$	0.0	0.0	0.0	0.2	0.0	0.0
Θ_{16}	$\theta_3 \leq \theta_2 \leq \theta_4 \leq \theta_1$	0.0	0.0	0.0	0.1	0.0	0.0
Θ_{18}	$\theta_3 \leq \theta_4 \leq \theta_2 \leq \theta_1$	0.0	0.0	0.0	0.1	0.0	0.0
Θ_{19}	$\theta_4 \leq \theta_1 \leq \theta_2 \leq \theta_3$	0.0	0.0	0.0	0.1	0.0	0.0
Θ_{23}	$\theta_4 \leq \theta_3 \leq \theta_1 \leq \theta_2$	0.0	0.0	0.0	0.1	0.0	0.0
Θ_{24}	$\theta_4 \leq \theta_3 \leq \theta_2 \leq \theta_1$	0.0	0.0	0.0	0.1	0.0	0.0

In particular the region where the trend is monotonically nondecreasing will be represented as $\Theta_1 = \{\boldsymbol{\theta} : \theta_1 \leq \theta_2 \leq \theta_3 \leq \theta_4\}$. The remaining orderings of the mean components will be represented s regions $\Theta_2, \ldots, \Theta_{24}$. These regions will only be specifically defined as they appear in this analysis. The observed confidence level measures $\hat{\alpha}^*_{stud}$, $\hat{\alpha}^*_{hyb}$ and $\hat{\alpha}^*_{perc}$ where computed on each of the regions Θ_1-Θ_{24} using $b = 1000$ resamples. Methods for computing these levels are discussed in Section 3.6. The entire set of calculations was repeated for the female subjects in the study as well. The results of these calculations are given in Table 3.2.

It is clear from Table 3.2 that all of the methods show a clear preference for the ordering $\Theta_1 = \{\boldsymbol{\theta} : \theta_1 \leq \theta_2 \leq \theta_3 \leq \theta_4\}$ for both genders, though the observed confidence level given by the studentized method is substantially lower than that given by the hybrid and percentile methods. The studentized method is preferred from an asymptotic viewpoint, and in the cases studied in the empirical study of Section 3.4. It was also observed in the empirical study that the hybrid and percentile methods tended to overstate the confidence level for a similarly shaped region. However, as reported in Section 3.4, the difference between the methods was generally slight for the cases studied there.

The difference could also be due to the added variability of estimating the six parameters in the covariance matrix based on a sample of size $n = 16$, whereas the empirical study in Section 3.4 only required the estimation of three parameters in the covariance matrix, with a minimum sample size of $n = 10$. These examples may indicate that the studentized method should be used for moderate sample sizes, or with some additional stabilization or smoothing technique. The methods of Polansky (2000), for example, could be extended from the univariate case for this purpose.

As indicated in Section 1.3.3, another problem of interest with this data is to compare the two mean growth curves. For example, to compute the observed confidence level that the mean growth curve for females is always less that the mean growth curve for males. The general setup for this problem can once again be studied within the framework of the smooth function model as indicated above. In this case the parameter vector is expanded to $\boldsymbol{\theta}' = (\theta_1 \cdots \theta_8)$ where the first four components represent the mean values for the four measurements of the male subjects, and the second four components represent the mean values of the four measurements of the female subjects. The remaining representation is similar to discussed above, except that the male and female subjects can be assumed to be independent of one another so that the covariance matrix would be a block-diagonal matrix. The inequality of the sample sizes for the two groups is also being tacitly ignored in this representation, though the effect of the asymptotic theory is minimal as long as neither sample size dominates in the limit.

The main region of interest for this example is that the mean male response at each observed time point is exceeds the female response. That is $\Theta_1 = \{\boldsymbol{\theta} : \theta_1 \geq \theta_5, \theta_2 \geq \theta_6, \theta_3 \geq \theta_7, \theta_4 \geq \theta_8\}$. Regions corresponding to crossings at each time point will also be considered. These regions are given by $\Theta_2 = \{\boldsymbol{\theta} : \theta_1 \leq \theta_5, \theta_2 \geq \theta_6, \theta_3 \geq \theta_7, \theta_4 \geq \theta_8\}$, $\Theta_3 = \{\boldsymbol{\theta} : \theta_1 \leq \theta_5, \theta_2 \leq \theta_6, \theta_3 \geq \theta_7, \theta_4 \geq \theta_8\}$, $\Theta_4 = \{\boldsymbol{\theta} : \theta_1 \leq \theta_5, \theta_2 \leq \theta_6, \theta_3 \leq \theta_7, \theta_4 \geq \theta_8\}$. Finally, the region Θ_5 represents the female response at each observed time point exceeds the male response. This region is given by $\Theta_5 = \{\boldsymbol{\theta} : \theta_1 \leq \theta_5, \theta_2 \leq \theta_6, \theta_3 \leq \theta_7, \theta_4 \leq \theta_8\}$.

The observed confidence level measures $\hat{\alpha}^*_{\text{stud}}$, $\hat{\alpha}^*_{\text{hyb}}$, and $\hat{\alpha}^*_{\text{perc}}$ where computed on each of the regions Θ_1-Θ_5 using $b = 1000$ resamples. For the studentized observed confidence level the studentized vector was computed for each gender separately. The results of these calculations are given in Table 3.3. It is clear from the table that there is a great degree of confidence that the mean male responses exceed the female mean responses at each observed time point. As with the calculations displayed above, the studentized levels are smaller than the hybrid and percentile observed confidence levels, with the region with the next greatest confidence being the one that models are crossing between the first and second observations.

Table 3.3 *Observed confidence levels (in percent) for the relationship between the mean function for the male and female subjects in the data of Potthoff and Roy (1964).*

Region	$\hat{\alpha}^*_{\text{stud}}$	$\hat{\alpha}^*_{\text{hyb}}$	$\hat{\alpha}^*_{\text{perc}}$
Θ_1	78.7	96.3	96.9
Θ_2	8.1	1.6	1.6
Θ_3	2.5	0.6	0.1
Θ_4	1.1	0.1	0.0
Θ_5	2.1	0.0	0.0

3.5.2 A Problem in Statistical Quality Control

Consider the quality assurance problem described in Section 1.3.2. Recall that the capability of a process whose quality characteristic X has mean μ and variance σ^2 was defined using the C_{pk} process capability index as

$$C_{pk} = \min\left\{\frac{U-\mu}{3\sigma}, \frac{\mu-L}{3\sigma}\right\},$$

where U and L are the upper and lower specification limits of the quality characteristic, respectively. While the C_{pk} index can be written as a function of a vector mean, the form of the index implies that the parameter of interest in this case does not fit fully within the framework of the smooth function model described in Section 3.2. This is due to the fact that the derivative of the C_{pk} index with respect to μ does not exist when $\mu = \frac{1}{2}(L+U)$. A plot of the value of the C_{pk} index as a function of μ for the special case when $\sigma = 1$, $L = 0$ and $U = 1$ is given in Figure 3.6, where the problem is easily visualized. This does not imply that observed confidence levels cannot, or should not, be computed in this case. However, it does mean that the asymptotic comparisons of Section 3.3 may not be completely valid for this problem when $\mu = \frac{1}{2}(U+L)$, and care must be taken in concluding that one level may be more accurate than another. The general intuitive idea that an observed confidence level that is derived from an accurate confidence interval should provide a reasonable method will be the guiding principle in this case.

To formalize the problem, let μ_i and σ_i be the mean and standard deviation of the quality characteristic for the ith supplier, where $i = 1, \ldots, 4$. The parameter vector of interest is

$$\boldsymbol{\theta}' = [C_{pk(1)}, C_{pk(2)}, C_{pk(3)}, C_{pk(4)}],$$

where

$$C_{pk(i)} = \min\left\{\frac{U-\mu_i}{3\sigma_i}, \frac{\mu_i-L}{3\sigma_i}\right\},$$

Figure 3.6 *The C_{pk} process capability index as a function of μ when $\sigma = 1$, $L = 0$ and $U = 1$.*

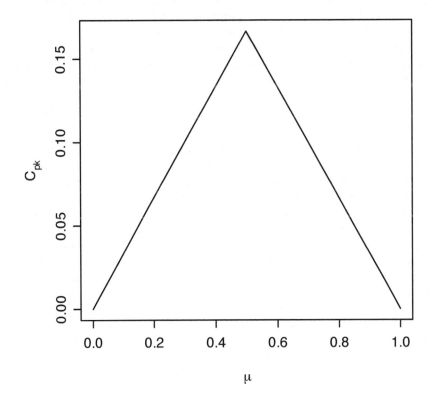

for $i = 1,\ldots,4$. Denote the sample of the quality characteristic from the manufacturing process of the ith supplier as $X_{i1},\ldots,X_{in_i}$, where n_i is the sample size from the ith supplier. The estimate C_{pk} index for Supplier i is then given by

$$\hat{C}_{pk(i)} = \min \left\{ \frac{U - \hat{\mu}_i}{3\hat{\sigma}_i}, \frac{\hat{\mu}_i - L}{3\hat{\sigma}_i} \right\},$$

where $\hat{\mu}_i$ and $\hat{\sigma}_i$ are the usual sample mean and standard deviation computed on $X_{i1},\ldots,X_{in_i}$. The main issue of interest in this example is to use the observed sample from each supplier to decide which supplier has the most capable process, which equivalently consists of deciding which supplier has the manufacturing process with the largest capability index. That is, we are interested in regions of the form

$$\Theta_i = \{\boldsymbol{\theta} : C_{pk(i)} \geq C_{pk(j)} \text{ for } j = 1,\ldots,4\}, \tag{3.23}$$

for $i = 1,\ldots,4$.

Exact confidence intervals for the C_{pk} process capability index are difficult to obtain and have no closed form even in the case when the process has a quality characteristic that follows a normal distribution. See Chou, Owen, and Borrego (1990). Quite accurate approximate confidence intervals do exist for normally distributed quality characteristics. See, for example, Franklin and Wasserman (1992a) and Kushler and Hurley (1991). Using the assumption of independence between the suppliers, it is relatively simple to construct a confidence region for a vector of the four C_{pk} indices corresponding to the four suppliers. Unfortunately, such a region will always be rectangular and would unsuitable for computing and observed confidence level of the regions of interest.

Franklin and Wasserman (1991, 1992b) explore the use of the bootstrap percentile method and the bias-corrected percentile to construct bootstrap confidence intervals for a single C_{pk} index. In general they find that percentile method confidence intervals do not attain the specified coverage level, except for the case when the data are normal and the sample size is 50 or greater. The histograms plotted in Figure 1.2 suggest approximate normality and the sample sizes are sufficiently large enough to consider the percentile method as a reasonable approach to computing the observed confidence levels for this problem. Therefore, observed confidence levels for the regions described in Equation (3.23) were computed using the nonparametric bootstrap and $b = 10,000$ resamples. To explore the possible effect of the normality assumption on this problem the analysis was repeated using the parametric bootstrap where the distribution of the quality characteristic of each supplier was modeled using a normal distribution with the same mean and variance and the observed sample data from the supplier. The parametric bootstrap works in the same way as the nonparametric bootstrap with the exception that the resamples are samples from the fitted parametric distribution instead of the empirical distribution. Note that the assumption of independence between the samples from each supplier allows one to resample separately from the sample of each supplier. For further details on the calculation of the observed confidence levels for this example consult Section 3.6.

The results of these calculations are given in Table 3.4. Note that there is little difference between the observed confidence levels computed using the parametric and the nonparametric bootstrap due to the fact that the normal model for the data fit quite well. The observed confidence levels clearly indicate that there is little confidence in Suppliers 2 and 3. The suppliers with the highest level of confidence are Suppliers 1 and 4, with a slight edge given to Supplier 4.

3.5.3 A Problem from Environmental Science

The use of the beta distribution has been recently suggested as an appropriate distribution to model human exposure to airborne contaminants. An example

Table 3.4 *Observed confidence levels for the four suppliers of piston rings. The observed confidence level represents the amount of confidence there is that the indicated supplier has the largest C_{pk} process capability index calculated using the normal parametric and nonparametric bootstrap.*

Supplier	1	2	3	4
Sample Size	50	75	70	75
Estimated C_{pk}	1.5302	1.1303	1.3309	1.5664
Parametric Bootstrap	41.25%	0.10%	4.08%	54.57%
Nonparametric Bootstrap	43.59%	0.07%	5.68%	50.66%

that considers exposure to the chemical naphthalene is considered by Flynn (2004). Flynn (2004) suggests that exposure measures of airborne concentration measurements should be divided by a maximum concentration so that the resulting concentrations are mapped to the interval $[0, 1]$. The selection of this value is specific to the chemical under consideration, its phase, and other considerations. For a vapor, such as in the naphthalene example, this maximum concentration is equal to $p_{sat}/760$, where p_{sat} is the saturation vapor pressure in mm Hg. For the data considered by Flynn (2004), this maximum concentration is 551,772 $\mu g/m^3$. The data considered by Flynn (2004) have a mean of 562.86 $\mu g/m^3$, a standard deviation of 644.90 $\mu g/m^3$, a minimum value of 0.67 $\mu g/m^3$, and a maximum value of 3911.00 $\mu g/m^3$. This results is a transformed values that roughly vary between 0.000 and 0.008. In many situations the theoretical maximum concentration value described by Flynn (2004) may exceed the usual observations in normal circumstances by a considerable amount. This results in a fitted beta distribution that has almost all of its density concentrated near 0. It may be more informative to fit a beta distribution of data that spans more of the unit interval. The example in this section will therefore fit a three parameter beta distribution, where the extra parameter is an upper limit of the range of the exposure data.

The beta distribution fit in this example has the form

$$f(x) = \begin{cases} \frac{(x/\theta_3)^{\theta_1-1}(1-x/\theta_3)^{\theta_2-1}}{\theta_3 B(\theta_1,\theta_2)}, & 0 \le x \le \theta_3, \\ 0 & \text{otherwise}, \end{cases}$$

where $B(\theta_1, \theta_2)$ is the beta function. For simplicity, parameter estimation for this density will be based on a hybrid algorithm of method of moments for θ_1 and θ_2, and conditional maximum likelihood for θ_3. The parameter θ_3 is estimated by $X_{(n)}$, the sample maximum. To estimate θ_1 and θ_2, define $Y_i = X_i/X_{(n)}$ for $i = 1, \ldots, n$. Estimates for θ_1 and θ_2 based on $Y_1, \ldots, Y_n$ given by

$$\hat{\theta}_1 = \frac{(1 - \bar{Y}) - (S_Y^2/\bar{Y})}{(S_Y/\bar{Y})^2},$$

Table 3.5 *Observed confidence levels, in percent, for the shape of the fitted beta distribution for the naphthalene concentration data. The region Θ_1 indicates that the distribution is unimodal. The region Θ_2 indicates that the distribution is "U"-shaped. The region Θ_3 indicates that the distribution is "J"-shaped. All bootstrap estimates are based on 1000 bootstrap resamples.*

	Θ_1	Θ_2	Θ_3
$\hat{\alpha}_{\text{stud}}$	0.00	0.07	99.93
$\hat{\alpha}^*_{\text{stud}}$	0.02	0.00	99.98
$\hat{\alpha}^*_{\text{hyb}}$	0.00	0.09	99.91
$\hat{\alpha}^*_{\text{perc}}$	0.01	0.00	99.99

and $\hat{\theta}_2 = \hat{\theta}_1/\bar{Y} - \hat{\theta}_1$, where $\bar{Y}$ and S_Y are the sample mean, standard deviation of $Y_1, \ldots, Y_n$, respectively.

The parameters θ_1 and θ_2 control the shape of the distribution. In particular if θ_1 and θ_2 exceed 1 then the distribution is unimodal. If one of the parameters is greater than one and the other is less than one then the distribution has the so-called "J"shape. If both parameters are less than 1 then the distribution is "U"-shaped. The shape of the distribution of the concentration data may have importance in environmental statistics. The parameter space of $\boldsymbol{\theta} = (\theta_1, \theta_2, \theta_3)'$ is $\Theta = \mathbb{R}^+ \times \mathbb{R}^+ \times \mathbb{R}^+$. Regions corresponding to each of the basic shapes of the distribution can then be defined as $\Theta_1 = \{\boldsymbol{\theta} : \theta_1 > 1, \theta_2 > 1\}$, $\Theta_2 = \{\boldsymbol{\theta} : \theta_1 < 1, \theta_2 < 1\}$, and $\Theta_3 = \{\boldsymbol{\theta} : \theta_1 > 1, \theta_2 < 1\} \cup \{\boldsymbol{\theta} : \theta_1 < 1, \theta_2 > 1\}$.

To demonstrate this application consider the data used by Flynn (2004). A histogram of 126 observed naphthalene is given in Figure 3.7. The parameter estimates for the data are $\hat{\theta}_1 = 0.5083$, $\hat{\theta}_2 = 3.0235$ and $\hat{\theta}_3 = 3911.0000$, which indicates a "J" shaped density. The parameters θ_1 and θ_2, conditional on θ_3 follows the smooth function model, so each of the observed confidence level methods will be considered. Table 3.5 gives observed confidence levels for the three regions of interest in fitting beta distributions described above. All of the methods yield a very high degree of confidence that the distribution is "J"-shaped.

3.6 Computation Using R

As mentioned in Section 2, the bootstrap estimates $\hat{g}_n(\mathbf{t})$ and $\hat{h}_n(\mathbf{t})$ rarely exist in a closed analytic form, and must be approximated using computer based simulations. This section will demonstrate how to perform these calculations using the R statistical computing environment. Following the development of Section 2.7, consider generating b resamples of size n where each sample is generated by randomly drawing n times, with replacement, from the original

Figure 3.7 *Frequency histogram of 126 observed concentrations of naphthalene* ($\mu g/m^3$). *The data were made available by Dr. Stephen Rappaport and Dr. Peter Egeghy. The support for the collection of the data was provided by grant from the U. S. Air Force and by NIEHS through Training Grant T32ES07018 and Project Grant P42ES05948.*

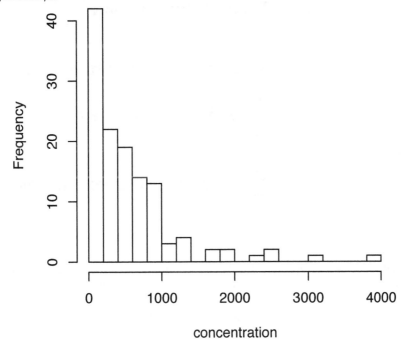

sample $\mathbf{X}_1, \ldots, \mathbf{X}_n$. Let $\bar{\mathbf{X}}_i^*$ be the sample mean of the ith resample, $\hat{\boldsymbol{\theta}}_i^* = g(\bar{\mathbf{X}}_i^*)$ and $\hat{\boldsymbol{\Sigma}}_i^* = h(\bar{\mathbf{X}}_i^*)$, for $i = 1, \ldots, b$. The bootstrap estimates of $g_n(\mathbf{t})$ and $h_n(\mathbf{t})$ can then be approximated by

$$\hat{g}_n(\mathbf{t}) \simeq b^{-1} \sum_{i=1}^{b} \delta[\mathbf{t}; n^{1/2} \hat{\boldsymbol{\Sigma}}^{-1/2} (\hat{\boldsymbol{\theta}}_i^* - \hat{\boldsymbol{\theta}})],$$

and

$$\hat{h}_n(\mathbf{t}) \simeq b^{-1} \sum_{i=1}^{b} \delta[\mathbf{t}; n^{1/2} \hat{\boldsymbol{\Sigma}}_i^{*-1/2} (\hat{\boldsymbol{\theta}}_i^* - \hat{\boldsymbol{\theta}})],$$

where $n^{1/2} \hat{\boldsymbol{\Sigma}}^{-1/2} (\hat{\boldsymbol{\theta}}_i^* - \hat{\boldsymbol{\theta}})$ and $n^{1/2} \hat{\boldsymbol{\Sigma}}_i^{*-1/2} (\hat{\boldsymbol{\theta}}_i^* - \hat{\boldsymbol{\theta}})$ are taken to be singleton sets. It then follows that the bootstrap estimates of the observed confidence level of a region Ψ can be approximated by

$$\hat{\alpha}^*_{\text{ord}}(\Psi) \;\simeq\; b^{-1} \sum_{i=1}^{b} \delta[n^{1/2}\hat{\Sigma}^{-1/2}(\hat{\theta}^*_i - \hat{\theta}); n^{1/2}\Sigma^{-1/2}(\hat{\theta} - \Psi)]$$

$$= \; b^{-1} \sum_{i=1}^{b} \delta[\hat{\theta} - \Sigma^{1/2}\hat{\Sigma}^{-1/2}(\hat{\theta}^*_i - \hat{\theta}); \Psi],$$

$$\hat{\alpha}^*_{\text{stud}}(\Psi) \;\simeq\; b^{-1} \sum_{i=1}^{b} \delta[n^{1/2}\hat{\Sigma}^{*-1/2}_i(\hat{\theta}^*_i - \hat{\theta}); n^{1/2}\Sigma^{-1/2}(\hat{\theta} - \Psi)]$$

$$= \; b^{-1} \sum_{i=1}^{b} \delta[\hat{\theta} - \hat{\Sigma}^{1/2}\hat{\Sigma}^{*-1/2}_i(\hat{\theta}^*_i - \hat{\theta}) \in \Psi],$$

and

$$\hat{\alpha}^*_{\text{hyb}}(\Psi) \;\simeq\; b^{-1} \sum_{i=1}^{b} \delta[n^{1/2}\hat{\Sigma}^{-1/2}(\hat{\theta}^*_i - \hat{\theta}); n^{1/2}\hat{\Sigma}^{-1/2}(\hat{\theta} - \Psi)]$$

$$= \; b^{-1} \sum_{i=1}^{b} \delta[2\hat{\theta} - \hat{\theta}^*_i; \Psi].$$

Finally, following Equation (3.11), the percentile observed confidence level of Ψ can be approximated using the resampling algorithm as

$$\hat{\alpha}^*_{\text{perc}}(\Psi) \simeq b^{-1} \sum_{i=1}^{b} \delta(\hat{\theta}^*_i; \Psi).$$

These expressions have been written in terms of the original region Ψ to simplify practical computation.

First consider the profile analysis problem studied in Sections 1.3.3 and 3.5.1. For such an example an R function for computing the observed confidence level of $\Theta_1 = \{\theta : \theta_1 \leq \theta_2 \leq \theta_3 \leq \theta_4\}$ is required. It is most efficient to write the function to return $\delta(\hat{\theta}^*_i; \Theta_1)$ for the ith resample, which is equivalent to returning 1 if $\hat{\theta}^*_{i1} \leq \hat{\theta}^*_{i2} \leq \hat{\theta}^*_{i3} \leq \hat{\theta}^*_{i4}$, and 0 otherwise, where we assume that $\hat{\theta}^*_i = (\hat{\theta}^*_{i1}, \hat{\theta}^*_{i2}, \hat{\theta}^*_{i3}, \hat{\theta}^*_{i4})'$. Such a function can be written as

```
perc.fun <- function(data,i) {
    m <- apply(data[i,],2,mean)
    if((m[1]<=m[2])&&(m[2]<=m[3])&&(m[3]<=m[4])) return(1)
    else return(0) }
```

The observed confidence level for Θ_1 can the be computed using the command

```
mean(boot(data=x,statistic=perc.fun,R=1000)$t)
```

where the data are stored in the matrix x and R=1000 specifies that $b = 1000$

resamples should be used in the calculation. If the confidence levels for all of
the regions studied in Section 5 are desired, then the function is somewhat
more tedious, but no less complicated, to write. This function can be written
as

```
perc.fun.all <- function(data,i) {
    m <- apply(data[i,],2,mean)
    if((m[1]<=m[2])&&(m[2]<=m[3])&&(m[3]<=m[4])) return(1)
    if((m[1]<=m[2])&&(m[2]<=m[4])&&(m[4]<=m[3])) return(2)
    if((m[1]<=m[3])&&(m[3]<=m[2])&&(m[2]<=m[4])) return(3)
    if((m[1]<=m[3])&&(m[3]<=m[4])&&(m[4]<=m[2])) return(4)
    if((m[1]<=m[4])&&(m[4]<=m[2])&&(m[2]<=m[3])) return(5)
    if((m[1]<=m[4])&&(m[4]<=m[2])&&(m[2]<=m[3])) return(6)

    if((m[2]<=m[1])&&(m[1]<=m[3])&&(m[3]<=m[4])) return(7)
    if((m[2]<=m[1])&&(m[1]<=m[4])&&(m[4]<=m[3])) return(8)
    if((m[2]<=m[3])&&(m[3]<=m[1])&&(m[1]<=m[4])) return(9)
    if((m[2]<=m[3])&&(m[3]<=m[4])&&(m[4]<=m[1])) return(10)
    if((m[2]<=m[4])&&(m[4]<=m[1])&&(m[1]<=m[3])) return(11)
    if((m[2]<=m[4])&&(m[4]<=m[3])&&(m[3]<=m[1])) return(12)

    if((m[3]<=m[1])&&(m[1]<=m[2])&&(m[2]<=m[4])) return(13)
    if((m[3]<=m[1])&&(m[1]<=m[4])&&(m[4]<=m[2])) return(14)
    if((m[3]<=m[2])&&(m[2]<=m[1])&&(m[1]<=m[4])) return(15)
    if((m[3]<=m[2])&&(m[2]<=m[4])&&(m[4]<=m[1])) return(16)
    if((m[3]<=m[4])&&(m[4]<=m[1])&&(m[1]<=m[2])) return(17)
    if((m[3]<=m[4])&&(m[4]<=m[2])&&(m[2]<=m[1])) return(18)

    if((m[4]<=m[1])&&(m[1]<=m[2])&&(m[2]<=m[3])) return(19)
    if((m[4]<=m[1])&&(m[1]<=m[3])&&(m[3]<=m[2])) return(20)
    if((m[4]<=m[2])&&(m[2]<=m[1])&&(m[1]<=m[3])) return(21)
    if((m[4]<=m[2])&&(m[2]<=m[3])&&(m[3]<=m[1])) return(22)
    if((m[4]<=m[3])&&(m[3]<=m[1])&&(m[1]<=m[2])) return(23)
    if((m[4]<=m[3])&&(m[3]<=m[2])&&(m[2]<=m[1])) return(24) }
```

The observed confidence levels can then be computed using the command

```
table(boot(data=x,statistic=perc.fun.all,R=1000)$t)/1000
```

The hybrid and studentized observed confidence levels can be computed using
the functions

```
hyb.fun.all <- function(data,i) {
    m1 <- apply(data[i,],2,mean)
    m0 <- apply(data,2,mean)
    m <- 2*m0-m1
```

```
        if((m[1]<=m[2])&&(m[2]<=m[3])&&(m[3]<=m[4])) return(1)
        if((m[1]<=m[2])&&(m[2]<=m[4])&&(m[4]<=m[3])) return(2)
        if((m[1]<=m[3])&&(m[3]<=m[2])&&(m[2]<=m[4])) return(3)
        if((m[1]<=m[3])&&(m[3]<=m[4])&&(m[4]<=m[2])) return(4)
        if((m[1]<=m[4])&&(m[4]<=m[2])&&(m[2]<=m[3])) return(5)
        if((m[1]<=m[4])&&(m[4]<=m[2])&&(m[2]<=m[3])) return(6)

        if((m[2]<=m[1])&&(m[1]<=m[3])&&(m[3]<=m[4])) return(7)
        if((m[2]<=m[1])&&(m[1]<=m[4])&&(m[4]<=m[3])) return(8)
        if((m[2]<=m[3])&&(m[3]<=m[1])&&(m[1]<=m[4])) return(9)
        if((m[2]<=m[3])&&(m[3]<=m[4])&&(m[4]<=m[1])) return(10)
        if((m[2]<=m[4])&&(m[4]<=m[1])&&(m[1]<=m[3])) return(11)
        if((m[2]<=m[4])&&(m[4]<=m[3])&&(m[3]<=m[1])) return(12)

        if((m[3]<=m[1])&&(m[1]<=m[2])&&(m[2]<=m[4])) return(13)
        if((m[3]<=m[1])&&(m[1]<=m[4])&&(m[4]<=m[2])) return(14)
        if((m[3]<=m[2])&&(m[2]<=m[1])&&(m[1]<=m[4])) return(15)
        if((m[3]<=m[2])&&(m[2]<=m[4])&&(m[4]<=m[1])) return(16)
        if((m[3]<=m[4])&&(m[4]<=m[1])&&(m[1]<=m[2])) return(17)
        if((m[3]<=m[4])&&(m[4]<=m[2])&&(m[2]<=m[1])) return(18)

        if((m[4]<=m[1])&&(m[1]<=m[2])&&(m[2]<=m[3])) return(19)
        if((m[4]<=m[1])&&(m[1]<=m[3])&&(m[3]<=m[2])) return(20)
        if((m[4]<=m[2])&&(m[2]<=m[1])&&(m[1]<=m[3])) return(21)
        if((m[4]<=m[2])&&(m[2]<=m[3])&&(m[3]<=m[1])) return(22)
        if((m[4]<=m[3])&&(m[3]<=m[1])&&(m[1]<=m[2])) return(23)
        if((m[4]<=m[3])&&(m[3]<=m[2])&&(m[2]<=m[1])) return(24) }
```

and

```
stud.fun.all <- function(data,i) {
    m1 <- apply(data[i,],2,mean)
    m0 <- apply(data,2,mean)
    s1d <- svd(cov(data[i,]))
    s0d <- svd(cov(data))
    s1 <- s1d$u%*%sqrt(diag(s1d$d))%*%s1d$u
    s0 <- s0d$u%*%sqrt(diag(s0d$d))%*%s0d$u
    m <- m0 - s0%*%solve(s1)%*%(m1-m0)
        if((m[1]<=m[2])&&(m[2]<=m[3])&&(m[3]<=m[4])) return(1)
    if((m[1]<=m[2])&&(m[2]<=m[4])&&(m[4]<=m[3])) return(2)
    if((m[1]<=m[3])&&(m[3]<=m[2])&&(m[2]<=m[4])) return(3)
    if((m[1]<=m[3])&&(m[3]<=m[4])&&(m[4]<=m[2])) return(4)
    if((m[1]<=m[4])&&(m[4]<=m[2])&&(m[2]<=m[3])) return(5)
    if((m[1]<=m[4])&&(m[4]<=m[2])&&(m[2]<=m[3])) return(6)

    if((m[2]<=m[1])&&(m[1]<=m[3])&&(m[3]<=m[4])) return(7)
```

```
    if((m[2]<=m[1])&&(m[1]<=m[4])&&(m[4]<=m[3])) return(8)
    if((m[2]<=m[3])&&(m[3]<=m[1])&&(m[1]<=m[4])) return(9)
    if((m[2]<=m[3])&&(m[3]<=m[4])&&(m[4]<=m[1])) return(10)
    if((m[2]<=m[4])&&(m[4]<=m[1])&&(m[1]<=m[3])) return(11)
    if((m[2]<=m[4])&&(m[4]<=m[3])&&(m[3]<=m[1])) return(12)

    if((m[3]<=m[1])&&(m[1]<=m[2])&&(m[2]<=m[4])) return(13)
    if((m[3]<=m[1])&&(m[1]<=m[4])&&(m[4]<=m[2])) return(14)
    if((m[3]<=m[2])&&(m[2]<=m[1])&&(m[1]<=m[4])) return(15)
    if((m[3]<=m[2])&&(m[2]<=m[4])&&(m[4]<=m[1])) return(16)
    if((m[3]<=m[4])&&(m[4]<=m[1])&&(m[1]<=m[2])) return(17)
    if((m[3]<=m[4])&&(m[4]<=m[2])&&(m[2]<=m[1])) return(18)

    if((m[4]<=m[1])&&(m[1]<=m[2])&&(m[2]<=m[3])) return(19)
    if((m[4]<=m[1])&&(m[1]<=m[3])&&(m[3]<=m[2])) return(20)
    if((m[4]<=m[2])&&(m[2]<=m[1])&&(m[1]<=m[3])) return(21)
    if((m[4]<=m[2])&&(m[2]<=m[3])&&(m[3]<=m[1])) return(22)
    if((m[4]<=m[3])&&(m[3]<=m[1])&&(m[1]<=m[2])) return(23)
    if((m[4]<=m[3])&&(m[3]<=m[2])&&(m[2]<=m[1])) return(24) }
```

respectively.

The quality assurance example studied in Section 3.5.2 requires a somewhat different strategy to compute the observed confidence levels in R. Because the four samples are independent of one another and the sample size are different for each supplier, resamples will be generated separately for each supplier. The R function to be used in conjunction with the bootstrap function is given by

```
cpk.perc <- function(data,i,UL,LL) {
mu <- mean(data[i])
sg <- sd(data[i])
ret <- min(c((UL-mu)/(3.0*sg),(mu-LL)/(3.0*sg)))
ret }
```

Suppose the data from the four suppliers 1–4 are stored in the objects S1, S2, S3, and S4, respectively and that the specification limits are stored in the objects U and L. Nonparametric bootstrap resamples for each supplier can be generated and stored in the objects RS1, RS2, RS3, and RS4 using the commands

```
RS1 <- boot(data=S1,statistic=cpk.perc,R=1000,UL=U,LL=L)$t
RS2 <- boot(data=S2,statistic=cpk.perc,R=1000,UL=U,LL=L)$t
RS3 <- boot(data=S3,statistic=cpk.perc,R=1000,UL=U,LL=L)$t
RS4 <- boot(data=S4,statistic=cpk.perc,R=1000,UL=U,LL=L)$t
```

The observed confidence level that the first supplier has the largest process capability can then be computed using the commands

```
RS <- cbind(RS1,RS2,RS3,RS4)
OLC1 <- matrix(0,1000,1)
for(i in 1:1000) if(RS1[i]=max(RS[i,])) OCL1[i] <- 1
mean(OLC1)
```

Similar commands can also be used to compute the observed confidence levels
for the remaining suppliers. The **boot** command also supports the parametric
bootstrap as well by specifying **sim="parametric"** and the model for generating the resamples using the **ran.gen** parameter and an appropriate function.
For normal parametric sampling sampling the function

```
ran.norm <- function(data,mle) {
out <- rnorm(length(data),mle[1],mle[2])
out }
```

will generate the required normal variates. The function used by the bootstrap
also must be changed to reflect the fact that the index variable is no longer
used. For the current example the function

```
cpk.perc <- function(data,UL,LL) {
mu <- mean(data)
sg <- sd(data)
ret <- min(c((UL-mu)/(3.0*sg),(mu-LL)/(3.0*sg)))
ret }
```

will suffice. The resamples for the first supplier can then be generated using
the commands

```
S1mle <- c(mean(S1),sd(S1))
PRS1 <- boot(data=S1,statistic=cpk.perc.norm,R=1000,
      UL=U,LL=L,sim="parametric",ran.gen=ran.norm,
      mle=S1mle)$t
```

with similar commands to generate PRS2, PRS3, and PRS4. The observed confidence levels for the remaining suppliers can then be computed using similar
commands to those presented earlier.

3.7 Exercises

3.7.1 Theoretical Exercises

1. Let $\mathbf{W}_1,\ldots,\mathbf{W}_n$ be a set of independent and identically distributed random
   variables from a bivariate distribution J. Denote the elements of each vector
   as $\mathbf{W}_i' = (W_{i1}, W_{i2})$ and the corresponding combined moments as $\mu_{jk} = E(W_{i1}^j W_{i2}^k)$. Assume that the covariance between W_{i1} and W_{i2} is zero.
   Define $\boldsymbol{\theta}' = (\mu_{20} - \mu_{10}^2, \mu_{02} - \mu_{01}^2)$, the vector of variances. Define a vector

$\mathbf{X}_i$ that is a function of the elements of $\mathbf{W}_i$ that is sufficient to define $\boldsymbol{\theta}$ and the asymptotic covariance matrix of $\boldsymbol{\theta}$ in terms of a smooth function of $E(\mathbf{X}_i)$. Define these functions so that the smooth function model is defined for this problem and find the plug-in estimates of $\boldsymbol{\theta}$ and $\boldsymbol{\Sigma}$.

2. Suppose that $\mathcal{G}_\alpha$ is a circle centered at the origin in the real plane $\mathbb{R}^2$ with radius r. Consider the ordinary confidence region

$$\mathbf{C}_{\mathrm{ord}}(\alpha;\mathbf{X}) = \{\hat{\boldsymbol{\theta}} - n^{-1/2}\boldsymbol{\Sigma}^{1/2}\mathbf{t} : \mathbf{t} \in \mathcal{G}_\alpha\},$$

where $\boldsymbol{\Sigma}$ and $\hat{\boldsymbol{\theta}}$ are known. Take $n = 1$ for simplicity.

(a) Suppose $\boldsymbol{\Sigma}$ is the identity matrix. Describe the geometric properties of the region $\mathbf{C}_{\mathrm{ord}}(\alpha;\mathbf{X})$.

(b) Suppose $\boldsymbol{\Sigma}$ is a diagonal matrix, but is otherwise arbitrary. Describe the geometric properties of the region $\mathbf{C}_{\mathrm{ord}}(\alpha;\mathbf{X})$.

(c) Suppose $\boldsymbol{\Sigma}$ is completely arbtrary. Describe the geometric properties of the region $\mathbf{C}_{\mathrm{ord}}(\alpha;\mathbf{X})$.

3. Prove that $\hat{\alpha}_{\mathrm{stud}}$ is first-order accurate.

4. Prove that $\hat{\alpha}^*_{\mathrm{stud}}$ is second-order accurate in probability.

5. Prove that Equation (3.13) implies that

$$\Delta_0(\mathcal{H}_\alpha) = \Lambda_0(\mathcal{H}_\alpha) = \alpha - n^{-1/2}\Lambda_1(\mathcal{H}_\alpha) + O(n^{-1}).$$

3.7.2 Applied Exercises

1. The data given in the table correspond to the waiting times for patients at a medical clinic that has five doctors. Each patient was randomly selected from a list of patients for each doctor during a one month period. A total of ten patients where selected for each doctor. The reported value is the number of minutes the patient was required to wait after the time of their appointment. If a patient was see before their appointment time then the values was accordingly recorded as negative.

Doctor	Observed Waiting Times									
A	11.0	5.8	10.9	8.2	8.2	10.2	10.7	6.9	12.6	8.3
B	5.4	7.8	6.5	6.3	5.7	6.9	6.3	7.7	6.6	6.4
C	6.3	8.5	5.6	7.3	5.8	6.3	5.5	5.9	9.4	8.4
D	5.0	0.9	6.1	5.6	17.5	10.3	-0.1	4.0	6.2	-1.0
E	7.3	9.4	10.9	-4.9	9.2	4.5	3.7	0.1	19.5	13.9

(a) Compute the average waiting time for each doctor. Does it appear that one of the doctors has a substantially larger waiting time than the rest? The goal of the clinic is to have a mean waiting time for patients that does not exceed 10 minutes, regardless of which doctor the patient has an appointment with. Based on these calculations how many doctors appear to have mean waiting times that do not exceed 10 minutes?

(b) Using the bootstrap with $b = 1000$ resamples, compute observed confidence levels for regions of the parameter space that correspond to all doctors having mean waiting times less than 15 minutes, all doctors having mean waiting times between 10 and 15 minutes, and all doctors having mean waiting times less than 10 minutes. Based on this analysis what conclusion can be made about the waiting times with respect to the goal of the clinic? Use the studentized, hybrid, and percentile methods. Repeat the analysis using the normal approximation. Does the normal approximation appear to be valid for this data?

(c) Using the bootstrap with $b = 1000$ resamples, compute observed confidence levels for regions of the parameter space that correspond to each doctor having the largest mean waiting time. Use the studentized, hybrid, and percentile methods. Use this analysis to suggest which doctors should be counseled about improving their patient waiting times?

(d) Consistency of service between patients and doctors is also important to the clinic. Compute the sample standard deviation for the waiting times for each doctor. Does it appear that the variability in the waiting times for each of the doctors is the same?

(e) Using the bootstrap with $b = 1000$ resamples, compute observed confidence levels for regions of the parameter space that correspond to each doctor having the largest variance in their waiting times. Use the studentized, hybrid, and percentile methods. Use this analysis to suggest which doctors should be counseled about improving the variance of their patient waiting times?

(f) Combine the two previous analyses as follows. Using the bootstrap with $b = 1000$ resamples, compute observed confidence levels for regions of the parameter space that correspond to each doctor having the largest mean waiting time *and* variance. Use the studentized, hybrid, and percentile methods. Does this analysis suggest that only one doctor is the source of both highest rankings?

2. Three suppliers of parts to a manufacturer are known to have manufacturing processes for the parts that have quality characteristics that follow distributions that are centered within the specification interval. This implies that the relative performance of the suppliers manufacturing processes can be determined by comparing the variability of the quality characteristics from each process. To make this comparison twenty-five samples of the parts manufactured by each supplier are obtained and the quality characteristic of each is measured. These observed quality characteristics are given in Table 3.6.

(a) Compute the sample standard deviation for the data from each supplier.

Table 3.6 *Data from three suppliers of parts.*

Supplier A

38.3	50.5	49.3	51.5	59.3
33.8	52.5	55.1	53.3	50.5
55.0	51.3	45.1	53.1	39.4
58.4	48.1	51.1	53.8	52.6
41.8	37.0	50.8	50.5	66.0

Supplier B

44.1	44.4	46.5	51.4	47.6
43.6	47.9	48.7	53.1	53.6
55.2	55.5	40.1	40.4	61.0
53.6	54.6	54.0	67.7	42.0
50.8	45.9	39.0	55.8	37.0

Supplier C

40.7	63.0	41.9	50.8	43.5
44.0	43.0	57.4	48.5	44.3
40.8	49.7	53.9	61.5	53.6
56.3	48.9	49.7	55.8	76.7
47.8	51.0	51.9	54.8	49.4

Does it appear that there is a supplier with a superior process, that is, one that has much lower variability than the others?

(b) Using the bootstrap with $b = 1000$ resamples, compute observed confidence levels based on the studentized, hybrid, and percentile methods for regions of the parameter space corresponding to the size possible orderings of the three suppliers based on the standard deviations of their processes. Based upon this analysis, what supplier or suppliers would you choose?

3. A military targeting system is designed to deliver ordinance within a circular target region defined in the real plane as $x^2 + y^2 \leq 1$. The coordinates are reported using a standardized coordinate system. The system is tested 27 times. The observed delivery points from these tests are reported in the table below. The system can be approved at three levels. The rating is excellent when the ordinance is within the circular region with a probability of at least 0.99. The rating is good when the ordinance is within the circular region with a probability of at least 0.95. The rating is fair when the ordinance is within the circular region with a probability of at least 0.90.

i	1	2	3	4	5	6	7	8	9
X_i	-0.63	0.92	-0.38	-0.99	-0.15	0.34	-0.52	0.14	-1.04
Y_i	1.05	-0.33	-0.18	-0.11	-1.10	0.45	-0.66	0.73	-1.20

i	10	11	12	13	14	15	16	17	18
X_i	-0.57	-0.10	-0.29	0.02	-0.18	-0.90	-0.60	0.50	1.14
Y_i	0.82	0.40	-0.21	0.00	0.87	0.19	0.68	0.67	-0.76

i	19	20	21	22	23	24	25	26	27
X_i	1.81	-0.74	0.18	-0.57	0.87	-1.64	0.39	-0.10	0.25
Y_i	0.26	0.06	-0.18	0.70	0.17	0.00	0.01	0.59	-0.59

(a) Assume that the data follow a multivariate normal distribution. Estimate the mean vector and the covariance matrix based on the observed
data. Using these estimates, estimate the probability that the ordinance
is within the target region by integrating the multivariate normal distribution with the estimated parameters over the target region.

(b) Compute observed confidence levels for each of the ratings possible for
this system using the percentile method and the parametric bootstrap
using 1000 samples. To implement the parametric bootstrap, simulate
1000 samples from a multivariate normal distribution whose parameters
are equal to the estimates based on the data. For each simulated sample,
compute the estimate of the probability that the ordinance is within the
target region based on the multivariate normal density as described in
part (a) above. The estimated observed confidence level for the excellent
rating will then be the proportion of times the estimate exceeds 0.99.
Observed confidence levels for the remaining two ratings are computed
using a similar method.

(c) Now assume that the distribution of the data is unknown. Estimate the
probability that the ordinance is within the target region by computing
the proportion of observations that are within the target region. Compare this estimate to that based on the multivariate normal assumption.

(d) Compute observed confidence levels for each of the ratings possible for
this system using the percentile method and the nonparametric bootstrap using 1000 samples. Compare these observed confidence levels to
those computed using the multivariate normal assumption.

4. A quality engineer is interested in determining the model for the distribution of gain of telecommunications amplifiers in decibels. A random sample
of seventy such amplifiers were tested. The observed gains of the amplifiers
are given in the table below.

7.65	10.28	8.65	8.32	7.41	9.43	8.93
8.62	8.38	9.52	7.48	9.41	7.98	8.41
8.07	7.31	8.60	8.19	8.36	8.64	8.54
7.24	8.98	8.95	7.75	9.85	8.99	8.10
7.28	8.48	7.43	8.47	8.37	7.02	7.94
9.97	8.97	7.43	7.68	9.07	8.38	7.67
9.65	7.95	7.97	7.81	8.66	8.54	9.20
8.52	10.79	7.95	8.14	8.23	8.65	8.92
8.07	7.26	9.37	7.68	7.23	7.07	8.75
7.90	8.31	9.25	8.99	10.53	7.94	8.13

(a) Consider a four parameter beta density of the form

$$f(x) = \begin{cases} \frac{[(x-\theta_3)/\theta_4]^{\theta_1-1}[1-(x-\theta_3)/\theta_4]^{\theta_2-1}}{\theta_4 B(\theta_1,\theta_2)}, & \theta_3 \le x \le \theta_3 + \theta_4 \\ 0 & \text{otherwise,} \end{cases}$$

which translates the beta density to the interval $(\theta_3, \theta_3 + \theta_4)$. Consider a hybrid method of estimating the parameters that begins with estimating $\hat{\theta}_3 = X_{(1)}$, the sample minimum and $\hat{\theta}_4 = X_{(n)} - X_{(1)}$, the sample range, where $X_1, \ldots, X_n$ represents the observed sample and $X_{(1)}, \ldots, X_{(n)}$ are the observed order statistics. Define $Y_i = (X_i - X_{(1)})/(X_{(n)} - X_{(1)})$ for $i = 1, \ldots, n$. Then $\hat{\theta}_1$ and $\hat{\theta}_2$ are estimated with

$$\hat{\theta}_1 = \frac{(1 - \bar{Y}) - (S_Y^2/\bar{Y})}{(S_Y/\bar{Y})^2},$$

and $\hat{\theta}_2 = \hat{\theta}_1/\bar{Y} - \hat{\theta}_1$, where $\bar{Y}$ and S_Y are the sample mean, standard deviation of $Y_1, \ldots, Y_n$, respectively. Compute these estimates for the data given in the table above.

(b) Compute observed confidence levels that the translated beta distribution has a unimodal shape, a "U" shape, or a "J" shape using the normal approximation, and the bootstrap estimates of the studentized, hybrid, and percentile methods.

5. One parameter of interest when evaluating the quality of a model railroad locomotive is the current draw, in amps, when the locomotive is running free under no load. The current draw is usually measured under several voltages applied to the locomotive motor. Consider a study of locomotives sampled from the stock of two locomotive manufacturers. A total of ten locomotives were sampled from each manufacturer, and the current draw for each was measured at 3, 6, 9 and 12 volts DC. The data are given in the table below.

Manufacturer	Volts (DC)			
	3	6	9	12
A	0.0445	0.0566	0.0711	0.1051
A	0.0289	0.0470	0.0704	0.1044
A	0.0547	0.0416	0.0717	0.1001
A	0.0497	0.0420	0.0761	0.0978
A	0.0472	0.0535	0.0729	0.0960
A	0.0379	0.0525	0.0826	0.0865
A	0.0615	0.0381	0.0747	0.0869
A	0.0522	0.0632	0.0743	0.0818
A	0.0489	0.0442	0.0761	0.0981
A	0.0399	0.0643	0.0793	0.0969
B	0.0695	0.0910	0.1602	0.2194
B	0.0721	0.1075	0.1588	0.2067
B	0.0823	0.0970	0.1488	0.2043
B	0.0796	0.1119	0.1820	0.2031
B	0.0696	0.0988	0.1640	0.2123
B	0.0572	0.1172	0.1626	0.2029
B	0.0747	0.1197	0.1569	0.1950
B	0.0720	0.1106	0.1381	0.1832
B	0.0759	0.1060	0.1630	0.1816
B	0.0816	0.1016	0.1687	0.2088

(a) Plot the current curves for each of the sample locomotives on a single axis for each manufacturer. On a third graph plot the mean current draw for each voltage for each manufacturer on a single set of axes. Does it appear that one current draw curve is always greater than the other?

(b) Compute observed confidence levels for the two regions corresponding to Manufacturer A having greater mean current draw than Manufacturer B at the four observed voltage levels, and Manufacturer B having greater mean current draw than Manufacturer A at the four observed voltage levels. Use the bootstrap estimates of the studentized, hybrid and percentile methods. What conclusions can be made from these observed confidence levels?

CHAPTER 4

Linear Models and Regression

4.1 Introduction

This chapter considers computing observed confidence levels for problems that occur in linear regression and linear models. The theory developed in the chapter focuses on regression models. Examples will demonstrate how problems from general linear models, including the one way layout, the randomized complete block design, and seemingly unrelated regressions can be considered within an extended regression framework, or within the general multiple parameter framework of Chapter 3.

The theoretical framework for this chapter closely resembles the framework used by Hall (1992a) to study the accuracy of bootstrap confidence regions for the problem of multivariate multiple regression. The development in this chapter will focus on univariate multiple regression. Extensions to the case of multivariate regression follow readily from the development in Hall (1992a). The approach and notation used in this chapter is necessarily nonstandard because more refined results are possible if the slope and intercept parameters are considered separately. As shown in Hall (1992a), asymptotically accurate confidence intervals for slope parameters are possible even with methods that are deficient in other problems. The results do transfer to the accuracy of observed confidence levels to a point. As is shown in Section 4.3, the bootstrap hybrid method is as asymptotically accurate as the bootstrap studentized method. However, the third-order accuracy that is realized for the bootstrap studentized confidence interval does not result in third-order accuracy for the corresponding observed confidence level. This is due to the fact that the bootstrap estimate of the studentized distribution is still only accurate to second-order. The accuracy of the observed confidence levels for the intercept parameter, as well as for the mean response, follow the accuracy results of the previous chapter.

Section 4.2 presents the framework within which the regression models will be studied. This section also includes several examples that develop similar frameworks for simple linear regression and for some general linear models. Several basic methods for computing observed confidence levels are developed in this section. The asymptotic accuracy of the methods developed in Section 4.2 are studied in Section 4.3. The presentation is slightly less formal as many of the details of the calculations follow from arguments given in Section 3.3. The proposed methods for computing observed confidence levels are studied empirically in Section 4.4 and several examples are considered in Section 4.5. Application of the calculations using R are studied in Section 4.7.

4.2 Statistical Framework

This chapter considers the standard linear multiple regression model where an n-dimensional random vector $\mathbf{Y}' = (Y_1, \ldots, Y_n)$ is observed conditional on an $n \times p$ matrix of constants $\mathbf{X}$ where $p < n$ and the rank of $\mathbf{X}$ is p. It is assumed that

$$E(\mathbf{Y}|\mathbf{X}) = \beta \mathbf{1}_n + \mathbf{X}\boldsymbol{\theta}, \tag{4.1}$$

for some p-dimensional parameter vector $\boldsymbol{\theta} \in \Theta$ where Θ is usually $\mathbb{R}^p$, and constant parameter $\beta \in \mathbb{R}$, where $\mathbf{1}_n$ is an $n \times 1$ vector with all elements equal to 1. The underlying stochastic mechanism for generating $\mathbf{Y}$ conditional on $\mathbf{X}$ is usually taken to be of the form $\mathbf{Y} = \beta \mathbf{1}_n + \mathbf{X}\boldsymbol{\theta} + \boldsymbol{\epsilon}$, where $\boldsymbol{\epsilon}$ is an n-dimensional random vector called the *error vector* such that $E(\boldsymbol{\epsilon}) = \mathbf{0}$ and $V(\boldsymbol{\epsilon}) = \boldsymbol{\Sigma}$ is an $n \times n$ diagonal matrix, usually of the form $\sigma^2 \mathbf{I}_n$ where $\sigma^2 < \infty$. Let $\mathbf{x}_i$ be the ith row of the matrix $\mathbf{X}$. The least squares estimators of β, $\boldsymbol{\theta}$, and σ are given by

$$\hat{\boldsymbol{\theta}} = n^{-1}\boldsymbol{\Sigma}_x^{-1} \sum_{i=1}^{n} (\mathbf{x}_i - \bar{\mathbf{x}})'(Y_i - \bar{Y}),$$

$\hat{\beta} = \bar{Y} - \bar{\mathbf{x}}\hat{\boldsymbol{\theta}}$, and $\hat{\sigma}^2 = n^{-1}\hat{\boldsymbol{\epsilon}}'\hat{\boldsymbol{\epsilon}}$ where

$$\bar{\mathbf{x}} = n^{-1} \sum_{i=1}^{n} \mathbf{x}_i,$$

$$\boldsymbol{\Sigma}_x = n^{-1} \sum_{i=1}^{n} (\mathbf{x}_i - \bar{\mathbf{x}})'(\mathbf{x}_i - \bar{\mathbf{x}}),$$

and $\hat{\boldsymbol{\epsilon}} = \mathbf{Y} - \hat{\beta}\mathbf{1}_n - \mathbf{X}\hat{\boldsymbol{\theta}}$. The main focus of most important problems in regression is on the slope parameter vector $\boldsymbol{\theta}$. Therefore define

$$G_n(\mathbf{t}) = P[n^{1/2}\boldsymbol{\Sigma}_x^{1/2}(\hat{\boldsymbol{\theta}} - \boldsymbol{\theta})/\sigma \leq \mathbf{t} | \epsilon_1, \ldots, \epsilon_n \sim F, \mathbf{X}],$$

and

$$H_n(\mathbf{t}) = P[n^{1/2}\boldsymbol{\Sigma}_x^{1/2}(\hat{\boldsymbol{\theta}} - \boldsymbol{\theta})/\hat{\sigma} \leq \mathbf{t} | \epsilon_1, \ldots, \epsilon_n \sim F, \mathbf{X}].$$

As with previous chapters, we assume that G_n and H_n are absolutely continuous with p-dimensional densities $g_n(\mathbf{t})$ and $h_n(\mathbf{t})$ corresponding to G_n and H_n respectively. Define $\mathcal{G}_\alpha$ and $\mathcal{H}_\alpha$ to be regions of $\mathbb{R}^p$ that respectively satisfy

$$P[n^{1/2}\boldsymbol{\Sigma}_x^{1/2}(\hat{\boldsymbol{\theta}} - \boldsymbol{\theta})/\sigma \in \mathcal{G}_\alpha | \epsilon_1, \ldots, \epsilon_n \sim F, \mathbf{X}] = \alpha,$$

and

$$P[n^{1/2}\boldsymbol{\Sigma}_x^{1/2}(\hat{\boldsymbol{\theta}} - \boldsymbol{\theta})/\hat{\sigma} \in \mathcal{H}_\alpha | \epsilon_1, \ldots, \epsilon_n \sim F, \mathbf{X}] = \alpha.$$

Confidence regions for $\boldsymbol{\theta}$ closely resemble those developed in Section 3.2. If σ is known then an ordinary confidence region for $\boldsymbol{\theta}$ is given by

$$\mathbf{C}_{\mathrm{ord}}(\alpha;\mathbf{X},\mathbf{Y}) = \{\hat{\boldsymbol{\theta}} - n^{-1/2}\sigma\boldsymbol{\Sigma}_x^{-1/2}\mathcal{G}_\alpha\},$$

and if σ is unknown then a studentized confidence region for $\boldsymbol{\theta}$ is given by

$$\mathbf{C}_{\mathrm{stud}}(\alpha;\mathbf{X},\mathbf{Y}) = \{\hat{\boldsymbol{\theta}} - n^{-1/2}\hat{\sigma}\boldsymbol{\Sigma}_x^{-1/2}\mathcal{H}_\alpha\}.$$

Taking the approximation $\mathcal{G}_\alpha \simeq \mathcal{H}_\alpha$ when σ is unknown in the studentized confidence region yields the hybrid confidence region given by

$$\mathbf{C}_{\mathrm{hyb}}(\alpha;\mathbf{X},\mathbf{Y}) = \{\hat{\boldsymbol{\theta}} - n^{-1/2}\hat{\sigma}\boldsymbol{\Sigma}_x^{-1/2}\mathcal{G}_\alpha\}.$$

Eicker (1963) and Jennrich (1969) provide conditions under which the least squares estimate of $\boldsymbol{\theta}$ is asymptotically normal. Under these conditions the normal approximation can be used to obtain approximate confidence regions for $\boldsymbol{\theta}$. The normal approximation of the ordinary and studentized confidence regions are given by

$$\hat{\mathbf{C}}_{\mathrm{ord}}(\alpha;\mathbf{X},\mathbf{Y}) = \{\hat{\boldsymbol{\theta}} - n^{-1/2}\sigma\boldsymbol{\Sigma}_x^{-1/2}\mathcal{N}_\alpha\},$$

and

$$\hat{\mathbf{C}}_{\mathrm{stud}}(\alpha;\mathbf{X},\mathbf{Y}) = \{\hat{\boldsymbol{\theta}} - n^{-1/2}\hat{\sigma}\boldsymbol{\Sigma}_x^{-1/2}\mathcal{N}_\alpha\},$$

where the region $\mathcal{N}_\alpha \subset \mathbb{R}^p$ satisfies $P(\mathbf{Z} \in \mathcal{N}_\alpha) = \alpha$ where $\mathbf{Z} \sim N_p(\mathbf{0}, I)$.

Observed confidence levels for the slope parameter vector $\boldsymbol{\theta}$ are computed by inverting the corresponding confidence regions as demonstrated in Section 3.2. Therefore, if $\Psi \subset \Theta$ is an arbitrary region and σ and $g_n(\mathbf{t})$ are known, the ordinary observed confidence level for Ψ is given by

$$\alpha_{\mathrm{ord}}(\Psi) = \int_{n^{1/2}\boldsymbol{\Sigma}_x^{1/2}(\hat{\boldsymbol{\theta}}-\Psi)/\sigma} g_n(\mathbf{t})d\mathbf{t}.$$

If $g_n(\mathbf{t})$ is unknown, then the normal approximation to the ordinary observed confidence level can be computed as

$$\hat{\alpha}_{\mathrm{ord}}(\Psi) = \int_{n^{1/2}\boldsymbol{\Sigma}_x^{1/2}(\hat{\boldsymbol{\theta}}-\Psi)/\sigma} \phi_p(\mathbf{t})d\mathbf{t}.$$

If σ is unknown then an observed confidence level for Ψ can be computed as

$$\alpha_{\mathrm{stud}}(\Psi) = \int_{n^{1/2}\boldsymbol{\Sigma}_x^{1/2}(\hat{\boldsymbol{\theta}}-\Psi)/\hat{\sigma}} h_n(\mathbf{t})d\mathbf{t},$$

when $h_n(\mathbf{t})$ is known. If $h_n(\mathbf{t})$ is unknown but $g_n(\mathbf{t})$ is known, then an observed confidence level for Ψ can be computed using the hybrid approximation $g_n(\mathbf{t}) \simeq h_n(\mathbf{t})$ as

$$\alpha_{\mathrm{hyb}}(\Psi) = \int_{n^{1/2}\boldsymbol{\Sigma}_x^{1/2}(\hat{\boldsymbol{\theta}}-\Psi)/\hat{\sigma}} g_n(\mathbf{t})d\mathbf{t}.$$

Finally if σ and both $g_n(\mathbf{t})$ and $h_n(\mathbf{t})$ are unknown then the normal approximation can be used to approximate the observed confidence level of Ψ as

$$\hat{\alpha}_{\mathrm{stud}}(\Psi) = \int_{n^{1/2}\mathbf{\Sigma}_x^{1/2}(\hat{\boldsymbol{\theta}}-\Psi)/\hat{\sigma}} \phi_p(\mathbf{t})d\mathbf{t}.$$

See Poliak (2007). As in previous chapters, if $g_n(\mathbf{t})$ and $h_n(\mathbf{t})$ are unknown and the normal approximation is not considered accurate enough, the bootstrap method of Efron (1979) is suggested as a method for approximating the observed confidence levels. The resampling algorithm is quite different for regression models, though the use of the bootstrap in regression models has been studied extensively, particularly for the construction of confidence sets for $\boldsymbol{\theta}$. See Adkins and Hill (1990), Freedman (1981), Freedman and Peters (1984), Hall (1989), Chapter 4 of Hall (1992a), Wu (1986), and Section 4.7. In the case of the regression model, the unknown distribution F corresponds to the distribution of the components of the error vector $\boldsymbol{\epsilon}$. Because $\boldsymbol{\epsilon}$ is not observed in the standard regression context, F is typically estimated using the empirical distribution of the elements of the residual vector $\hat{\boldsymbol{\epsilon}}$. Therefore, the bootstrap estimates of G_n and H_n are given by

$$\hat{G}_n(\mathbf{t}) = P^*[n^{1/2}\mathbf{\Sigma}_x^{1/2}(\hat{\boldsymbol{\theta}}^* - \hat{\boldsymbol{\theta}})/\hat{\sigma} \le \mathbf{t}|\epsilon_1^*,\dots,\epsilon_n^* \sim \hat{F}_n, \mathbf{X}], \qquad (4.2)$$

and

$$\hat{H}_n(\mathbf{t}) = P^*[n^{1/2}\mathbf{\Sigma}_x^{1/2}(\hat{\boldsymbol{\theta}}^* - \hat{\boldsymbol{\theta}})/\hat{\sigma}^{*2} \le \mathbf{t}|\epsilon_1^*,\dots,\epsilon_n^* \sim \hat{F}_n, \mathbf{X}],$$

where $P^*(A) = P(A|\mathbf{Y},\mathbf{X})$. To compute $\hat{\boldsymbol{\theta}}^*$, a new observed vector is generated as $\mathbf{Y}^* = \hat{\beta}\mathbf{1}_n + \mathbf{X}\hat{\boldsymbol{\theta}} + \hat{\boldsymbol{\epsilon}}^*$, then

$$\hat{\boldsymbol{\theta}}^* = n^{-1}\mathbf{\Sigma}_x^{-1} \sum_{i=1}^{n}(\mathbf{x}_i - \bar{\mathbf{x}})'(Y_i^* - \bar{Y}^*),$$

and $\hat{\sigma}^{*2} = n^{-1}\hat{\boldsymbol{\epsilon}}^{'*}\hat{\boldsymbol{\epsilon}}^*$. Taking $\hat{g}_n(\mathbf{t})$ and $\hat{h}_n(\mathbf{t})$ to be the mass functions corresponding to the distributions $\hat{G}_n(\mathbf{t})$ and $\hat{H}_n(\mathbf{t})$, respectively, allows one to compute observed confidence levels based on the bootstrap estimates. The bootstrap estimates of the ordinary, studentized and hybrid observed confidence levels for an arbitrary region $\Psi \subset \Theta$ are given by

$$\hat{\alpha}_{\mathrm{ord}}^*(\Psi) = \int_{n^{1/2}\mathbf{\Sigma}_x^{1/2}(\hat{\boldsymbol{\theta}}-\Psi)/\sigma} \hat{g}_n(\mathbf{t})d\mathbf{t},$$

$$\hat{\alpha}_{\mathrm{stud}}^*(\Psi) = \int_{n^{1/2}\mathbf{\Sigma}_x^{1/2}(\hat{\boldsymbol{\theta}}-\Psi)/\hat{\sigma}} \hat{h}_n(\mathbf{t})d\mathbf{t},$$

and

$$\hat{\alpha}_{\mathrm{hyb}}^*(\Psi) = \int_{n^{1/2}\mathbf{\Sigma}_x^{1/2}(\hat{\boldsymbol{\theta}}-\Psi)/\hat{\sigma}} \hat{g}_n(\mathbf{t})d\mathbf{t}.$$

As with most methods based on bootstrap estimates, $\hat{g}_n(\mathbf{t})$ and $\hat{h}_n(\mathbf{t})$ rarely exist in a closed analytic form so that simulation based methods must be usually employed to compute $\hat{\alpha}_{\mathrm{ord}}^*$, $\hat{\alpha}_{\mathrm{stud}}^*$ and $\hat{\alpha}_{\mathrm{hyb}}^*$. These issues will be discussed in Section 4.7.

It is also possible to compute observed confidence levels based on the percentile method. In this case let

$$V_n(\mathbf{t}) = P(\hat{\boldsymbol{\theta}} \leq \mathbf{t}|\epsilon_1, \ldots, \epsilon_n \sim F, \mathbf{X}, \mathbf{Y}),$$

with corresponding density $v_n(\mathbf{t})$. The bootstrap estimate of $V_n(\mathbf{t})$ is

$$\hat{V}_n(\mathbf{t}) = P^*(\hat{\boldsymbol{\theta}} \leq \mathbf{t}|\epsilon_1^*, \ldots, \epsilon_n^* \sim \hat{F}_n),$$

where, once again, $\hat{F}_n$ is the empirical distribution of $\hat{\epsilon}_1, \ldots, \hat{\epsilon}_n$. Let $\hat{v}_n(\mathbf{t})$ be the mass function corresponding to to $\hat{V}_n(\mathbf{t})$, and let $\mathcal{V}_\alpha$ be defined as in Equation (3.10). Then the observed confidence level for an arbitrary region $\Psi \subset \Theta$ is given by

$$\hat{\alpha}_{\text{perc}}^*(\Psi) = \int_\Psi \hat{v}_n(\mathbf{t})dt.$$

The reliability of all the methods for computing observed confidence levels in this section will be discussed in Section 4.3.

4.2.1 Simple Linear Regression

The simple linear regression model is a special case of the model proposed in Equation (4.1) with $p = 1$. That is, $\mathbf{Y}$ is an $n \times 1$ random vector that is observed conditionally on an $n \times 1$ vector of constants $\mathbf{x}$, such that $E(\mathbf{Y}|\mathbf{x}) = \theta\mathbf{x} + \beta\mathbf{1}$, where θ and β are unknown parameters. In this case the underlying mechanism for generating $\mathbf{Y}$ is of the form $\mathbf{Y} = \theta\mathbf{x} + \beta\mathbf{1} + \boldsymbol{\epsilon}$, where $\boldsymbol{\epsilon}$ follows the assumptions outlined above. The least squares estimates of β, θ and σ^2 are given by

$$\hat{\theta} = \sigma_x^{-2} n^{-1} \sum_{i=1}^{n} (x_i - \bar{x})(Y_i - \bar{Y}),$$

$\hat{\beta} = \bar{Y} - \bar{x}\hat{\theta}$ and $\hat{\sigma}^2 = n^{-1}\hat{\boldsymbol{\epsilon}}'\hat{\boldsymbol{\epsilon}}$, where

$$\sigma_x^2 = n^{-1} \sum_{i=1}^{n} (x_i - \bar{x})^2,$$

and $\hat{\boldsymbol{\epsilon}} = \mathbf{Y} - \hat{\theta}\mathbf{x} - \hat{\beta}\mathbf{1}$. For the slope parameter $\hat{\theta}$ define

$$G_n(t) = P[n^{1/2}(\hat{\theta} - \theta)\sigma_x/\sigma \leq t|\epsilon_1, \ldots, \epsilon_n \sim F, \mathbf{x}],$$

and

$$H_n(t) = P[n^{1/2}(\hat{\theta} - \theta)\sigma_x/\hat{\sigma} \leq t|\epsilon_1, \ldots, \epsilon_n \sim F, \mathbf{x}],$$

with corresponding absolutely continuous densities $g_n(t)$ and $h_n(t)$. Let g_α and h_α be the αth percentiles of $G_n(t)$ and $H_n(t)$, respectively. Confidence intervals for θ are constructed as in Section 2.3 using the four critical points defined in that chapter. If σ is known then the ordinary critical point can be used, and is given by

$$\hat{\theta}_{\text{ord}}(\alpha) = \hat{\theta} - n^{-1/2}\sigma_x^{-1}\sigma g_{1-\alpha}.$$

If σ is unknown, then the studentized critical point can be used, and is given by

$$\hat{\theta}_{\text{stud}}(\alpha) = \hat{\theta} - n^{-1/2}\sigma_x^{-1}\hat{\sigma}h_{1-\alpha}.$$

If the percentile $h_{1-\alpha}$ is unknown, but $g_{1-\alpha}$ is known then the approximation $h_{1-\alpha} \simeq g_{1-\alpha}$ can be used to yield the hybrid critical point given by

$$\hat{\theta}_{\text{hyb}}(\alpha) = \hat{\theta} - n^{-1/2}\sigma_x^{-1}\hat{\sigma}g_{1-\alpha}.$$

Finally, if the approximation $h_{1-\alpha} \simeq -g_\alpha$ is used then the percentile critical point results, given by

$$\hat{\theta}_{\text{perc}}(\alpha) = \hat{\theta} + n^{-1/2}\sigma_x^{-1}\hat{\sigma}g_\alpha.$$

Let $\Psi = (t_L, t_U)$ be an arbitrary interval subset of Θ, which in this case is usually $\mathbb{R}$. The observed confidence levels associated with these critical points are given by Poliak (2007) as

$$\alpha_{\text{ord}}(\Psi) = G_n\left[\frac{n^{1/2}\sigma_x(\hat{\theta} - t_L)}{\sigma}\right] - G_n\left[\frac{n^{1/2}\sigma_x(\hat{\theta} - t_U)}{\sigma}\right],$$

$$\alpha_{\text{stud}}(\Psi) = H_n\left[\frac{n^{1/2}\sigma_x(\hat{\theta} - t_L)}{\hat{\sigma}}\right] - H_n\left[\frac{n^{1/2}\sigma_x(\hat{\theta} - t_U)}{\hat{\sigma}}\right],$$

$$\alpha_{\text{hyb}}(\Psi) = G_n\left[\frac{n^{1/2}\sigma_x(\hat{\theta} - t_L)}{\hat{\sigma}}\right] - G_n\left[\frac{n^{1/2}\sigma_x(\hat{\theta} - t_U)}{\hat{\sigma}}\right],$$

and

$$\alpha_{\text{perc}}(\Psi) = G_n\left[\frac{n^{1/2}\sigma_x(t_U - \hat{\theta})}{\hat{\sigma}}\right] - G_n\left[\frac{n^{1/2}\sigma_x(t_L - \hat{\theta})}{\hat{\sigma}}\right].$$

In the case where G_n and H_n are unknown the normal approximations

$$\hat{\alpha}_{\text{ord}}(\Psi) = \Phi\left[\frac{n^{1/2}\sigma_x(\hat{\theta} - t_L)}{\sigma}\right] - \Phi\left[\frac{n^{1/2}\sigma_x(\hat{\theta} - t_U)}{\sigma}\right],$$

$$\hat{\alpha}_{\text{stud}}(\Psi) = \Phi\left[\frac{n^{1/2}\sigma_x(\hat{\theta} - t_L)}{\hat{\sigma}}\right] - \Phi\left[\frac{n^{1/2}\sigma_x(\hat{\theta} - t_U)}{\hat{\sigma}}\right], \qquad (4.3)$$

can be used for the case when σ is known, and unknown, respectively. As usual, if the normal approximation is not accurate enough a bootstrap estimate can be used. Let

$$\hat{G}_n(t) = P^*[n^{1/2}\sigma_x(\hat{\theta}^* - \hat{\theta})/\hat{\sigma} \le t | \epsilon_1^*, \ldots, \epsilon_n^* \sim \hat{F}_n],$$

and

$$\hat{H}_n(t) = P^*[n^{1/2}\sigma_x(\hat{\theta}^* - \hat{\theta})/\hat{\sigma}^* \le t | \epsilon_1^*, \ldots, \epsilon_n^* \sim \hat{F}_n].$$

The corresponding bootstrap estimates of the observed confidence levels for an arbitrary interval $\Psi = (t_L, t_U)$ are given by

$$\hat{\alpha}_{\text{ord}}^*(\Psi) = \hat{G}_n\left[\frac{n^{1/2}\sigma_x(\hat{\theta} - t_L)}{\sigma}\right] - \hat{G}_n\left[\frac{n^{1/2}\sigma_x(\hat{\theta} - t_U)}{\sigma}\right],$$

$$\hat{\alpha}^*_{\text{stud}}(\Psi) = \hat{H}_n \left[\frac{n^{1/2}\sigma_x(\hat{\theta} - t_L)}{\hat{\sigma}} \right] - \hat{H}_n \left[\frac{n^{1/2}\sigma_x(\hat{\theta} - t_U)}{\hat{\sigma}} \right],$$

$$\hat{\alpha}^*_{\text{hyb}}(\Psi) = \hat{G}_n \left[\frac{n^{1/2}\sigma_x(\hat{\theta} - t_L)}{\hat{\sigma}} \right] - \hat{G}_n \left[\frac{n^{1/2}\sigma_x(\hat{\theta} - t_U)}{\hat{\sigma}} \right],$$

and

$$\hat{\alpha}^*_{\text{perc}}(\Psi) = \hat{G}_n \left[\frac{n^{1/2}\sigma_x(t_U - \hat{\theta})}{\hat{\sigma}} \right] - \hat{G}_n \left[\frac{n^{1/2}\sigma_x(t_L - \hat{\theta})}{\hat{\sigma}} \right],$$

respectively.

The conditional expectation of Y for a given value of $x = x_0$ within the range of the observed $\mathbf{x}$ vector is given by $y_0 = E(Y|x = x_0) = \theta x_0 + \beta$ and can be estimated as $\hat{y}_0 = \hat{\theta}x_0 + \hat{\beta}$. Note that setting $x_0 = 0$ yields $\hat{y}_0 = \hat{\beta}$ so that a study of estimating these conditional expectations will also automatically include the problem of estimating the intercept of the regression model. The asymptotic variance of $n^{1/2}\hat{y}_0$ is $\sigma^2\sigma_y^2$ where $\sigma_y^2 = 1 + \sigma_x^{-2}(x_0 - \bar{x})^2$. Hence, define the standardized and studentized distributions of $\hat{y}_0$ as

$$G_n^0(t) = P[n^{1/2}(\hat{y}_0 - y_0)/(\sigma\sigma_y) \le t | \epsilon_1, \ldots, \epsilon_n \sim F, \mathbf{x}],$$

and

$$H_n^0(t) = P[n^{1/2}(\hat{y}_0 - y_0)/(\hat{\sigma}\sigma_y) \le t | \epsilon_1, \ldots, \epsilon_n \sim F, \mathbf{x}].$$

The development of methods for computing observed confidence levels now proceeds as outlined above for the slope parameter. For example, the studentized observed confidence level for $\hat{y}_0$ for arbitrary interval $\Psi = (t_L, t_U)$ is given by

$$\alpha_{\text{stud}}^0(\Psi) = H_n \left[\frac{n^{1/2}(\hat{y}_0 - t_L)}{\hat{\sigma}\sigma_y} \right] - H_n \left[\frac{n^{1/2}(\hat{y}_0 - t_U)}{\hat{\sigma}\sigma_y} \right].$$

If the normal approximation is appropriate then the studentized observed confidence level can be approximated with

$$\hat{\alpha}_{\text{stud}}^0(\Psi) = \Phi \left[\frac{n^{1/2}(\hat{y}_0 - t_L)}{\hat{\sigma}\sigma_y} \right] - \Phi \left[\frac{n^{1/2}(\hat{y}_0 - t_U)}{\hat{\sigma}\sigma_y} \right].$$

If the normal approximation is not accurate enough then the bootstrap can be used to estimate G_n^0 as

$$H_n^0(t) = P^*[n^{1/2}(\hat{y}_0^* - \hat{y}_0)/(\hat{\sigma}^*\sigma_y) \le t | \epsilon_1^*, \ldots, \epsilon_n^* \sim F],$$

which yields a bootstrap estimate of the studentized observed confidence level given by

$$\hat{\alpha}_{\text{stud}}^{0*}(\Psi) = \hat{H}_n \left[\frac{n^{1/2}(\hat{y}_0 - t_L)}{\hat{\sigma}\sigma_y} \right] - \hat{H}_n \left[\frac{n^{1/2}(\hat{y}_0 - t_U)}{\hat{\sigma}\sigma_y} \right].$$

Observed confidence levels based on the hybrid, percentile, and ordinary confidence intervals are developed in a similar way.

4.2.2 One-Way Layout

Some general linear models can also be studied under the framework developed in this chapter, though the results of Chapter 3 can often be used as well, depending on the structure of the model. As an example consider the one-way layout, which corresponds to an analysis of variance of a single factor.

In a one-way layout with k treatment levels an n-dimensional random vector $\mathbf{Y}$ is observed conditional on a $n \times (k+1)$ matrix of constants $\mathbf{D}$ where $k + 1 < n$ and the rank of $\mathbf{D}$ is k. To simplify the exposition of this example, a balanced model will be considered where each treatment level is observed an equal number of times. In this case $m = n/k$, where m is assumed to be a positive integer. It is convenient to index the vector $\mathbf{Y}$ in this case as

$$\mathbf{Y} = [Y_{11}, \ldots, Y_{1m}, Y_{21}, \ldots, Y_{2m}, \ldots, Y_{k1}, \ldots, Y_{km}], \qquad (4.4)$$

where Y_{ij} represents the jth observation or replication observed under treatment i for $i = 1, \ldots, k$ and $j = 1, \ldots, m$. The design matrix then has the general form

$$\mathbf{D} = \begin{bmatrix} \mathbf{1}_m & \mathbf{1}_m & \mathbf{0}_m & \mathbf{0}_m & \cdots & \mathbf{0}_m \\ \mathbf{1}_m & \mathbf{0}_m & \mathbf{1}_m & \mathbf{0}_m & \cdots & \mathbf{0}_m \\ \mathbf{1}_m & \mathbf{0}_m & \mathbf{0}_m & \mathbf{1}_m & \cdots & \mathbf{0}_m \\ \vdots & \vdots & \vdots & \vdots & \ddots & \vdots \\ \mathbf{1}_m & \mathbf{0}_m & \mathbf{0}_m & \mathbf{0}_m & \cdots & \mathbf{1}_m \end{bmatrix}$$

where $\mathbf{1}_m$ represents an $m \times 1$ vector of 1's and $\mathbf{0}_m$ represents a $m \times 1$ vector of 0's. The underlying mechanism for generating $\mathbf{Y}$ conditional of $\mathbf{D}$ is taken to be of the form $\mathbf{Y} = \mathbf{D}\gamma + \epsilon$ where $\gamma' = (\gamma_0, \ldots, \gamma_k)$ is a $(k+1)$ vector of constant parameters and ϵ is an n-dimensional error vector such that $E(\epsilon) = \mathbf{0}_n$ and $V(\epsilon) = \sigma^2 \mathbf{I}$ where $\sigma^2 < \infty$. This can be written in terms of each Y_{ij} as

$$Y_{ij} = \gamma_0 + \gamma_i + \epsilon_{ij}, \qquad (4.5)$$

for $i = 1, \ldots, k$ and $j = 1, \ldots, m$ where ϵ is assumed to be indexed in the same way as the vector $\mathbf{Y}$ in Equation (4.4).

As parameterized this model does not fall within the regression framework as described in this section as $\mathbf{D}$ is not a full rank matrix. This results in the fact that the parameters identified in Equation (4.5) are not unique. Linear models such as this are usually studied through estimable functions of the parameter γ, which are unique. A unique set of parameters can also be obtained by adding side conditions or constraints to the parameter vector γ. A common condition, which will be used in this section, will be to set $\gamma_0 = 0$ and define $\theta_i = \gamma_i$ for $i = 1, \ldots, k$. In this case the model can be written as $\mathbf{Y} = \mathbf{X}\theta + \epsilon$

where $\mathbf{X}$ has the form

$$
\mathbf{X} = \begin{bmatrix}
\mathbf{1}_m & \mathbf{0}_m & \mathbf{0}_m & \cdots & \mathbf{0}_m \\
\mathbf{0}_m & \mathbf{1}_m & \mathbf{0}_m & \cdots & \mathbf{0}_m \\
\mathbf{0}_m & \mathbf{0}_m & \mathbf{1}_m & \cdots & \mathbf{0}_m \\
\vdots & \vdots & \vdots & \ddots & \vdots \\
\mathbf{0}_m & \mathbf{0}_m & \mathbf{0}_m & \cdots & \mathbf{1}_m
\end{bmatrix}
$$

which is a full rank $n \times k$ matrix and $\boldsymbol{\theta}' = (\theta_1, \ldots, \theta_k)$. This model now fits within the regression framework outlined in this Section. While the asymptotic properties of this model can be studied under the regression context of this chapter, the simplified structure of the model allows for a simpler analysis using the multiple parameter context of the Chapter 3.

For the k-way layout consider a sequence of $k \times 1$ random vectors $\mathbf{W}_1, \ldots, \mathbf{W}_m$ where $\mathbf{W}_j' = (Y_{1j}, \ldots, Y_{kj})$. Under the assumption that the elements of the error vector are independent of one another and are identically distributed following a univariate distribution F, the sequence $\mathbf{W}_1, \ldots, \mathbf{W}_m$ can be taken to be a set of independent and identically distributed random vectors from a k-dimensional distribution $\bar{F}$. The distribution $\bar{F}$ is related to F through a linear transformation. Suppose $\epsilon_{1j}, \ldots, \epsilon_{kj} \sim F$, then $\bar{F}$ would correspond to the joint distribution of the random variable $\boldsymbol{\theta} + \boldsymbol{\epsilon}_j$ where $\boldsymbol{\epsilon}_j' = (\epsilon_{1j}, \ldots, \epsilon_{kj})$. Therefore, in the context of this model, $E(\mathbf{W}_j) = \boldsymbol{\theta}$. Under the assumption that $\bar{F}$ follows the conditions outlined in Chapter 3, observed confidence levels can be developed under the multiple parameter framework of that chapter.

4.3 Asymptotic Accuracy

The asymptotic behavior of the methods for computing observed confidence levels in the regression problem is similar to the asymptotic study of Section 3.3. When σ is known, the ordinary confidence region $\mathbf{C}_{\text{ord}}(\alpha; \mathbf{X}, \mathbf{Y})$ is used as the standard confidence region where an arbitrary method for computing an observed confidence level $\tilde{\alpha}$ is k^{th}-order accurate if $\tilde{\alpha}[\mathbf{C}_{\text{ord}}(\alpha; \mathbf{X}, \mathbf{Y})] = \alpha + O(n^{-k/2})$. For the case when σ is unknown, the confidence region $\mathbf{C}_{\text{stud}}(\alpha; \mathbf{X}, \mathbf{Y})$ is used as the standard confidence region with a similar definition for k^{th}-order accuracy. As in Section 3.3, it is clear that α_{ord} and α_{stud} are accurate.

The asymptotic theory for the approximate methods for computing observed confidence levels follows the multivariate Edgeworth expansion theory used in Section 3.3. Under suitable conditions on F and $\boldsymbol{\theta}$, Hall (1992a, Section 4.3.6) argues that for $\mathcal{S} \subset \mathbb{R}^p$, a finite union of convex sets,

$$
\begin{aligned}
P[n^{1/2} &\boldsymbol{\Sigma}_x^{1/2}(\hat{\boldsymbol{\theta}} - \boldsymbol{\theta})/\sigma \in \mathcal{S}] \\
&= \int_{\mathcal{S}} \left[1 + \sum_{i=1}^{k} n^{-i/2} r_i(\mathbf{t}) \right] \phi_p(\mathbf{t})d\mathbf{t} + o(n^{-k/2}) \\
&= \sum_{i=1}^{k} n^{-i/2} \Delta_i(\mathcal{S}) + o(n^{-k/2}),
\end{aligned} \tag{4.6}
$$

and

$$
\begin{aligned}
P[n^{1/2} &\boldsymbol{\Sigma}_x^{1/2}(\hat{\boldsymbol{\theta}} - \boldsymbol{\theta})/\hat{\sigma} \in \mathcal{S}] \\
&= \int_{\mathcal{S}} \left[1 + \sum_{i=1}^{k} n^{-i/2} s_i(\mathbf{t}) \right] \phi_p(\mathbf{t})d\mathbf{t} + o(n^{-k/2}), \\
&= \sum_{i=1}^{k} n^{-i/2} \Lambda_i(\mathcal{S}) + o(n^{-k/2}),
\end{aligned} \tag{4.7}
$$

where the polynomials $r_i(\mathbf{t})$ and $s_i(\mathbf{t})$ are defined in Section 4.3.6 of Hall (1992a), and $\Delta_i(\mathcal{S})$ and $\Lambda(\mathcal{S})$ are defined as in Section 3.3. The general form of the polynomials $r_i(\mathbf{t})$ and $s_i(\mathbf{t})$ are the same as the polynomials that appear in Section 3.3, though the coefficients will obviously differ. The polynomials $r_1(\mathbf{t})$ and $s_1(\mathbf{t})$ are a special case as they match the polynomials in Section 3.3, but Hall (1992a, Section 4.3.6) shows that $r_1(\mathbf{t}) = s_1(\mathbf{t})$ for all $\mathbf{t} \in \mathbb{R}^p$.

We begin by considering the case when σ is known. As discussed above, α_{ord} is correct, and is useful as long as the distribution $G_n(\mathbf{t})$ is known. If $G_n(\mathbf{t})$ is unknown then the observed confidence level based on the normal approximation, $\hat{\alpha}_{\text{ord}}$, can be used. Following the development of Section 3.3 is can be shown that

$$
\hat{\alpha}_{\text{ord}}[\mathbf{C}_{\text{ord}}(\alpha; \mathbf{X}, \mathbf{Y})] = \int_{\mathcal{G}_\alpha} \phi_p(\mathbf{t})d\mathbf{t} = \Delta_0(\mathcal{G}_\alpha),
$$

as long as $\mathcal{G}_\alpha$ is a finite union of convex sets. Noting that

$$
P[n^{1/2} \boldsymbol{\Sigma}_x^{1/2}(\hat{\boldsymbol{\theta}} - \boldsymbol{\theta})/\sigma \in \mathcal{G}_\alpha] = \alpha,
$$

it follows from Equation (4.6) that

$$
\alpha = \hat{\alpha}_{\text{ord}}[\mathbf{C}_{\text{ord}}(\alpha; \mathbf{X}, \mathbf{Y})] + o(n^{-1/2}),
$$

so that $\hat{\alpha}_{\text{ord}}$ is first-order correct. The bootstrap estimate $\hat{\alpha}_{\text{ord}}^*$ will be discussed later in this section. A similar argument can be used to show that $\hat{\alpha}_{\text{stud}}$ is first-order correct when σ is unknown. The analysis of the hybrid observed confidence level follows in a manner similar to the methods used to establish Equation (3.17), so that it follows that

$$
\alpha_{\text{hyb}}[\mathbf{C}(\alpha; \mathbf{X}, \mathbf{Y})] = \alpha + n^{-1/2}[\Delta_1(\mathcal{H}_\alpha) - \Lambda_1(\mathcal{H}_\alpha)] + O(n^{-1}).
$$

In the case of Equation (3.17) $\Delta_1(\mathcal{H}_\alpha) = \Lambda_1(\mathcal{H}_\alpha)$ only in special cases, such as when $\mathcal{H}_\alpha$ is a sphere centered at the origin. In the regression setting it is known that $r_1(\mathbf{t}) = s_1(\mathbf{t})$ and therefore $\Delta_1(\mathcal{S}) = \Lambda_1(\mathcal{S})$ for all $\mathcal{S} \subset \mathbb{R}^p$. Therefore, it follows that

$$\alpha_{\text{hyb}}[\mathbf{C}_{\text{stud}}(\alpha; \mathbf{X}, \mathbf{Y})] = \alpha + O(n^{-1}),$$

and therefore α_{hyb} is second-order correct. The increased accuracy of bootstrap methods in regression was first discussed by Hall (1992a). However, as will be shown later in this section, this increased accuracy does not always translate to more accurate methods for computing observed confidence levels.

As with the multiparameter case discussed in Section 3.3, Edgeworth expansions for the bootstrap estimates of $g_n(\mathbf{t})$ and $h_n(\mathbf{t})$ will allows us to analyze the accuracy of the bootstrap methods for computing observed confidence levels. In particular, following Equations (3.20) and (3.21) it follows in the regression setting that

$$P^*[n^{1/2}\mathbf{\Sigma}_x^{1/2}(\hat{\boldsymbol{\theta}} - \boldsymbol{\theta})/\hat{\sigma} \in \mathcal{S}|\epsilon_1^*, \ldots \epsilon_n^* \sim \hat{F}_n]$$
$$= \Delta_0(\mathcal{S}) + n^{-1/2}\Delta_1(\mathcal{S}) + O_p(n^{-1}),$$

and

$$P^*[n^{1/2}\mathbf{\Sigma}_x^{1/2}(\hat{\boldsymbol{\theta}} - \boldsymbol{\theta})/\hat{\sigma}^* \in \mathcal{S}|\epsilon_1^*, \ldots \epsilon_n^* \sim \hat{F}_n]$$
$$= \Lambda_0(\mathcal{S}) + n^{-1/2}\Lambda_1(\mathcal{S}) + O_p(n^{-1}).$$

As in Section 3.3, these results can be used to establish that $\hat{\alpha}_{\text{ord}}^*$, $\hat{\alpha}_{\text{stud}}^*$, and $\hat{\alpha}_{\text{hyb}}^*$ are second order accurate in probability. The accuracy of the percentile method is still first-order. Following the arguments used to establish Equation (3.22) it can be shown that in the regression case

$$\hat{\alpha}_{\text{perc}}^*[\mathbf{C}_{\text{stud}}(\alpha; \mathbf{X}, \mathbf{Y})] = \alpha - 2n^{-1/2}\Delta_1(\mathcal{H}_\alpha) + O_p(n^{-1}).$$

The asymptotic accuracy of observed confidence levels in the case of simple linear regression follows closely to the development in Section 2.4. Of course α_{ord} and α_{stud} are correct. To analyze $\hat{\alpha}_{\text{ord}}$ note that

$$\hat{\alpha}_{\text{ord}}[C_{\text{ord}}(\alpha, \boldsymbol{\omega}; \mathbf{X}, \mathbf{Y})] = \Phi\left\{\frac{n^{1/2}\sigma_x[\hat{\theta} - \hat{\theta}_{\text{ord}}(\omega_L)]}{\sigma}\right\} -$$
$$\Phi\left\{\frac{n^{1/2}\sigma_x[\hat{\theta} - \hat{\theta}_{\text{ord}}(\omega_U)]}{\sigma}\right\}$$
$$= \Phi(g_{1-\omega_L}) - \Phi(g_{1-\omega_U}).$$

which exactly matches Equation (2.33). Following the analysis of Section 2.4 it follows that $\hat{\alpha}_{\text{ord}}$ is first-order accurate for the case when σ is known. A similar analysis can be used to show that $\hat{\alpha}_{\text{stud}}$ is also first-order accurate for the case when σ is unknown. As in Section 2.4, it can be established that

$$\alpha_{\text{hyb}}[C_{\text{stud}}(\alpha, \boldsymbol{\omega}; \mathbf{X}, \mathbf{Y})] = \alpha + n^{-1/2}\Lambda(\omega_L, \omega_U) - n^{-1/2}\Delta(\omega_L, \omega_U) + O(n^{-1}).$$

In this case

$$
\begin{aligned}
\Lambda(\omega_L, \omega_U) - \Delta(\omega_L, \omega_U) &= q_1(z_{\omega_U})\phi(z_{\omega_U}) - p_1(z_{\omega_U})\phi(z_{\omega_U}) \\
&\quad - q_1(z_{\omega_L})\phi(z_{\omega_L}) + p_1(z_{\omega_L})\phi(z_{\omega_L}) \\
&= 0
\end{aligned}
$$

due to the fact that $p_1(t) = q_1(t)$ is the regression setting. Therefore α_{hyb} is second-order correct. As in the case of multiple regression, α_{perc} remains first-order correct as

$$
\alpha_{\text{perc}}[C_{\text{stud}}(\alpha, \boldsymbol{\omega}; \mathbf{X}, \mathbf{Y})] = \alpha + 2n^{-1/2}\Lambda(\omega_L, \omega_U) + O(n^{-1}),
$$

in the regression setting. For the bootstrap estimates of the observed confidence levels, the analysis is virtually the same because the asymptotic expansions of the confidence levels match their nonbootstrap counterparts to second order. Therefore $\hat{\alpha}^*_{\text{ord}}$, $\hat{\alpha}^*_{\text{hyb}}$, and $\hat{\alpha}^*_{\text{stud}}$ are all second-order accurate in probability, while $\hat{\alpha}^*_{\text{perc}}$ is first-order accurate in probability. Hall (1992a, pages 169–170) shows that for the regression problem the bootstrap studentized confidence interval for θ is actually third-order accurate. However, this property does not hold for the bootstrap estimate of the studentized observed confidence level due to the fact that the bootstrap estimate of $H_n(t)$ remains second-order accurate.

As developed in Chapter 3, the bias-corrected and accelerated bias-corrected confidence intervals can be used to construct corresponding methods for observed confidence levels. The methods have a very similar form to those developed in Chapter 3. See Poliak (2007) for further details.

4.4 Empirical Comparisons

The asymptotic comparisons of Section 4.3 indicate that the bootstrap methods used for computing observed confidence levels in the regression setting are equally accurate for large samples with the exception of the percentile method. If these trends hold for finite samples the result is important as the hybrid observed confidence level is typically easier and quicker to compute. This section empirically investigates the finite sample accuracy of these methods in the case of simple linear regression.

In particular, for a sample of size n consider an equally spaced grid of points on the interval $[0, 1]$ given by $x_i = i/n$ for $i = 1, \ldots, n$ for the elements of the vector $\mathbf{x}$, and a simple linear regression model of the form $\mathbf{Y} = \beta\mathbf{x} + \boldsymbol{\epsilon}$. The values of β considered in this study are $\beta = \frac{1}{4}, \frac{1}{2}, 1$ and 2 with samples of size $n = 25, 50$, and 100. The error vectors for the study are generated from three of the six normal mixtures first studied in Section 2.5 and plotted in Figure 2.1. The distributions studied are the standard normal, skewed unimodal, and kurtotic unimodal distributions. For each combination

Table 4.1 *Average observed confidence level (in percent) that $\beta \in \Psi = [0,1]$ from 100 simulated sets of regression data of the form $\mathbf{Y} = \mathbf{x}\beta + \boldsymbol{\epsilon}$. The measure whose average is closest to the average level of α_{stud} is italicized.*

Distribution	β	n	α_{stud}	$\hat{\alpha}_{\text{stud}}$	$\hat{\alpha}^*_{\text{stud}}$	$\hat{\alpha}^*_{\text{hyb}}$	$\hat{\alpha}^*_{\text{perc}}$
Standard	0.25	25	38.56	40.18	40.04	*39.98*	40.13
Normal		50	51.06	*51.18*	50.91	51.29	51.21
		100	63.09	63.20	63.30	*63.09*	63.26
	0.50	25	41.26	*41.27*	41.08	41.21	41.40
		50	55.46	*55.76*	55.81	55.85	55.84
		100	71.69	72.12	72.15	72.08	*71.84*
	1.00	25	34.09	34.20	*34.10*	34.29	34.24
		50	41.31	*40.88*	40.87	40.88	40.88
		100	51.60	47.23	49.93	50.19	46.99
	2.00	25	13.78	13.60	13.85	12.57	12.63
		50	6.04	5.97	8.74	6.44	8.00
		100	0.02	0.02	0.01	0.02	0.02
Skewed	0.25	25	43.33	51.33	44.72	*44.47*	49.04
Unimodal		50	58.53	60.39	*56.69*	60.51	54.10
		100	69.39	66.92	74.46	*69.51*	69.81
	0.50	25	48.56	*48.56*	49.18	49.44	49.48
		50	61.54	63.87	63.53	*63.23*	64.39
		100	81.41	79.44	76.12	78.92	*79.47*
	1.00	25	41.90	46.42	37.57	42.19	*41.66*
		50	47.82	45.71	45.56	*48.95*	44.52
		100	43.31	51.28	46.67	46.68	*46.23*
	2.00	25	9.43	9.55	*9.36*	9.18	7.84
		50	5.10	2.48	3.25	2.39	*3.38*
		100	0.27	0.78	0.80	*0.35*	1.37
Kurtotic	0.25	25	47.23	*47.35*	44.84	48.66	46.23
		50	62.75	60.60	60.81	59.43	*61.82*
		100	71.35	68.68	*71.77*	72.96	70.27
	0.50	25	48.64	50.09	46.51	*50.71*	51.99
		50	65.57	64.21	61.92	61.21	*65.13*
		100	82.76	77.03	76.31	79.66	*82.07*
	1.00	25	37.57	43.29	41.50	*39.85*	42.24
		50	42.71	44.56	44.85	*41.39*	51.40
		100	50.66	47.21	*48.45*	52.87	53.90
	2.00	25	10.45	8.34	9.35	*9.83*	9.44
		50	3.86	3.95	*3.82*	4.11	2.90
		100	*0.61*	0.66	0.78	0.86	0.35

of simulation parameters 100 **Y** vectors are simulated using the simple linear regression model given above.

The results of the simulation are given in Table 4.1. When interpreting the results of this study it should be remembered that the hybrid and studentized method are both second-order accurate. One can observed from Table 4.1 that, while there is no method that had the best accuracy for all of the situations studied, the studentized and hybrid methods do best in a majority of the situations. More extensive empirical studies have been performed by Poliak (2007). These more extensive studies also show that the studentized and hybrid methods do best in a majority of the situations.

4.5 Examples

4.5.1 A Problem from Analytical Chemistry

Finding alternate techniques to determine difficult to obtain measurements is a major avenue of research in the physical sciences. Chikae, et al. (2007) study the problem of determining compost maturity on the basis of three electrically measured parameters. An efficient compost system can aid in the recovery of waste products, such as from food processing and domestic refuse, as new materials or biosources for energy. A common application of compost products is in the field of agriculture. However, the application of immature compost materials can be harmful to crops. See, for example, Zucconi, et al. (1981). It is therefore important to be able to assess whether composting material has matured.

A common method of measuring the maturity of composting material is based on the germination index. To compute the germination index a water extract sample is taken from the compost material and seeds of a certain plant are left to germinate in the water for 48 hours. Another set of seeds of the same type are germinated in distilled water to serve as a control. The germination index is then a function of the number of seeds that germinated in each water sample along with the length of the roots of the germinated seeds. A germination index that exceeds 50% is considered an indication that the compost material is sufficiently mature. See Zucconi, et al. (1981). The 48-hour waiting period for this measurement makes it difficult to quickly analyze the maturity of compost material.

Chikae, et al. (2007) develop a quick and simple method for estimating the germination index based three electrochemical measurments: pH, NH_4^+ concentration, and phosphatase activity in a water extract a compost sample. The measurements of these three properties can be easily obtained using portable sensors and a laptop computer. The time required to take the measurements is approximately 2 minutes. To investigate the reliability of this system, Chikae

Figure 4.1 *The estimated germination index and the actual germination index computed from twenty-four samples of compost material. The solid line is the least squares linear regression line for the data. The dotted line is the identity function.*

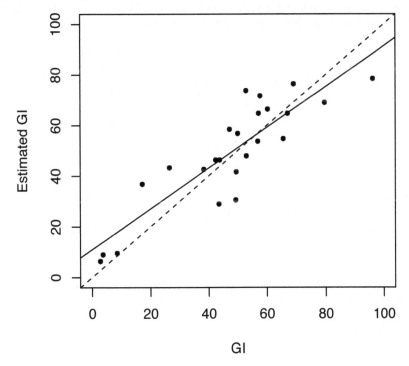

et al. (2007) obtained germination index measurements for twenty-four compost samples. The actual germination index, along with the estimate of the germination index obtained from the portable sensor where obtained for each sample. These measurements are plotted in Figure 4.1.

One can observe from Figure 4.1 that the estimation process is not a perfect predictor of the actual germination index. The fitted regression line indicates the mean estimated germination index tends to overestimate smaller values and underestimate larger values of the actual germination index. Given this observed behavior, how much confidence is there that the slope of the regression line is less than 1? This region will be defined as Θ_1. It is important in this problem to detect whether the compost is really mature or not based on the estimated germination index. Within this context it is important to determine how much confidence there is given that the actual germination index does not exceed 50%. This region will be defined as Θ_2.

This simple linear regression model fits within the statistical framework described in Section 4.2.1, and the question of the observed confidence level of the region corresponding to the slope being less than 1 is easily computed.

Table 4.2 *Estimated observed confidence levels (in percent) for the regions of interest in the germination index example. The region Θ_1 corresponds to the regression line having a slope less than 1. The region Θ_2 corresponds to the regression line staying below 50 when the actual germination index is below 50. The region Θ_3 corresponds to the regression line being below 50 when the actual germination index is exactly 50.*

| | Observed Confidence Levels | | | |
Region	$\hat{\alpha}_{\text{stud}}$	$\hat{\alpha}^*_{\text{stud}}$	$\hat{\alpha}^*_{\text{hyb}}$	$\hat{\alpha}^*_{\text{perc}}$
Θ_1	98.46	96.00	98.80	98.90
Θ_2	48.10	32.70	31.20	31.90
Θ_3	49.50	32.90	31.20	31.90

The least squares estimates for the regression line plotted in Figure 4.1 are $\hat{\beta} = 10.912$, $\hat{\theta} = 0.804$, and $\hat{\sigma} = 9.953$ with $\sigma_x = 22.389$. The normal approximation for the amount of confidence there is that the slope is less than 1 is

$$\hat{\alpha}_{\text{stud}}(\Theta_1) = 1 - \Phi(-2.160) = 0.9846.$$

The bootstrap estimates of this observed confidence level for this region are presented in Table 4.2. See Section 4.7 for the details of these calculations. It is clear from Table 4.2 that there is a large amount of confidence that the slope parameter is less than 1.

The computations for the region Θ_2 are slightly more complicated by the fact that the region jointly depends on the slope and intercept and that the region is a complicated function of those two parameters. Let $\boldsymbol{\theta}' = (\beta, \theta)$ where, using the notation established in Section 4.2, β is the intercept and θ is the slope of the linear regression model. The region Θ_2 is then given by

$$\Theta_2 = \{\boldsymbol{\theta} \in \mathbb{R}^2 : \beta \leq 50, \beta + 50\theta \leq 50\}.$$

To compute the observed confidence level for this region, the joint sampling distribution of $\boldsymbol{\theta}$ must be used. Let

$$G_n(\mathbf{t}) = P[n^{1/2}\tilde{\boldsymbol{\Sigma}}_x^{-1/2}(\hat{\boldsymbol{\theta}} - \boldsymbol{\theta})/\sigma \leq \mathbf{t}|\epsilon_1. \ldots, \epsilon_n \sim F, \mathbf{X}],$$

and

$$H_n(\mathbf{t}) = P[n^{1/2}\tilde{\boldsymbol{\Sigma}}_x^{-1/2}(\hat{\boldsymbol{\theta}} - \boldsymbol{\theta})/\hat{\sigma} \leq \mathbf{t}|\epsilon_1. \ldots, \epsilon_n \sim F, \mathbf{X}],$$

where

$$\tilde{\boldsymbol{\Sigma}}_x = \begin{bmatrix} n^{-1} + \mu_x^2 d_x^{-2} & -\mu_x d_x^{-2} \\ -\mu_x d_x^{-2} & d_x^{-2} \end{bmatrix},$$

$$d_x^2 = \sum_{i=1}^n (X_i - \bar{X})^2,$$

and

$$\mu_x = n^{-1} \sum_{i=1}^{n} x_i.$$

The corresponding observed confidence levels are then developed using these distributions as in Section 4.2. The observed confidence levels for this region are reported in Table 4.2. Details for computing the observed confidence levels using R are given in Section 4.7. It is apparent that there is around 30% confidence that the regression line stays below 50 when the actual germination index is below 50. Certainly for small values of the germination index there should be little problem with concluding that the compost is mature based on the estimated germination index. The problem lies, not surprisingly, when the germination index approaches 50. This can be verified by computing the observed confidence level of the region

$$\Theta_3 = \{\boldsymbol{\theta} \in \mathbb{R}^2 : \beta + 50\theta \le 50\}.$$

Note that $\Theta_2 \subset \Theta_3$ and that the region Θ_3 only requires that the regression line be less than or equal to 50 when the actual germination index is exactly 50. The observed confidence levels for Θ_3 are reported in Table 4.2. From Table 4.2 one can observe that the estimated observed confidence levels for Θ_3 are very close to that of Θ_2. This is an indication that it is the restriction that $\beta + 50\theta \le 50$, which is driving the observed confidence level down to around 30%, and not the restriction that $\beta \le 50$.

4.5.2 A Problem in Experimental Archaeology

In a study of the ability of flint axes to fell trees, Mathieu and Meyer (1997) consider data originally obtained by Jørgensen (1985) on the breast height diameter of trees in centimeters and the time to fell the tree in minutes for three types of trees: maple, oak, and birch. Data similar to what was reported are presented in Figure 4.2. Consider modeling the data using three separate linear regression lines: one for each tree type. The slopes of the regression lines then indicate how difficult each type of tree is to fell using a flint axe. Observed confidence levels can then be used to measure the amount of confidence there is that, for example, Maple is the most difficult type of tree to fell, followed in order by Oak and Birch.

Let Y_{ij} be the observed felling time for a tree with diameter x_{ij} of type i for $j = 1, \ldots, n_i$ and $i = 1, 2, 3$. The tree types are indexed as Maple (1), Oak (2) and Birch (3). The value n_i corresponds to the number of trees of type i used in the study. For the data in Figure 4.2, $n_1 = n_2 = n_3 = 20$. The value $n = n_1 + n_2 + n_3 = 60$ will denote the total sample size. To fit a separate regression line for each tree type, the model $E(\mathbf{Y}|\mathbf{X}) = \mathbf{W}\boldsymbol{\beta} + \mathbf{X}\boldsymbol{\theta}$ where

$$\mathbf{Y}' = (Y_{11}, \ldots, Y_{1n_1}, Y_{21}, \ldots, Y_{2n_2}, Y_{31}, \ldots, Y_{3n_3}),$$

Figure 4.2 *Data on felling time and breast height diameter for three tree types: Maple (○), Oak (△), and Birch (+). The plotted lines are individual least squares regression lines for the three tree types: Maple (solid line), Oak (dashed line) and Birch (dotted line).*

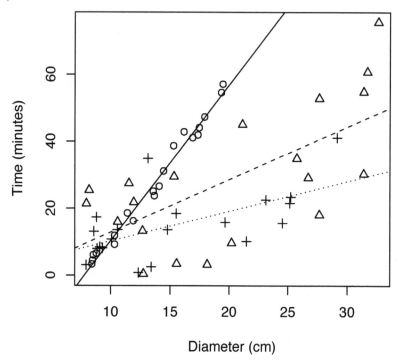

$\boldsymbol{\beta}' = (\beta_1, \beta_2, \beta_3)$, $\boldsymbol{\theta}' = (\theta_1, \theta_2, \theta_3)$,

$$\mathbf{W} = \begin{bmatrix} \mathbf{1}_{n_1} & \mathbf{0}_{n_1} & \mathbf{0}_{n_1} \\ \mathbf{0}_{n_2} & \mathbf{1}_{n_2} & \mathbf{0}_{n_2} \\ \mathbf{0}_{n_3} & \mathbf{0}_{n_3} & \mathbf{1}_{n_3} \end{bmatrix},$$

and

$$\mathbf{X} = \begin{bmatrix} \mathbf{x}_1 & \mathbf{0}_{n_1} & \mathbf{0}_{n_1} \\ \mathbf{0}_{n_2} & \mathbf{x}_2 & \mathbf{0}_{n_2} \\ \mathbf{0}_{n_3} & \mathbf{0}_{n_3} & \mathbf{x}_3 \end{bmatrix},$$

where $\mathbf{x}_i' = (x_{i1}, \ldots, x_{in_i})$.

The mechanism for generating $\mathbf{Y}$ conditional on $\mathbf{X}$ is assumed to have the form $\mathbf{Y} = \mathbf{W}\boldsymbol{\beta} + \mathbf{X}\boldsymbol{\theta} + \boldsymbol{\epsilon}$ where $E(\boldsymbol{\epsilon}) = \mathbf{0}_n$ and $V(\boldsymbol{\epsilon}) = \boldsymbol{\Sigma}$, where the covariance matrix has the form

$$\boldsymbol{\Sigma} = \begin{bmatrix} \sigma_1^2 \mathbf{I}_{n_1} & \mathbf{0}_{n_1, n_2} & \mathbf{0}_{n_1, n_3} \\ \mathbf{0}_{n_2, n_1} & \sigma_2^2 \mathbf{I}_{n_2} & \mathbf{0}_{n_2, n_3} \\ \mathbf{0}_{n_3, n_1} & \mathbf{0}_{n_3, n_2} & \sigma^2 \mathbf{I}_{n_3} \end{bmatrix}.$$

Table 4.3 *Observed confidence levels for the six possible orderings of the slope parameters in the tree felling example.*

Region	Ordering	$\hat{\alpha}_{\text{stud}}$	$\hat{\alpha}^*_{\text{stud}}$	$\hat{\alpha}^*_{\text{hyb}}$	$\hat{\alpha}^*_{\text{perc}}$
Θ_1	$\theta_1 < \theta_2 < \theta_3$	0	0	0	0
Θ_2	$\theta_1 < \theta_3 < \theta_2$	0	0	0	0
Θ_3	$\theta_2 < \theta_1 < \theta_3$	0	0	0	0
Θ_4	$\theta_2 < \theta_3 < \theta_1$	5.4	10.5	7.8	7.5
Θ_5	$\theta_3 < \theta_1 < \theta_2$	0	0	0	0
Θ_6	$\theta_3 < \theta_2 < \theta_1$	94.6	89.5	92.2	92.5

Table 4.4 *Least squares estimates for fitting a linear regression to each of the three tree types separately.*

i	Tree Type	$\hat{\beta}_i$	$\hat{\theta}_i$	$\hat{\sigma}_i$
1	Maple	-36.080	4.646	2.222
2	Oak	-2.893	1.573	14.767
3	Birch	1.335	0.889	7.842

Hence a different error variance will be used for each tree type. A close inspection of Figure 4.2 strongly indicates that the variability between tree types differs. It is also assumed that the three experiments are independent of one another. In this case it is efficient to fit the three regression lines separately to obtain estimates of θ and ϵ. See Zellner (1962). This model is an extension of the one given in Equation (4.1), though this type of model still falls within the general linear model, and the observed confidence levels defined in 4.2 can be adapted to this problem with only minor modifications as discussed below.

Of interest for this problem is computing levels of confidence for each of the six possible orderings of θ_1, θ_2, and θ_3. The corresponding regions of $\Theta = \mathbb{R}^3$ are defined in Table 4.3. The least squares estimates obtained by fitting a separate regression line to each type are given in Table 4.4.

By assumption the three estimates $\hat{\theta}_1$, $\hat{\theta}_2$, and $\hat{\theta}_3$ are independent of one another meaning that the asymptotic normality of $\hat{\theta}' = (\hat{\theta}_1, \hat{\theta}_2, \hat{\theta}_3)$ follows from the asymptotic normality of the individual estimators. This suggests that a reasonable method for computing observed confidence levels is given by the normal approximation of Equation (4.3), with suitable modifications for the current model. Due to the independence of the elements of $\hat{\theta}$, the asymptotic covariance matrix of $n^{1/2}\hat{\boldsymbol{\theta}}$ is given by

$$\boldsymbol{\Sigma}_T = \begin{bmatrix} \sigma_1^2 d_{x,1}^{-2} & 0 & 0 \\ 0 & \sigma_2^2 d_{x,2}^{-2} & 0 \\ 0 & 0 & \sigma_3^2 d_{x,3}^{-2} \end{bmatrix},$$

where

$$d_{x,i}^2 = \sum_{i=1}^{n_i} (x_{ij} - \bar{x}_i)^2,$$

and

$$\bar{x}_i = n_i^{-1} \sum_{i=1}^{n_1} x_{ij}.$$

The covariance matrix $\boldsymbol{\Sigma}_T$ is estimated by

$$\hat{\boldsymbol{\Sigma}}_T = \begin{bmatrix} \hat{\sigma}_1^2 d_{x,1}^{-2} & 0 & 0 \\ 0 & \hat{\sigma}_2^2 d_{x,2}^{-2} & 0 \\ 0 & 0 & \hat{\sigma}_3^2 d_{x,3}^{-2} \end{bmatrix}$$

where $\hat{\sigma}_i^2 = \hat{\boldsymbol{\epsilon}}_i' \hat{\boldsymbol{\epsilon}}_i$ where the residual vector $\hat{\boldsymbol{\epsilon}}' = (\hat{\boldsymbol{\epsilon}}_1, \hat{\boldsymbol{\epsilon}}_2, \hat{\boldsymbol{\epsilon}}_3)$, with each subvector $\hat{\boldsymbol{\epsilon}}_i$ corresponding to the residuals for the separate fit of each tree type. In the context of this problem, define

$$H_n(\mathbf{t}) = P[n^{1/2} \hat{\boldsymbol{\Sigma}}_T^{-1/2}(\hat{\boldsymbol{\theta}} - \boldsymbol{\theta}) \le \mathbf{t}|E]$$

where E is the event that $\epsilon_{i1}, \ldots, \epsilon_{in_i} \sim F_i$ for $i = 1, 2, 3$ where it is assumed that the vector $\boldsymbol{\epsilon}$ is indexed in the same way as $\mathbf{Y}$. It follows that an observed confidence level for an arbitrary region $\Psi \subset \Theta = \mathbb{R}^3$ based on a studentized confidence region for $\boldsymbol{\theta}$ is given by

$$\alpha_{\text{stud}}(\Psi) = \int_{n^{1/2} \hat{\boldsymbol{\Sigma}}_T^{-1/2}(\hat{\boldsymbol{\theta}} - \Psi)} h_n(\mathbf{t}) d\mathbf{t},$$

where $h_n(\mathbf{t})$ is the three-dimensional density corresponding to $H_n(\mathbf{t})$. For this example F_1, F_2, and F_3 are unknown so that $h_n(\mathbf{t})$ is also unknown. As stated above, the normal approximation can be applied to this case to obtain

$$\hat{\alpha}_{\text{stud}}(\Psi) = \int_{n^{1/2} \hat{\boldsymbol{\Sigma}}_T^{-1/2}(\hat{\boldsymbol{\theta}} - \Psi)} \phi_3(\mathbf{t}) d\mathbf{t}.$$

For comparison, estimates based on the bootstrap will also be implemented. The bootstrap estimate of $H_n(\mathbf{t})$ is given by

$$\hat{H}_n(\mathbf{t}) = P[n^{1/2} \hat{\boldsymbol{\Sigma}}_T^{*-1/2}(\hat{\boldsymbol{\theta}}^* - \hat{\boldsymbol{\theta}}) \le \mathbf{t}|E^*],$$

where E^* is the event that $\epsilon_{i1}^*, \ldots, \epsilon_{in_i}^* \sim \hat{F}_i$ for $i = 1, 2, 3$, conditional on the event E defined earlier. The distribution $\hat{F}_i$ corresponds to the empirical distribution of $\hat{\epsilon}_{i1}, \ldots, \hat{\epsilon}_{in_i}$, conditional on E, for $i = 1, 2, 3$. The bootstrap estimate can be used to compute the bootstrap estimate of $\alpha_{\text{stud}}(\Psi)$ given by

$$\hat{\alpha}_{\text{stud}}^*(\Psi) = \int_{n^{1/2} \hat{\boldsymbol{\Sigma}}_T^{-1/2}(\hat{\boldsymbol{\theta}} - \Psi)} \hat{h}_n(\mathbf{t}) d\mathbf{t}.$$

Bootstrap estimates of the hybrid and percentile method observed confidence levels are developed in an analogous way. The estimates for each of these methods and regions are reported in Table 4.3. It is clear from Table 4.3 that there is a great amount of confidence that maple is the most difficult tree type to fell. The remaining question arises as to whether oak or birch is more difficult to fell. The observed confidence levels reported in Table 4.3 that there is a great deal of confidence (90%-95%) that oak is more difficult to fell than birch.

4.5.3 A Problem from Biostatistics

Consider the example described in Section 1.3.5, which concerned selecting the best hyperactivity treatment from four possible treatments on the basis of a study which used a randomized complete block design. Let $\mathbf{Y}$ be a vector indexed as in Equation (4.4), where Y_{ij} represents the response for the ith treatment within the jth block for $i = 1, \ldots, 4$ and $j = 1, \ldots, 4$. The treatments have been labeled, in order, as A, B, C, and D. There are no replications in this experiment. This experiment can essentially be analyzed in terms of a two-way layout with no interaction. Such a model can then be set within the statistical framework of Chapter 3. See Theoretical Exercise 2. Of interest in this design is the effect of each treatment. Let θ_i be the estimable effect for treatment i for $i = 1, \ldots, 4$ with $\boldsymbol{\theta}' = (\theta_1, \theta_2, \theta_3, \theta_4)$, where as in Section 4.2.2, the overall mean of the model is taken to be 0. A secondary restriction that the effect of block 1 is 0 will also be used to create a unique model. Under these restrictions the effect of the ith treatment coincides with the mean response for treatment i so that the estimated effect of treatment i is then given by

$$\hat{\theta}_i = \frac{1}{4} \sum_{j=1}^{4} Y_{ij}.$$

Note that a different set of restrictions, such as the more common one given in Equation (5-19) of Montgomery (1997), could also be used without any modification. This is due to the fact that ordering the treatment means under those restrictions is equivalent to ordering the estimated treatment effects. The regions of interest correspond to each of the treatment effects having the lowest value without regard to the ordering of the remaining effects. That is

$$\Theta_i = \{\boldsymbol{\theta} \in \mathbb{R}^4 : \theta_i \geq \theta_j \text{ for } j = 1, \ldots 4\}.$$

To analyze this problem we use the definitions of $G_n(\mathbf{t})$ and $H_n(\mathbf{t})$ defined in Equations (3.1) and (3.2). The covariance matrix in this case is given by $\boldsymbol{\Sigma} = \sigma^2 \mathbf{I}_4$ where σ^2 is the error variance. The observed confidence levels for the four regions $\Theta_1, \ldots, \Theta_4$ are then developed as in Chapter 3. The observed confidence levels for these regions are reported in Table 4.5. One can observe from Table 4.5 that Treatment D has the highest observed confidence level for being most effective followed by Treatment A. One can also note that

Table 4.5 *Observed confidence levels for the four each of the four treatments in the hyperactivity example being the most effective.*

Region	$\hat{\alpha}_{stud}$	$\hat{\alpha}^*_{stud}$	$\hat{\alpha}^*_{hyb}$	$\hat{\alpha}^*_{perc}$
Θ_1	17.3	32.7	30.9	29.6
Θ_2	0	0	0	0
Θ_3	0	0	0	0
Θ_4	82.7	67.3	69.1	70.4

Treatment A has less than half the confidence of Treatment D, and that there is essentially no confidence that either Treatment C or Treatment D is the most effective.

4.5.4 A Problem from Biostatistics

The LRC-CPRT study was a large scale investigation of the effect of the drug cholostyramine on blood cholesterol levels. Efron and Feldman (1991) and Efron and Tibshirani (1996, 1998) consider modeling the decrease in the total blood cholesterol level for each participant in the study as a polynomial model of each participant's compliance with the intended dose of the drug. The data considered by Efron and Tibshirani (1996, 1998) consist of 201 observations from participants of the study who where in the control group and therefore received a placebo instead of the active treatment. A scatterplot of this data along with a cubic polynomial fit is displayed in Figure 4.3. There is clear curvature in the polynomial fit, but how much confidence is there in the shape of the regression line. For example, how much confidence is there that the curve is actually concave down over the range of the data dependent data?

Let Y_i be the decrease in total blood cholesterol level for participant i, and let x_i be the corresponding compliance level of the same participant. The regression model fit in Figure 4.3 then has the form

$$E(Y_i|x_i) = \theta_0 + \theta_1 x_i + \theta_2 x_i^2 + \theta_3 x_i^3.$$

Let $\boldsymbol{\theta}' = (\theta_0, \theta_1, \theta_2, \theta_3)$ be the parameter vector of interest in this problem with parameter space $\Theta = \mathbb{R}^4$. The region corresponding to the regression function being concave down has the form

$$\Theta_d = \{\boldsymbol{\theta} : 2\theta_2 + 6x\theta_3 < 0 \text{ for all } x \in [0, 100]\},$$

where the interval $[0, 100]$ corresponds to the range of participant compliance. Similarly, the region were the regression function would be concave up has the form

$$\Theta_u = \{\boldsymbol{\theta} : 2\theta_2 + 6x\theta_3 > 0 \text{ for all } x \in [0, 100]\}.$$

Figure 4.3 *Cholesterol decrease versus compliance for the participants in the control group for the Minnesota arm of the LRC-CPRT study. The solid line corresponds to a cubic regression fit.*

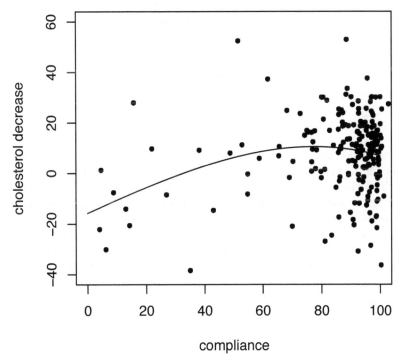

The region Θ_n will represent the remaining possibility that the regression line is neither strictly concave up or concave down in the range of compliance between 0 and 100.

Table 4.6 presents observed confidence levels for these regions using the methods described in this chapter. The contention that the curve is actually concave down, as observed in the fit of the data, only has observed confidence levels of about 30%. There is little or no confidence that the curve is actually concave up. Most of the observed confidence is that the curve is neither concave up or concave down.

4.6 Further Issues in Linear Regression

4.6.1 Model Selection

A major potential application of observed confidence levels in linear models and regression is in model selection. In the usual framework of model selection

Table 4.6 *Observed confidence levels in percent that the actual regression line is concave down (Θ_d), concave up (Θ_u) or neither (Θ_n). The bootstrap calculations are based on 10,000 resamples.*

Method	Regions		
	Θ_d	Θ_u	Θ_n
$\hat{\alpha}^*_{\text{stud}}$	27.2	0.00	72.8
$\hat{\alpha}^*_{\text{hyb}}$	30.15	0.02	69.83
$\hat{\alpha}^*_{\text{perc}}$	29.10	0.01	70.89

it is assumed that all of the relevant independent variables have been observed to model the dependent variable, and the problem of interest is then to discover which of the variables do not have a significant effect and can be removed from the model. Model selection was mentioned Efron and Gong (1983) as a potential application of a bootstrap method where the proportion of time each model was selected in a series of bootstrap resamples was suggested as an indication of which variables where relatively important in the model. Model selection in polynomial regression is also an example of significant interest in Efron and Tibshirani (1996, 1998). This section will explore these methods and will discuss the possible application of observed confidence levels to these types of problems.

In direct application observed confidence levels are of little help to the regression model selection problem. This is because the region corresponding to a particular variable not having a significant effect on the dependent variable usually has measure 0 with respect to the probability measure that provides the observed confidence levels for each region. As an example consider the example of fitting a polynomial regression model to the cholesterol data studied in Section 4.5.4. The interest in this example is determining the correct order of the polynomial for the regression function. Suppose a fifth-order polynomial is considered to be the absolute highest order to be considered. Therefore, the regression model to be fit has the form

$$E(Y|x) = \sum_{i=0}^{5} x^i \theta_i,$$

where the parameter vector $\boldsymbol{\theta}' = (\theta_0, \dots, \theta_5)$ has parameter space $\Theta = \mathbb{R}^6$. Consider the region corresponding to a fourth-order polynomial being sufficient for the model, that is $\Theta_4 = \{\boldsymbol{\theta} : \theta_5 = 0\}$. The region is a surface in $\mathbb{R}^5$ running along one of the axes in $\mathbb{R}^6$. Using the studentized method for computing observed confidence levels it follows that

$$\alpha_{\text{stud}}(\Theta_4) = \int_{n^{1/2} \boldsymbol{\Sigma}_x^{1/2} (\hat{\boldsymbol{\theta}} - \Theta_4)} h_n(\mathbf{t}) dt = 0,$$

whenever $h_n(\mathbf{t})$ is a continuous density in $\mathbb{R}^6$. This will be true for any model selection problem where one of the regression parameters must equal 0 for a smaller model to be true.

Efron and Tibshirani (1996, 1998) compute confidence levels for this example that are nonzero. The question of interest then is how the levels are computed, and what they mean. For the cholesterol data, Efron and Tibshirani (1996, 1998) first divide the cholesterol decrease values by 15.09 so that the variance of the data is approximately 1 and then consider a transformation of this standardized response vector $\mathbf{Y}$ so that the square of the successive coordinates correspond to the square root of the reduction in error sums of squares realized by adding one to the polynomial order of the model. Efron and Tibshirani (1998) report that the first three coordinates of this vector are 7.20, 2.74, and -2.55. The uncorrected total sum of squares of the standardized data for the model is

$$\sum_{i=1}^{201} Y_i^2 = 260.4234,$$

and the sums of squares, corrected for fitting a constant model is

$$\sum_{i=1}^{201} (Y_i - \bar{Y})^2 = 208.6101.$$

Therefore the square root of the decrease in sums of squares for fitting a constant model is $(260.4234 - 208.601)^{1/2} = 7.1979$, yielding the first element of the transformed vector. The remaining elements follow in a similar manner.

To choose the order of the polynomial model Efron and Tibshirani (1998) use Mallow's C_p criteria, which has the sample form

$$\hat{C}_p(j) = \sum_{i=j+1}^{201} \tilde{Y}_i^2 + 2(j+1), \tag{4.8}$$

where $\tilde{Y}_i$ is the ith component of the transformed response vector for the cholesterol data and j is the order of polynomial under consideration. The usual criteria for implementing the $\hat{C}_p$ is to compute $\hat{C}_p(j)$ for $j = 0, \ldots, k$ where k is the highest order polynomial under consideration. The jth-order polynomial is the indicated as the proper order when $\hat{C}_p(j)$ is the smallest of $\hat{C}_p(0), \ldots, \hat{C}_p(k)$. See Mallows (1973). In this context, Efron and Tibshirani (1998) define $\boldsymbol{\theta} = E(\tilde{\mathbf{Y}})$ with $\hat{\boldsymbol{\theta}} = \tilde{\mathbf{Y}}$. The corresponding parameter space is $\Theta = \mathbb{R}^{201}$. Assuming that the correct order of the model is p, the estimator in Equation (4.8) is an estimator of

$$\Gamma(j) = \sigma^{-2} E(\text{SSE}_j) - (n - 2p),$$

where SSE_j is the error sum of squares for the jth order model. When $j = p$, the correct order of the model, $E(\text{SSE}_j) = \sigma^2(n-p)$ and $\Gamma(p) = p$. Therefore the population selection criteria for the jth order model is that $\Gamma(j) = j$. Using the parameter space of the transformed regression vector defined above, the jth order polynomial is correct when $\boldsymbol{\theta} \in \Theta_j$ where $\Theta_j = \{\boldsymbol{\theta} : \theta_j = \theta_{j+1} = \cdots = \theta_k = 0\}$, which will have measure zero with respect to the sampling distribution of $\boldsymbol{\theta}$, for example. This differs from the application of the sample case implemented by Efron and Tibshirani (1998), which selects the jth order model when θ_j is the minimum of $\theta_1, \ldots, \theta_k$. Therefore, the method used by Efron and Tibshirani (1998) is actually equivalent to choosing the jth order model when $\boldsymbol{\theta}$ is in a neighborhood of Θ_j. The transformation implemented by Efron and Tibshirani (1998) does not solve the fundamental problem that the region of interest has measure 0 with respect to the probability measure of interest.

The possibility that a transformation of the parameter space will solve this problem is remote in most cases. To see this consider the simple example where $\mathbf{X} \sim F_{\boldsymbol{\theta}}$ is an observed d-dimensional random vector where Θ is the parameter space of $\boldsymbol{\theta}$. Take $\hat{\boldsymbol{\theta}} = \mathbf{X}$ and let $\Psi \subset \Theta = \mathbb{R}^d$ be any region such that

$$\int_{\Psi} dF_{\boldsymbol{\theta}} = 0,$$

for all $\boldsymbol{\theta} \in \Theta$. Now consider computing the percentile method observed confidence level of Ψ, where $\hat{F}$ is estimated by the parametric estimate $\hat{F} = F_{\hat{\boldsymbol{\theta}}}$. This observed confidence level is then given by

$$\alpha_{\text{perc}}(\Psi) = P(\hat{\boldsymbol{\theta}}^* \in \Psi | \hat{\boldsymbol{\theta}}^* \sim F_{\hat{\boldsymbol{\theta}}}, \mathbf{X}) = \int_{\Psi} dF_{\hat{\boldsymbol{\theta}}} = 0,$$

as long as $\hat{\boldsymbol{\theta}} \in \Theta$.

Can a transformation of the data vector $\mathbf{X}$, and therefore the parameter space Θ, provide a nonzero observed confidence level for Ψ? Let $\mathbf{Y} = L(\mathbf{X})$ for some real function L so that the parameter space of the problem is transformed to Λ_{Θ}. That is

$$\Lambda_{\Theta} = \{\boldsymbol{\lambda} : \boldsymbol{\lambda} = E(\mathbf{Y}), \mathbf{Y} = L(\mathbf{X}), \mathbf{X} \sim F_{\boldsymbol{\theta}}\}.$$

Let Λ_{Ψ} be the region of Λ_{Θ} that is *equivalent* to Ψ. This means we should get the same probability model using either representation, or that $\boldsymbol{\theta} \in \Psi$ if and only if $\boldsymbol{\lambda} \in \Lambda_{\Theta}$. Similarly, it will be assumed that $\hat{\boldsymbol{\theta}} \in \Psi$ if and only if $\hat{\boldsymbol{\lambda}} \in \Lambda_{\Theta}$. Note that if the regions are not equivalent the confidence level of Λ_{Ψ} will be computing a confidence level for a different model than is specified by Ψ. Now compute the percentile method observed confidence level for Λ_{Ψ}. Let G be the distribution of $L(\mathbf{X})$ and let $\hat{G}$ be an estimate of G. Then the percentile method observed confidence level for Λ_{Ψ} is

$$\alpha_{\text{perc}}(\Lambda_{\Psi}) = P(\hat{\boldsymbol{\lambda}}^* \in \Lambda_{\Psi} | \hat{\boldsymbol{\lambda}}^* \sim \hat{G}, \mathbf{X}).$$

The probability model $\hat{G}$ has not been specified here, and it could be chosen so that $\alpha_{\text{perc}}(\Lambda_{\Psi}) > 0$, but under the assumptions of this section it must follow that

$$P(\hat{\boldsymbol{\lambda}}^* \in \Lambda_{\Psi} | \hat{\boldsymbol{\lambda}}^* \sim \hat{G}, \mathbf{X}) = P(\hat{\boldsymbol{\theta}}^* \in \Psi | \hat{\boldsymbol{\theta}}^* \sim F_{\hat{\boldsymbol{\theta}}}, \mathbf{X}) = 0,$$

as before. It is clear then that two equivalent regions must have the same observed confidence level.

4.6.2 Observed Prediction Levels

Consider once again the simple linear regression model $\mathbf{Y} = \theta\mathbf{x} + \beta\mathbf{1} + \epsilon$ where the basic framework of Section 4.2.1 will be assumed along with the additional assumption that ϵ has a normal distribution with mean vector $\mathbf{0}$ and covariance matrix $\sigma^2\mathbf{I}$. Suppose that one wishes to predict the value of the dependent variable for a specified value of the independent variable x_p. In the usual case where θ and β are unknown, this predicted value will be given by $\hat{Y}_p = \hat{\theta}x_p + \hat{\beta}$ where $\hat{\theta}$ and $\hat{\beta}$ are the least squares estimates of θ and β computed from the observed data. To account for the variability inherent in making such a prediction, a $100\alpha\%$ prediction interval can be computed for $\hat{Y}_p$ as

$$D(\alpha, \boldsymbol{\omega}; \mathbf{x}, \mathbf{Y}) = [\hat{Y}_p - t_{1-\omega_L, n-2}\hat{\sigma}_p, \hat{Y}_p - t_{1-\omega_U, n-2}\hat{\sigma}_p], \tag{4.9}$$

where $\alpha = \omega_U - \omega_L$ and $\hat{\sigma}_p^2$ is an estimate of the variance of the predicted value given by

$$\hat{\sigma}_p^2 = \frac{\sum_{i=1}^n (Y_i - \hat{\beta} - \hat{\theta}x_i)^2}{n-2} \left[1 + n^{-1} + \frac{(x_p - \bar{x})^2}{\sum_{i=1}^n (x_i - \bar{x})^2} \right].$$

As with confidence intervals, one may be interested in how much confidence there is that the predicted value will be within a specified region, say $\Psi = (t_L, t_U)$. In this case the prediction analog of an observed confidence level can be computed by setting $\Psi = D(\alpha, \boldsymbol{\omega}; \mathbf{x}, \mathbf{Y})$ and solving for α. The resulting measure can then be called the *observed prediction level* of the region Ψ. For example, for the prediction interval given in Equation (4.9) the observed prediction level for $\Psi = (t_L, t_U)$ is given by

$$\alpha(\Psi) = T_{n-2}\left[\frac{\hat{Y}_p - t_L}{\hat{\sigma}_p}\right] - T_{n-2}\left[\frac{\hat{Y}_p - t_U}{\hat{\sigma}_p}\right].$$

4.7 Computation Using R

The fundamental framework for calculating observed confidence levels in R in the regression setting is very similar to what was developed for the multiparameter case in Section 3.6. The main difference is that the algorithm for generating resamples relies on an empirical model approach in which the residuals from the original fit of the linear model are resampled. These resamples are then filtered through the fitted model to produce resampled data. The boot function is general enough to handle this case with the only modifications occuring in the way that the resampling function is developed.

To motive the method that is used to generate the resamples, note that in

Equation 4.2, $\hat{\boldsymbol{\theta}}^*$ is the least squares estimate of the slope parameter vector fitted from resampled data $\mathbf{Y}^*$ with the same constant matrix $\mathbf{X}$. The error vector $\boldsymbol{\epsilon}^*$ is generated by resampling from the residuals from the original fit of the model given by $\hat{\boldsymbol{\epsilon}}$, that is, by taking a sample of size n from the empirical distribution of $\hat{\boldsymbol{\epsilon}}$ conditional on $\mathbf{Y}$ and $\mathbf{X}$. The resampled vector $\mathbf{Y}$ is then generated as $\mathbf{Y}^* = \hat{\beta}\mathbf{1}_n + \mathbf{X}\hat{\boldsymbol{\theta}} + \boldsymbol{\epsilon}^*$. This method of generating resamples in the regression model, usually known as *residual resampling*, is generally considered to be most appropriate for regression models where $\mathbf{X}$ is a fixed matrix of constants. The alternate method of resampling the rows of $\mathbf{Y}$ along with the corresponding rows of $\mathbf{X}$ is equivalent to resampling from the joint empirical distribution of $\mathbf{Y}$ and $\mathbf{X}$ taken together. This method is more appropriate when $\mathbf{X}$ as well as $\mathbf{Y}$ are stochastic. See Chapter 7 of Shao and Tu (1995).

Suppose $\hat{\boldsymbol{\theta}}_i^*$ and $\hat{\sigma}_i^*$ are the least squares estimates of $\boldsymbol{\theta}$ and σ computed on the ith set of resampled data given by $\mathbf{Y}_i^*$ and $\mathbf{X}$ for $i = 1, \ldots, b$. The bootstrap estimates of $g_n(\mathbf{t})$ and $h_n(\mathbf{t})$ can then be approximated by

$$\hat{g}_n(\mathbf{t}) \simeq b^{-1} \sum_{i=1}^{b} \delta[\mathbf{t} : n^{1/2}\boldsymbol{\Sigma}_x^{1/2}(\hat{\boldsymbol{\theta}}_i^* - \hat{\boldsymbol{\theta}})/\hat{\sigma}],$$

and

$$\hat{h}_n(\mathbf{t}) \simeq b^{-1} \sum_{i=1}^{b} \delta[\mathbf{t} : n^{1/2}\boldsymbol{\Sigma}_x^{1/2}(\hat{\boldsymbol{\theta}}_i^* - \hat{\boldsymbol{\theta}})/\hat{\sigma}_i^*],$$

where $n^{1/2}\boldsymbol{\Sigma}_x^{1/2}(\hat{\boldsymbol{\theta}}_i^* - \hat{\boldsymbol{\theta}})/\hat{\sigma}$ and $n^{1/2}\boldsymbol{\Sigma}_x^{1/2}(\hat{\boldsymbol{\theta}}_i^* - \hat{\boldsymbol{\theta}})/\hat{\sigma}_i^*$ are taken to be singleton sets. The bootstrap estimates of the observed confidence levels for a region Ψ can then be approximated by

$$\hat{\alpha}_{\text{ord}}^*(\Psi) \simeq b^{-1} \sum_{i=1}^{b} \delta[\hat{\boldsymbol{\theta}} - \sigma(\hat{\boldsymbol{\theta}}_i^* - \hat{\boldsymbol{\theta}})/\hat{\sigma}; \Psi],$$

$$\hat{\alpha}_{\text{stud}}^*(\Psi) \simeq b^{-1} \sum_{i=1}^{b} \delta[\hat{\boldsymbol{\theta}} - \hat{\sigma}(\hat{\boldsymbol{\theta}}_i^* - \hat{\boldsymbol{\theta}})/\hat{\sigma}_i^*; \Psi],$$

$$\hat{\alpha}_{\text{hyb}}^*(\Psi) \simeq b^{-1} \sum_{i=1}^{b} \delta[2\hat{\boldsymbol{\theta}} - \hat{\boldsymbol{\theta}}_i^*; \Psi],$$

and

$$\hat{\alpha}_{\text{perc}}^*(\Psi) \simeq b^{-1} \sum_{i=1}^{b} \delta[\hat{\boldsymbol{\theta}}_i^*; \Psi],$$

following the parallel development in Section 3.6.

The calculations performed for the example from Section 4.5.1 will demonstrate many of the principles used to compute observed confidence levels for regression models in R. It will be assumed that the data are stored in an objects called GI and EGI. The object GI corresponds to the actual germination index values and the object EGI corresponds to the estimated germination index values. A simple linear regression model can be fit in R using the command GI.fit <- lm(EGI~GI) where the information from the fit of the model is stored in the object GI.fit. Information about the fit can be obtained by applying commands such as anova and summary to the object. For example, the summary command yields

```
> summary(GI.fit)

Call:
lm(formula = EGI ~ GI)

Residuals:
     Min        1Q    Median        3Q       Max
-20.04915  -7.20928   0.00631   8.23430  20.25649

Coefficients:
            Estimate Std. Error t value Pr(>|t|)
(Intercept) 10.91242    4.97057   2.195    0.039 *
GI           0.80399    0.09478   8.483  2.2e-08 ***
---
Signif. codes:  0 *** 0.001 ** 0.01 * 0.05 . 0.1   1

Residual standard error: 10.4 on 22 degrees of freedom
Multiple R-Squared: 0.7659,Adjusted R-squared: 0.7552
F-statistic: 71.96 on 1 and 22 DF,  p-value: 2.201e-08
```

which provides the least squares estimates of β and θ. Using the estimator defined in Section 4.2.1, the estimate of σ is can be obtained by extracting the residuals from the object GI.fit using the command

```
sqrt(sum(GI.fit$residuals^2)/length(GI.fit$residuals))
```

The first region of interest in this example is the region Θ_1 which corresponds to all values of the slope parameter that are less than 1. The function used by the boot that is suitable for computing $\hat{\alpha}_{hyb}^*(\Theta_1)$ is given by

```
giboothyb <- function(data,i,tval,fit) {
    Ystar <- fit$coefficients[1] + fit$coefficients[2]*data
        + fit$residuals[i]
    fitstar <- lm(Ystar~data)
    if(2*fit$coefficients[2]-fitstar$coefficients[2]<tval)
        delta <- 1
    else delta <- 0
    return(delta) }
```

In this case it is assumed that the only data specified in the boot command is the GI object. It is also assumed that the objects GI.fit and tval are passed to this function through the boot command. This allows the giboothyb function to resample from the residuals of the original fit for without having to compute the least squares fit each time. The command

```
mean(boot(GI,statistic=giboothyb,R=1000,tval=1,fit=GI.fit)$t)
```

will the return the estimated observed confidence level for Θ_1 based on the hybrid bootstrap method. To compute the $\hat{\alpha}^*_{\text{perc}}(\Theta_1)$ the function

```
gibootperc <- function(data,i,tval,fit) {
    Ystar <- fit$coefficients[1] + fit$coefficients[2]*data
        + fit$residuals[i]
    fitstar <- lm(Ystar~data)
    if(fitstar$coefficients[2]<tval) delta <- 1
    else delta <- 0
    return(delta) }
```

is used. To compute $\hat{\alpha}^*_{\text{stud}}(\Theta_1)$ we add the minor complication that $\hat{\sigma}^*_i$ must be computed with each resample. In this case this does not require an added level of resampling since the estimate has a closed form. A suitable function to compute $\hat{\alpha}^*_{\text{stud}}(\Theta_1)$ with the boot command is

```
gibootstud <- function(data,i,tval,fit) {
    Ystar <- fit$coefficients[1] + fit$coefficients[2]*data
        + fit$residuals[i]
    fitstar <- lm(Ystar~data)
    shat <- sum(fit$residuals^2)/length(fit$residuals)
    shatstar <- sum(fitstar$residuals^2)
        /length(fitstar$residuals)
    if(fit$coefficients[2]-shat*(fitstar$coefficients[2]
        -fit$coefficients[2])/shatstar<tval) delta <- 1
    else delta <- 0
    return(delta) }
```

Note that a slightly more efficient version of this function can be created by sending shat directly to the function so that it need not be recalculated for each resample.

To compute the observed confidence levels for Θ_2 similar functions are used. For example, a suitable function to compute $\hat{\alpha}^*_{\text{stud}}(\Theta_2)$ with the boot command is given by

```
gibootstud2 <- function(data,i,fit) {
    Ystar <- fit$coefficients[1] + fit$coefficients[2]*data
        + fit$residuals[i]
    fitstar <- lm(Ystar~data)
    shat <- sum(fit$residuals^2)/length(fit$residuals)
```

```
shatstar <- sum(fitstar$residuals^2)
    /length(fitstar$residuals)
tbeta <- fit$coefficients[1] - shat*
    (fitstar$coefficients[1]
    -fit$coefficients[1])/shatstar
ttheta <- fit$coefficients[2] - shat*
    (fitstar$coefficients[2]
    -fit$coefficients[2])/shatstar
if((tbeta<=50)&&(tbeta+50*ttheta<=50)) delta <- 1
    else delta <- 0
return(delta) }
```

The command

```
mean(boot(GI,statistic=gibootstud2,R=1000,fit=GI.fit)$t)
```

will the return the estimated observed confidence level for Θ_2 based on the studentized bootstrap method.

The normal approximation $\hat{\alpha}_{\text{stud}}$ can be computed using numerical integration of the bivariate standard normal density, or by using simulation techniques as follows. Let $\mathbf{Z}_1^*, \ldots, \mathbf{Z}_b^*$ be a set of independent and identically distributed random variables from a bivariate normal density with mean vector $\mathbf{0}$ and covariance matrix $\mathbf{I}$. Then

$$
\begin{aligned}
\hat{\alpha}_{\text{stud}}(\Psi) &= \int_{n^{1/2}\tilde{\boldsymbol{\Sigma}}_x^{1/2}(\hat{\boldsymbol{\theta}}-\Psi)/\hat{\sigma}} \phi(\mathbf{t})d\mathbf{t} \\
&\simeq b^{-1}\sum_{i=1}^{b}\delta[\mathbf{Z}_i^*; n^{1/2}\tilde{\boldsymbol{\Sigma}}_x^{1/2}(\hat{\boldsymbol{\theta}}-\Psi)/\hat{\sigma}] \\
&= b^{-1}\sum_{i=1}^{b}\delta[\hat{\boldsymbol{\theta}} - \hat{\sigma}n^{-1/2}\tilde{\boldsymbol{\Sigma}}_x^{-1/2}\mathbf{Z}_i^*; \Psi].
\end{aligned}
$$

This calculation is easily implemented in R using the following commands

```
b <- 1000
n <- length(GI)
kount <- 0
that <- GI.fit$coefficients
shat <- sqrt(sum(GI.fit$residuals^2)/n)
mx <- mean(GI)
dx2 <- sum((GI-mx)^2)
sx <- sqrt(dx2/n)
Sx <- matrix(c(1/n+mx^2/(dx2),-1.0*mx/(dx2),
    -1.0*mx/(dx^2),1/dx2),2,2)
svdSx <- svd(Sx)
srSx <- svdSx$u%*%sqrt(diag(svdSx$d))%*%svdSx$u
```

```
Z <- matrix(rnorm(2*b),b,2)
for(i in 1:b)
{
    Zt <- that - shat/sqrt(n)*solve(srSx)%*%Z[i,]
    if((Zt[1]<=50)&&(Zt[1]+50*Zt[2]<=50)) kount <- kount + 1
}
```

4.8 Exercises

4.8.1 Theoretical Exercises

1. Consider the simple linear regression model described in Section 4.2.1.

   (a) Derive the hybrid critical point for estimating the intercept parameter β. Use this critical point to write out a $100\alpha\%$ confidence interval for β.

   (b) Using the confidence interval derived in Part (a), find the hybrid observed confidence level for β for an arbitrary interval region $\Psi = (t_L, t_U)$.

   (c) Find the normal and bootstrap approximations to the hybrid observed confidence level for β for an arbitrary interval region $\Psi = (t_L, t_U)$.

2. Consider a two-factor analysis of variance without interaction. In this case a random vector $\mathbf{Y}$ indexed as Y_{ijk} is observed where

$$Y_{ijk} = \mu + \gamma_i + \delta_j + \epsilon_{ijk}$$

   for $i = 1, \ldots, p$, $j = 1, \ldots, q$ and $k = 1, \ldots, n$. The parameters $\mu, \gamma_1, \ldots, \gamma_p$ and $\delta_1, \ldots, \delta_q$ are unknown constants.

   (a) Write out the general form of a matrix $\mathbf{D}$ that would allow this model to be written as $\mathbf{Y} = \mathbf{D}\boldsymbol{\lambda} + \boldsymbol{\epsilon}$ where

$$\boldsymbol{\lambda}' = (\mu, \gamma_1, \ldots, \gamma_p, \delta_1, \ldots, \delta_q)$$

   is the parameter vector and $\boldsymbol{\epsilon}$ is a vector containing the error random variables ϵ_{ijk} and is indexed the same as the vector $\mathbf{Y}$ above. Find the rank of $\mathbf{D}$.

   (b) Reparameterize the model given in part (a) so that it is a full rank model with unique parameters.

   (c) Write the model given above in a form so that it can be analyzed using the multiple parameter methods of Chapter 3.

3. Consider the percentile method for computing observed confidence levels for the slope parameter θ of a simple linear regression model. Following the development in Chapter 3, develop a bias-corrected observed confidence level for this parameter, and perform an asymptotic accuracy analysis to determine the order of accuracy of the method (Poliak, 2007).

Table 4.7 *Drilling data.*

Brand A

Depth	71.0	92.2	87.4	51.6	70.2	65.1	50.3	91.4
Time	34.9	45.7	43.7	29.1	35.9	33.0	25.1	47.4

Depth	86.2	97.0	30.7	43.5	48.9	96.2	74.9	19.9
Time	44.7	45.9	16.0	23.3	23.9	45.6	35.6	12.0

Brand B

Depth	30.7	67.1	47.7	47.3	40.7	41.9	22.5	20.3
Time	18.5	27.6	27.5	20.4	16.2	20.0	11.1	8.8

Depth	36.0	27.1	87.3	30.9	90.4	38.4	90.1	40.6
Time	21.3	11.8	31.9	19.2	36.5	20.7	38.6	15.9

Brand C

Depth	20.7	98.6	11.2	49.2	52.2	53.1	64.1	53.7
Time	21.9	74.1	15.8	45.3	42.4	42.8	51.1	36.0

Depth	80.8	82.9	88.2	23.1	96.6	17.8	34.2	29.3
Time	62.4	64.0	62.4	20.5	70.3	11.0	25.5	18.7

4. Consider the percentile method for computing observed confidence levels for the slope parameter θ of a simple linear regression model. Following the development in Chapter 3, develop an accelerated bias-corrected observed confidence level for this parameter, and perform an asymptotic accuracy analysis to determine the order of accuracy of the method. (Poliak, 2007).

5. Following the framework used in this chapter, show that the hybrid method produces second-order accurate observed confidence levels for slope parameters in the linear regression model.

4.8.2 Applied Exercises

1. A mining company is considering three possible brands of drill bits that are designed to penetrate a certain type of granite. In order to study the three brands, the company purchases sixteen of each type of bit and observes the amount of time (in seconds) each bit takes to drill holes of varying depths (in centimeters) throughout the work week. The observed data from this study are given in Table 4.7. Assume that the data are independent between brands.

   (a) Fit a linear regression model for each brand separately. Compare the estimates of the slopes of each fit. Which brand appears to be most efficient in drilling through the granite, that is, which brand has the largest slope?

(b) Compute observed confidence levels that each brand has the largest slope using the normal approximation, and bootstrap estimates of the studentized, hybrid and percentile methods. Which brand, or brands, would you recommend to the company?

2. An automotive engine design firm is considering two new types of software for their electronic engine control systems. They wish to compare these systems to a standard system which is currently installed in a certain model of vehicle. A set of twenty-one automobiles of this model are obtained and are randomly divided into three groups of seven each. Each group is then randomly assigned to one of the software systems. The software systems are then installed in the automobiles and the gasoline mileage for each vehicle was determined using a standard testing routine. The data are given in the table below.

Standard Software System						
27.4	28.3	27.2	28.3	27.7	27.9	26.9

New Software System I						
28.2	28.5	28.9	28.8	27.2	29.4	28.8

New Software System II						
31.0	30.3	31.4	29.0	31.4	30.8	30.6

(a) Fit a one-way analysis of variance on the data given in the table. Is there a significant difference in the mean gasoline mileage for the three systems?

(b) Using the bootstrap with $b = 1000$ resamples compute observed confidence levels using the studentized, hybrid, and percentile methods for regions of the parameter space corresponding to each possible ordering of the software systems with respect to mean gasoline mileage. What conclusions can be made about the gasoline mileage of the three systems?

3. The printing time for a page (in seconds) of grayscale graphics is being studied in terms of the pixel density of the page being printed and the type of printer being used. The pixel density of a page is a measure of how dark the entire area of the page is that uses both the number of nonwhite pixels used on the page as well as the grayscale darkness of the pixel. The pixel density is scaled to be between 0 and 100 where 0 represents a blank page and 100 represents a totally black page. Ten different graphics, each with varying levels of pixel density, were printed using three different printers. The observed data is given in Table 4.8.

Table 4.8 *Printing time data.*

Graphic	Pixel Density	Printing Time Printer I	Printer II	Printer III
1	89	18.1	17.3	7.3
2	26	9.3	5.0	8.8
3	34	8.6	7.6	18.3
4	12	4.4	5.2	8.7
5	24	7.8	8.7	4.2
6	52	9.0	10.9	12.4
7	30	8.9	8.1	8.0
8	31	7.7	9.3	11.2
9	71	12.8	13.1	8.5
10	50	11.0	8.7	10.6

(a) Fit a separate regression model for each printer using printing time as the dependent variable and pixel density as the independent variable. Does it appear that the slopes of the three fits, which is a measure of how many seconds are spent printing per unit of pixel density, are the same? Examining the plots, does it appear that there are other factors besides the slopes that might affect the printing speed of the printers?

(b) Compute observed confidence levels for each of the six possible orderings of the three slopes using the normal approximation and the bootstrap estimates of the studentized, hybrid, and percentile methods. For the bootstrap calculations use $b = 1000$. The printer with the smallest slope is considered the fastest per unit of pixel density. Do the observed confidence levels suggest which printer or printers are the best in terms of this measure?

(c) Consider only the data from Printer I. Consider the regions for the slope parameter equal to $\Theta_1 = (0, 0.05]$, $\Theta_2 = (0.5, 0.10]$ $\Theta_3 = (0.10, 0.15]$, $\Theta_3 = (0.15, 0.20]$ and $\Theta_3 = (0.20, \infty)$ corresponding to different classifications of the speed of the printer. Compute observed confidence levels that the slope parameter for Printer I is within each of these regions using the normal approximation and the bootstrap estimates of the studentized, hybrid, and percentile methods. For the bootstrap calculations use $b = 1000$. Is there a region that appears most likely?

(d) Consider again only the data from Printer I. Using the data construct a 95% prediction interval for the printing time of a page that has a pixel density equal to 75 using the usual normal theory based prediction interval. Now consider regions of printing times equal to $(0, 5]$, $(5, 10]$ and $(10, 20]$. Compute observed prediction levels for each of these regions using the usual normal theory based prediction interval as a basis for computing the observed prediction levels.

Table 4.9 *Viscosity data.*

Sample	1	2	3	4	5
Standard Method	0.5142	0.5027	0.4867	0.4938	0.5230
New Method	0.5017	0.4838	0.4703	0.4806	0.5128
Sample	6	7	8	9	10
Standard Method	0.4984	0.4842	0.5062	0.4837	0.5108
New Method	0.4827	0.4666	0.4911	0.4723	0.4988
Sample	11	12	13	14	15
Standard Method	0.5136	0.4846	0.5145	0.4912	0.4970
New Method	0.4969	0.4729	0.4955	0.4803	0.4844
Sample	16	17	18	19	20
Standard Method	0.5017	0.5067	0.5168	0.4857	0.5015
New Method	0.4896	0.4901	0.5003	0.4709	0.4870

4. The viscosity (measured in centipoise, or cP) of a chemical used to produce paint is an important factor that determines how well the finished paint product will perform. The company purchasing the chemical currently uses a standard test procedure to determine the viscosity of a shipment of the chemical, but the procedure is time consuming. A new procedure, which is much faster, is tested to determine if it is a reliable substitute for the standard procedure. Each method was used to test the viscosity of twenty shipments of the chemical. The data are given in Table 4.9.

   (a) Fit a linear regression line using the new method as the independent variable and the standard method as the dependent variable. Using this line, what assessment would you give as to the relationship between the two measurements? Does the new method tend to systematically under or over estimate the viscosity? What correction would you suggest making to the new method observations to make them agree more with the standard method observations? Discuss the possible hazards of making such corrections.

   (b) The company decides to accept the new method as valid if there is a high degree of confidence that the intercept of the true regression line is in the range $(-0.05, 0.05)$ and the slope of the true regression line is between 0.95 and 1.05. Compute an observed confidence level for this region using normal approximation and the bootstrap estimates of the studentized, hybrid, and percentile methods. What recommendation should be made to the company?

   (c) Compute a 95% confidence interval for the mean measurement of the standard method when the new method equals 0.5. Compute observed confidence levels that this mean value is within 0.01, 0.05, and 0.10 of 0.5. Compare these results to the confidence interval.

(d) Compute a normal theory 95% prediction interval for the measurement of the standard method when the new method equals 0.5. Compute normal theory observed prediction levels that this value is within 0.01, 0.05, and 0.10 of 0.5. Compare these results to the prediction interval.

5. Fischer and Polansky (2006) study the effect of normalized remote tensile stress on the number of joints, or cracks, in bedded layers of sedimentary rock. Some representative data of such a process is given in the table below.

Stress	8.1	2.9	5.0	6.7	6.9	7.8
Number of Joints	116	98	113	114	113	115
Stress	3.2	1.3	3.7	8.8	2.0	4.2
Number of Joints	98	56	105	117	83	107
Stress	4.9	5.8	3.2	3.7	5.2	3.3
Number of Joints	111	115	100	108	113	102
Stress	5.3	9.7	8.1	2.8	4.9	6.3
Number of Joints	111	119	118	100	114	115
Stress	7.8	8.5	5.5	7.5	4.6	1.1
Number of Joints	118	117	112	115	111	38

(a) Fit a polynomial regression model to the data using a method such as Mallow's C_p to determine the order of the polynomial. Perform a residual analysis on the fit. Does the fit appear to be reasonable?

(b) Consider a general k^{th} order polynomial of the form

$$f(x) = \sum_{i=0}^{k+1} \theta_i x^i,$$

where k is the order of polynomial determined in Part (a). Let $\theta = (\theta_0, \ldots, \theta_{k+1})$ and describe regions Θ_U and Θ_D that specify parameters of the polynomial model where f is concave up, and down, respectively, within a specified interval (a, b).

(c) Using the representation from Part (b), compute observed confidence levels that the $(k + 1)$-order polynomial regression model that predicts the number of joints based on the remote tensile stress is concave down, concave up, or neither on the range of stress between 1 and 10. Use the normal approximation, and the bootstrap estimates of the studentized, hybrid and percentile methods.

(d) Compute a 99% confidence interval for the mean number of joints when the remote tensile stress is 3. Compute observed confidence levels that the mean number of joints when the tensile stress if 3 is in the regions $[95, 100)$ and $[100, 105)$. Compare the observed confidence levels to the confidence interval. For both calculations use the normal approximation, and the bootstrap estimates of the studentized, hybrid and percentile methods.

(e) Compute a normal theory 99% prediction interval for the mean number of joints when the remote tensile stress is 3. Compute normal theory observed prediction levels that the mean number of joints when the tensile stress of 3 is in the regions $[95, 100)$ and $[100, 105)$. Compare the observed prediction levels to the prediction interval. For both calculations use the normal-based theory.

CHAPTER 5

Nonparametric Smoothing Problems

5.1 Introduction

Many modern problems in nonparametric statistics have focused on the estimation of smooth functions. Nonparametric density estimation considers estimating an unknown smooth density using only assumptions that address the smoothness of the unknown density and the weight of its tails. Nonparametric regression techniques focus on the estimation of an unknown smooth conditional expectation function under similar assumptions. This chapter addresses the issues of computing and assessing the accuracy of observed confidence levels for these types of problems. The regions of interest in these problems often consider the density or regression function itself as the parameter of interest. As such, the parameter space becomes a space of functions, and the regions of interest often correspond to subsets of the function space that contain functions with particular properties. For example, in nonparametric density estimation interest may lie in assessing the amount of confidence there is, based on the observed data, that the unknown population density has one, two or three modes. In nonparametric regression problems it may be of interest to assess the amount confidence there is that the regression function lies within a certain region.

Section 5.2 will consider computing observed confidence levels for problems in nonparametric density estimation. For simplicity, this section will focus on the method of kernel density estimation. This method is widely used and is comparatively simple to apply, but is sophisticated enough to provide a sound theoretical framework for the development of the observed confidence levels. Other methods can be used as well with the proper modifications to the corresponding theory. Section 5.2 also includes some empirical results on the behavior of observed confidence levels in these problems as well the application of the methods to several examples. Section 5.5 considers computing observed confidence levels for problems in nonparametric regression. As with section 5.2, Section 5.5 will focus on the particular method of local polynomial regression, also known as kernel regression. As with kernel density estimation, local polynomial regression is widely used, is simple to apply, and has a sophisticated enough theoretical framework to allow for the theoretical development of observed confidence levels to these problems. This section also includes some empirical results on the behavior of observed confidence levels in these problems as well as the application of these methods to two examples.

5.2 Nonparametric Density Estimation

5.2.1 Methods for Nonparametric Density Estimation

This section reviews the basic elements of kernel density estimation. Further details can be found in Chapter 2 of Wand and Jones (1995). Additional methods for nonparametric density estimation can be found in Chapters 1–2 of Bowman and Azzalini (1997) and Chapters 2–3 in Simonoff (1996).

Let $X_1, \ldots, X_n$ be a set of independent and identically distributed random variables from a distribution F that has continuous density f. Section 1.2 introduced the empirical distribution function given by

$$\hat{F}_n(t) = n^{-1} \sum_{i=1}^{n} \delta(t; [X_i, \infty)),$$

for all $t \in \mathbb{R}$ as a estimate of the distribution function F. No assumptions are necessary to show that this estimator is point-wise unbiased and consistent. Therefore, the empirical distribution function is considered a nonparametric estimate of F. Unfortunately, the empirical distribution function cannot directly provide a nonparametric estimate of the continuous density f. This is because $\hat{F}_n(t)$ is a step function with steps of size $n^{-1} \sum_{i=1}^{n} \delta(t; \{X_i\})$ for each $t \in \mathbb{R}$. Hence the step sizes correspond to the number of values in the observed sample that equal t. Therefore, the empirical distribution function corresponds to the discrete distribution

$$p(t) = n^{-1} \sum_{i=1}^{n} \delta(t; \{X_i\}),$$

for all $t \in \mathbb{R}$. Fix and Hodges (1951, 1989) considered using the proportion of X_i values within a neighborhood of t to estimate an unknown density $f(t)$. The width of the neighborhood was decreased as a function of the sample size to create a consistent and continuous estimate of $f(t)$. This is equivalent to taking the convolution between the empirical distribution function $\hat{F}_n$ and a uniform density whose range depends on the sample size n. Rosenblatt (1956) and Parzen (1962) considered the generalization of taking the convolution of $\hat{F}_n$ with a continuous weight function, that is not necessarily uniform, to develop what is called *kernel density estimator*. Therefore, a kernel density estimate of the continuous unknown density f has the form

$$\hat{f}_{n,h}(x) = (nh)^{-1} \sum_{i=1}^{n} K\left(\frac{x - X_i}{h}\right),$$

where K is a continuous function called the *kernel function* where

$$\int_{\mathbb{R}} K(x)dx = 1,$$

$$\int_{\mathbb{R}} x^2 K(x)dx = \sigma_K^2 < \infty,$$

and

$$\int_{\mathbb{R}} xK(x)dx = \int_{\mathbb{R}} x^3 K(x)dx = 0.$$

The parameter h is called the *bandwidth*, and controls how smooth the resulting estimate is. Both the kernel function K and the bandwidth h control the shape of the resulting estimate, and therefore both must be chosen in a way that produces a reasonable estimate. There are many ways to evaluate how close the kernel estimate $\hat{f}_{n,h}$ is to the true density f. The most common approach is to use the *mean integrated square error* (MISE) given by

$$\text{MISE}(\hat{f}_{n,h}, f) = \int_{\mathbb{R}} E[\hat{f}_{n,h}(x) - f(x)]^2 dx.$$

Rosenblatt (1956) and Parzen (1962) consider an asymptotic analysis of $\text{MISE}(\hat{f}_{n,h}, f)$. Such an analysis requires the following assumptions.

1. The density f has a continuous, square integrable and an ultimately monotone second derivative f''.

2. The bandwidth h is a sequence of constants depending on the sample size n such that $h \to 0$ and $nh \to \infty$ as $n \to \infty$.

3. The kernel function K is a symmetric and bounded probability density having at least four finite moments.

For further details on the first assumption see page 20 of Wand and Jones (1995). Under Assumptions 1–3 it can be shown that

$$\text{MISE}(\hat{f}_{n,h}, f) = (nh)^{-1}R(K) + \tfrac{1}{4}h^4 R(f'')\sigma_K^2 + o[(nh)^{-1} + h^4], \qquad (5.1)$$

where, for a real function g that is square integrable,

$$R(g) = \int_{\mathbb{R}} g^2(x)dx.$$

If the error term in Equation (5.1) is ignored and the remaining terms are minimized with respect to h for a fixed kernel function K then an asymptotically optimal value for the bandwidth h can be found as

$$h_{\text{opt}} = n^{-1/5}\left[\frac{R(K)}{\sigma_K^4 R(f'')}\right]^{1/5}. \qquad (5.2)$$

Epanechnikov (1969) shows that there also exists an optimal kernel function. Unfortunately, this choice of kernel function can result in a kernel density estimate that is not differentiable everywhere, which may affect the calculation of some functionals of f. A careful study of the asymptotic expansion in

Equation (5.1) reveals that the choice of the kernel function does not have an effect of the asymptotic order of $\mathrm{MISE}(\hat{f}_{n,h}, f)$. Further, the relative efficiency of alternate kernel functions, such as the normal kernel, is quite close to 1. Therefore, from an asymptotic viewpoint, the choice of kernel function is usually considered a secondary issue. In this book, unless otherwise specified, the normal kernel function will always be used in examples and simulations.

It is apparent from Equation (5.2) that the optimal value of the crucial bandwidth depends on the unknown density through the function $R(f'')$. That is, h_{opt} must be estimated based on the observed sample $X_1, \ldots, X_n$. A substantial amount of research has been devoted to the problem of estimating the asymptotically optimal bandwidth. See Silverman (1995), Simonoff (1996), and Wand and Jones (1995) for detailed reviews of the more common methods.

A simple method for estimating h_{opt} is given by the least squares cross validation method of Rudemo (1982) and Bowman (1984). This method is based on the idea that

$$\mathrm{MISE}(\hat{f}_{n,h}, f) = E\left[R(\hat{f}_{n,h}) - 2\int_{\mathbb{R}} \hat{f}_{n,h}(x)f(x)dx\right] + R(f).$$

It can be shown that an unbiased estimator of $\mathrm{MISE}(\hat{f}_{n,h}, f) - R(f)$ is given by

$$\mathrm{LSCV}(h) = R(\hat{f}_{h,n}) - 2n^{-1}\sum_{i=1}^{n}\hat{f}_{n,h,(-i)}(X_i),$$

where

$$\hat{f}_{n,h,(-i)}(X_i) = [(n-1)h]^{-1}\sum_{j\neq i}^{n}K\left(\frac{x - X_j}{h}\right),$$

is the kernel density estimate of f with bandwidth h, based on the sample where the observation X_i has been deleted. Because $R(f)$ does not depend on the bandwidth h, minimizing $\mathrm{MISE}(\hat{f}_{n,h}, f)$ is equivalent to minimizing $\mathrm{MISE}(\hat{f}_{n,h}, f) - R(f)$, that is $\hat{h}_{\mathrm{LSCV}} = \arg\min\{h : \mathrm{LSCV}(h)\}$. It can occur that the function $\mathrm{LSCV}(h)$ may have more than one local minimum. In such a case has been suggested that the largest local minimum be used as the estimate of h_{opt}. See Hall and Marron (1991) and Marron (1993). Theoretical and empirical studies of the least squares cross-validation estimator of h_{opt} show that $\hat{h}_{\mathrm{LSCV}}$ usually has a large variance. See Hall and Marron (1987a) and Park and Marron (1990). A biased version of this method, which reduces the variance, has been suggested by Scott and Terrell (1987).

An alternative method to least squares cross-validation is based on direct plug-in methods. Note that from Equation (5.2), estimating the asymptotically optimal bandwidth is equivalent to estimating $R(f'')$. It is convenient at this point to generalize the problem. Let

$$R(f^{(s)}) = \int_{\mathbb{R}} [f^{(s)}(x)]^2 dx.$$

Hall and Marron (1987b) and Jones and Sheather (1991) suggest a kernel estimator of $R(f^{(s)})$ of the form

$$\hat{R}_{n,g}(f^{(s)}) = (-1)^s g^{-(2s+1)} n^{-2} \sum_{i=1}^{n} \sum_{j=1}^{n} L^{(2s)}\left(\frac{X_i - X_j}{g}\right), \qquad (5.3)$$

where g is a bandwidth which may differ from h and L is a kernel function which may differ from K. Under suitable conditions it can be shown that the asymptotically optimal bandwidth g for estimating $R(f^{(s)})$ is given by

$$g_{\text{opt}} = n^{-(2s+3)} \left[\frac{2L^{(2s)}(0)}{-\sigma_L^2 R(f^{(s+1)})}\right]^{1/(2s+3)}, \qquad (5.4)$$

which obviously depends on $R(f^{(s+1)})$. This suggests an iterative process for estimating h_{opt}. To begin the process the bandwidth in Equation (5.4) is approximated by assuming that f is a normal density whose standard deviation is given by $\tilde{\sigma} = \min\{\hat{\sigma}, \text{IQR}/1.349\}$ as suggested by Silverman (1986, page 47), where $\hat{\sigma}$ is the sample standard deviation and IQR is the sample interquartile range. In this case it can be shown that the kernel estimate in Equation (5.3) reduces to

$$\hat{R}_{\phi}(f^{(s)}) = \frac{(-1)^s (2s)!}{(2\tilde{\sigma})^{2s+1} s! \pi^{1/2}},$$

where it has been assumed that $L = \phi$. This is known as a *normal scale rule*. The estimate $\hat{R}_{\phi}(f^{(s)})$ is used to estimate the optimal bandwidth for estimating $R(f^{(s-1)})$ using the kernel estimate in Equation (5.3). The estimate $\hat{R}_{n,g}(f^{(s-1)})$ is then computed using this bandwidth. The process continues until the estimate $\hat{R}_{n,g}(f'')$ is computed. This estimate is then used to estimate h_{opt}.

For example, the two-stage plug-in bandwidth selector of Sheather and Jones (1991) uses the following algorithm. Again, it is assumed that $L = \phi$.

1. Calculate the normal scale estimate

$$\hat{R}_{\phi}(f^{(4)}) = \frac{105}{32\pi^{1/2}\tilde{\sigma}^9}.$$

2. Calculate $\hat{R}_{n,g}(f^{(3)})$ using the bandwidth

$$\hat{g}_3 = n^{-1/9} \left[\frac{30}{(2\pi)^{1/2}\hat{R}_{\phi}(f^{(4)})}\right]^{1/9}.$$

3. Calculate $\hat{R}_{n,g}(f'')$ using the bandwidth

$$\hat{g}_2 = n^{-1/7} \left[\frac{-6}{(2\pi)^{1/2} \hat{R}_{n,\hat{g}_3}(f^{(3)})} \right]^{1/7}.$$

4. The estimate of h_{opt} is then calculated as

$$\hat{h}_{\text{pi},2} = n^{-1/5} \left[\frac{R(K)}{\sigma_K^2 \hat{R}_{n,\hat{g}_2}(f'')} \right]^{1/5}.$$

There are several more proposed methods for estimating the asymptotically optimal kernel smoothing bandwidth. The methods presented in this section were chosen because of their relative simplicity and reasonable performance. Other potential problems such a boundary bias have been ignored to keep the presentation as simple as possible. In effect, it will be assumed that the support of the densities studied in this chapter consist of the entire real line, or that the effect of boundary problems of the densities studied here are minimal.

Alternate bandwidth selection methods can be used in the following development with little modification as long as the bandwidth estimates are asymptotically equivalent to the methods presented here. However, specific finite sample results such as those presented in the emprirical studies and the examples may differ slightly. Alternate methods for nonparametrically estimating densities may also be used in principle, though the corresponding theory developed below is unlikely to still be valid.

5.2.2 The Search for Modes

The premier problem addressed in nonparametric density estimation is the estimation of the number and location of modes of an unknown continuous density. A mode of a smooth density is a local maximum, that is, a mode exists for a sufficiently smooth density f at a point t if $f'(t) = 0$ and $f''(t) < 0$. The possibility that a density has more than one mode may be an important indication that the population consists of two or more subpopulations, or that the density f is a mixture. See Cox (1966). It is a simple matter to inspect a kernel density estimate of f based on a sample $X_1, \ldots, X_n$ and locate and count the number of modes in the estimate. The central problem is concerned with accounting for the variability of such estimates. This is particularly true with the case of estimating the number of modes a density has. Outliers can cause small modes to appear in the tails of kernel density estimates. In most cases, kernel density estimates have a high degree of variation in the tails due to the sparsity of data in those regions. Therefore, it becomes crucial to determine whether these modes are indicative of some real underlying behavior, or just due to the variability in the data. This section will discuss methods for computing observed confidence levels on the number of modes of an unknown continuous density. As is discussed below, it is much more difficult to address this problem than the problems studied in the previous chapters.

A hypothesis testing approach to counting the number of modes in an unknown smooth density based on kernel density estimation was suggested by Silverman (1981). In this approach a hypothesis of unimodality is tested against a hypothesis of multimodality by finding the smallest bandwidth h_{crit} for which the kernel estimate of f is unimodal. This takes advantage of the fact that the number of modes in $\hat{f}_{n,h}$ decreases as the bandwidth h increases for certain kernel functions, most notably the normal kernel function. Because samples from multimodal densities generally require more smoothing (larger h) to obtain a unimodal kernel density estimate than samples from unimodal densities, the hypothesis of unimodality is rejected when h_{crit} is too large. Of course, determining what values of h_{crit} should result in rejecting the hypothesis of unimodality is the key difficulty in applying this test.

Silverman (1981, 1983) discusses the use of standard families to obtain the rejection region for this test and studies the asymptotic behavior of h_{crit}. A clearly more universal method is based on the smoothed bootstrap. To test unimodality one would obtain the distribution of h_{crit} assuming that $\hat{f}_{n,h_{\text{crit}}}$ is the true population. That is

$$P(h_{\text{crit}} \leq t) \simeq P^*(h^*_{\text{crit}} \leq t | X_1^*, \ldots, X_n^* \sim \hat{f}_{n,h_{\text{crit}}}), \qquad (5.5)$$

where h^*_{crit} is computed on the sample $X_1^*, \ldots, X_n^*$. Section 5.4 will discuss the computational issues concerned with computing this estimate. The null hypothesis is then rejected when h_{crit} exceeds the $(1 - \alpha)^{\text{th}}$ percentile of the estimated distribution in Equation (5.5). Further calibration of this method is discussed by Hall and York (2001).

The use of the bootstrap was adapted by Efron and Tibshirani (1998) as a method for computing confidence levels for the number of modes of a smooth density. In their approach they essentially implement the percentile method for computing observed confidence levels. Define θ to be a functional parameter corresponding to the number of modes of f. A simplistic mathematical definition of θ can be taken as

$$\theta = \theta(f) = \#\{t \in \mathbb{R} : f'(t) = 0, f''(t) < 0\}, \qquad (5.6)$$

where f is assumed to have at least two continuous derivatives. Donoho (1988) points out potential problems in defining the mode functional. For example, the uniform density may be defined to have 0, 1, or infinitely many modes depending on the definition of mode. An alternate definition of the mode functional used by Donoho (1988) is given by

$$\theta = \tilde{\theta}(f) = \lim_{h \to 0} \theta(f \oplus \phi_{0,h^2}),$$

where $\oplus$ is the convolution operator, and $\theta(\cdot)$ is the functional defined in Equation (5.6). Consider the bootstrap estimate of the distribution of the number of modes observed in a kernel estimate given by

$$\hat{v}(t) = P^*[\theta(\hat{f}_{n,\hat{h}^*}) = t | X_1^*, \ldots, X_n^* \sim \hat{F}_n],$$

where $\hat{f}_{n,\hat{h}^*}$ is a kernel estimate computed on $X_1^*, \ldots, X_n^*$ with bandwidth estimate also computed on $X_1^*, \ldots, X_n^*$. It is also possible, as suggested by Efron and Tibshirani (1998), to use the original bandwidth estimate computed on the original sample as the bandwidth for all of the resamples as well. In either case, Efron and Tibshirani (1998) use the cross-validation technique for their work, though any reasonable bandwidth selection method could be implemented. Note that $\hat{v}_n(t)$, as well as $v_n(t)$, are necessarily discrete. Following the general idea of the percentile method, the approximate observed confidence level for f having t modes is simply $\hat{\alpha}^*_{\text{perc}}(\Psi = \{t\}) = \hat{v}_n(t)$. Section 5.4 will discuss the practical application of this method.

Note that studentized and hybrid observed confidence levels can also be developed for this problem. For example, let

$$\hat{h}_n(t) = P^*\{[\theta(\hat{f}^*_{n,\hat{h}^*}) - \theta(\hat{f}_{n,\hat{h}})]/\hat{SE}^*[\theta(\hat{f}^*_{n,\hat{h}^*})] = t | X_1^*, \ldots, X_n^* \sim \hat{F}_n\},$$

where $\hat{SE}^*[\theta(\hat{f}^*_{n,\hat{h}^*})]$ is the bootstrap estimate of the standard error of $\theta(\hat{f}^*_{n,\hat{h}^*})$. Then the studentized observed confidence level for the density f having t modes is given by

$$\hat{\alpha}^*_{\text{stud}}(\Psi = \{t\}) = \hat{h}_n\{[\theta(\hat{f}_{n,\hat{h}}) - t]/\hat{SE}[\theta(\hat{f}_{n,\hat{h}})]\}.$$

The mode functional does not fall within the smooth function model, and therefore it is not clear whether $\hat{\alpha}^*_{\text{stud}}(\Psi = \{t\})$ yields any improvement over $\hat{\alpha}^*_{\text{perc}}(\Psi = \{t\})$ in this case, or whether these estimates are even consistent. Hall and Ooi (2004) address some of the issues of the development of asymptotic theory for this functional. Summarizing this research shows that there are two main issues to be dealt with. The first is that a bandwidth that will insure consistency in the usual sense when mode are clearly defined must converge to zero slower than the usual rate of $n^{-1/5}$ as implemented by Efron and Tibshirani (1998). Unfortunately no optimal rate or form for the bandwidth is known at this time. The second result states that when the density may not have clearly defined modes, such as when the density is allowed to have shoulders where $f'(t) = f''(t) = 0$, then an even larger bandwidth that converges to 0 slower than $n^{-1/7}$ must be employed. See Hall and Ooi (2004) for further details of these results. Related issues in the nonbootstrap case are studied by Mammen (1995), Mammen, Marron, and Fisher (1992), and Konakov and Mammen (1998). The general need for the larger bandwidths in both of these problems is to avoid the effect of small spurious modes that often occur in the tails of kernel density estimates. The larger bandwidth in the second case insures that shoulders are sufficiently smoothed so that they are not taken for modes either. This would generally indicate that the method of Efron and Tibshirani (1998) would tend to overestimate the likelihood of larger numbers of modes. The effect of these results is studied empirically in Section 5.2.3, where the method of Efron and Tibshirani (1998) is implemented using a two-stage plug-in bandwidth estimate which is of order $n^{-1/5}$. Further research is required to obtain optimal bandwidths for this problem, if they are available.

The rate at which the bandwidths converge to 0 in kernel density estimates is not the only fundamental problem in attempting to estimate the number of modes of an unknown smooth density. As the development below shows, certain topological properties of the problem complicate matters even further. The number of modes is a nonlinear functional that falls within a class of problems studied by Donoho (1988) that includes mixture complexity and various density norms. For these types of functionals Donoho (1988) argues that it is not possible to obtain two-sided confidence intervals in a completely nonparametric setting. The reasoning is quite simple. Considering the case of interest in this section where θ is the number of modes of f, it turns out that there are an infinite number of densities that are arbitrarily close to a specified density that have any number of modes more than is observed in f. The metric used by Donoho (1988) is the variation distance, where the distance between to distribution functions F and G is defined as

$$\sup_{m \in \mathcal{M}[0,1]} \left| \int_{\mathbb{R}} m(t)dF(t) - \int_{\mathbb{R}} m(t)dG(t) \right|, \qquad (5.7)$$

where $\mathcal{M}[0, 1]$ is the collection of all measurable functions on $\mathbb{R}$ with values in $[0, 1]$. For additional details on this distance measure see Le Cam (1973, 1986). An example of this effect can be visualized using Figure 5.1. The density indicated by the solid line in Figure 5.1 is a bimodal density. The density indicated by the dashed line is the same density with a small perturbation which causes the density to become trimodal. The two densities can be have a distance that is arbitrarily close using the measure in Equation (5.7) by simply making the small third mode arbitrarily small. Thus there is always a trimodal density whose samples are virtually indistinguishable from any specified bimodal density, regardless of the sample size. This result can be extended to add any countable number of modes. This effect makes it impossible to set an upper confidence bound on the number of modes in a density without further specific assumptions on the behavior of the density. It is possible, however, to set a lower confidence bound on the number of mode of an unknown density. This is because it is not always possible to find a density with *fewer* modes than a specified density that has an arbitrarily small distance from the density. Using these arguments, Donoho (1988) argues that only lower confidence bounds are possible for such functionals.

This obviously has a very profound effect on the construction of observed confidence levels for the number of models of an unknown density. Using these results, it appears to be only practical to obtain observed confidence levels for regions of the form $\Psi_k = \{\theta \in \mathbb{N} : \theta \geq k\}$. Donoho (1988) derives a conservative lower confidence bound for θ. Let $||F - G||$ denote the Kolomogorov

Figure 5.1 *A bimodal density (solid line) and the same density with a small pertur-bation that makes it a trimodal density.*

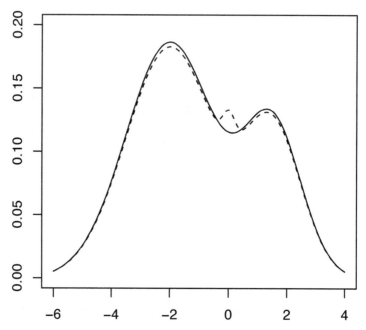

distance between two distribution functions F and G defined by

$$||F - G|| = \sup_{t \in \mathbb{R}} |F(t) - G(t)|.$$

A *lower ϵ-envelope* is established for the functional $\theta(F)$ using this distance as

$$L(F, \epsilon) = \inf_{G \in \mathcal{F}} \{\theta(G) : ||F - G|| \leq \epsilon\},$$

which for the functional in this section is the minimum number of modes found in all of the distributions that have a Kolmogorov distance within ϵ of F. The collection $\mathcal{F}$ will be taken to be all distributions with a continuous density that has at least one smooth derivative. It can be shown that $||\hat{F}_n - F||$ is distribution-free over $\mathcal{F}$. Therefore, let d_ξ be the ξth percentile of the distribution of $||\hat{U}_n - U||$ where $\hat{U}_n$ is the empirical distribution function computed on a set of n independent and identically distributed observations from a uniform distribution on $[0, 1]$ and $U(t) = t\delta(t; [0, 1]) + \delta(t; (1, \infty))$. The proposition in Section 3.2 of Donoho (1988) the implies that $L(\hat{F}_n, d_\alpha)$ is a lower confidence bound on the number of modes $\theta(F)$ with a coverage probability of at least α. Using the general method for computing observed confidence levels described in Section 2.2, the observed confidence level for Ψ_k is computed by setting

$\Psi_k = [L(\hat{F}_n, d_\alpha), \infty) \cap \mathbb{N}$ and solving for α, which is equivalent to setting $k = L(\hat{F}_n, d_\alpha)$ and solving for α. Equivalently this requires the value of ϵ such that

$$k = \inf_{G \in \mathcal{F}} \{\theta(G) : ||\hat{F}_n - G|| \le \epsilon\}.$$

For this value of ϵ there is at least one member of $\mathcal{F}$ that has k modes whose Kolmogorov distance to $\hat{F}_n$ is not greater than ϵ. Note that the other distributions in $\mathcal{F}$ with k modes may have Kolmogorov distances larger than, or less than or equal to ϵ. When focus is shifted to distributions in $\mathcal{F}$ that have $k - 1$ or fewer modes, none of these distributions can have a Kolmogorov distance to $\hat{F}_n$ that is less than or equal to ϵ. Therefore, ϵ is the largest Kolmogorov distance that does not include any distributions with $k - 1$ or fewer modes. Let ϵ_k be equal to this value of ϵ. The observed confidence level for Ψ_k is then computed as $P(||\hat{U}_n - U|| < \epsilon)$. Due to the conservative nature of the lower confidence bound, this observed confidence level is technically a lower bound on the observed confidence level. Further, this method does not require any smoothing and so the issue of bandwidth selection is eliminated. This measure is studied empirically in Section 5.2.3.

Müller and Sawitzki (1991) provide an alternative approach to considering statistical inference about the number of modes of a smooth density. They argue that a mode should be associated with a point that has a high probability in a neighborhood around the point. Therefore, small modes that are small perturbations of a smooth density are not statistically interesting. Their approach does not use the usual analytic definition of modes given above. Rather, modes are studied through a functional known as *excess mass*. The univariate case will be considered in this section. Extension to the multivariate case in the context of testing for modes is considered by Müller and Sawitzki (1991) and Polonik (1995). The excess mass functional of a smooth density f in one dimension is defined as

$$e(\lambda) = \int_{\mathbb{R}} [f(t) - \lambda]^+ dt = \int_C [f(t) - \lambda] dt, \tag{5.8}$$

where $\lambda \ge 0$, and for a real function m, $m^+(t) = m(t)\delta(m(t); [0, \infty))$. That is $e(\lambda)$ measures the amount of probability mass exceeding the density level λ. The set C is Equation (5.8) is given by $\{t \in \mathbb{R} : f(t) \ge \lambda\}$. The set C can be written as the union of a collection of connected, disjoint subsets in $\mathbb{R}$. That is, C is the union of a set of interval subsets of $\mathbb{R}$. The set C is usually written as the most efficient representation, that is the smallest collection of nonoverlapping intervals whose union is C. In this case the intervals are called λ-clusters and are denoted $C_1(\lambda), \ldots, C_m(\lambda)$. Figures 5.2–5.3 show the λ-clusters for a trimodal density for two different values of λ. Note that a distribution that has m modes has at most m λ-clusters.

Figure 5.2 *The excess mass $e(\lambda)$ and λ-clusters of a trimodal density for a specific value of λ. In this example the number of λ-clusters equals the number of modes.*

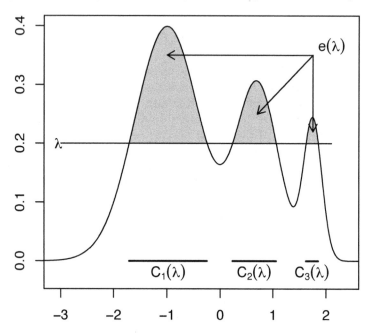

This leads Müller and Sawitzki (1991) to define

$$e_m(\lambda) = \sup_{\{C_1(\lambda),...,C_m(\lambda)\}\in \mathcal{I}_m} \sum_{j=1}^{m} \int_{C_j(\lambda)} (f(t) - \lambda)dt$$

where $\mathcal{I}_m$ is the collection of all sets of m disjoint interval subsets of $\mathbb{R}$. That is $\mathcal{I}_m$ can be taken to be the collection of all possibles sets of m λ-clusters. If f has fewer than m modes then some of the λ-clusters will be empty. Note that this can also occur when λ is below the "dip" between one or more modes, or when λ is above one of the modes, as is demonstrated in Figure 5.3. If f has more than m modes then the clusters will be concentrated on the m regions that contain the most excess mass at density level λ. The clusters again can contain more than one mode depending on the value of λ. This will be assumed in the discussion below. The difference $d_m(\lambda) = e_m(\lambda) - e_1(\lambda)$ therefore measures how much excess mass is accounted for between the single mode with the most excess mass above λ and the excess mass of the remaining $m - 1$ modes, if they exist. Note that is f is unimodal then $d_m(\lambda) = 0$ for all $\lambda \geq 0$, but that if f has more than one mode $d_m(\lambda) \geq 0$ for $\lambda \geq 0$. This motivates a test statistic for testing the hypothesis of a single mode versus more than one mode.

Figure 5.3 *The excess mass $e(\lambda)$ and λ-clusters of a trimodal density for a specific value of λ. In this example the number of λ-clusters is less than the number of modes.*

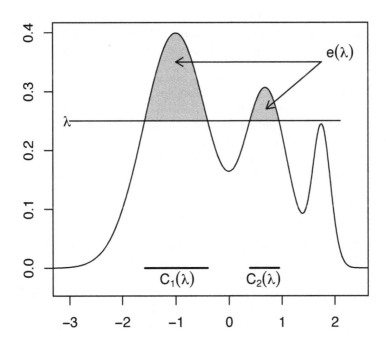

To develop the sample version of $d_m(\lambda)$ in the univariate case, consider a λ-cluster of the form $C_j(\lambda) = (a_j(\lambda), b_j(\lambda))$. Then

$$
\begin{aligned}
\int_{C_j(\lambda)} (f(t) - \lambda)dt &= \int_{a_j(\lambda)}^{b_j(\lambda)} f(t)dt - \lambda \int_{a_j(\lambda)}^{b_j(\lambda)} dt \\
&= F(b_j(\lambda)) - F(a_j(\lambda)) - \lambda(b_j(\lambda) - a_j(\lambda)). \quad (5.9)
\end{aligned}
$$

Replacing F with the empirical distribution function $\hat{F}_n$ yields an estimate of the integral given in Equation (5.9) as

$$
\hat{F}_n(b_j(\lambda)) - \hat{F}_n(a_j(\lambda)) - \lambda(b_j(\lambda) - a_j(\lambda)) = n^{-1} \sum_{i=1}^{n} \delta(X_i; C_j(\lambda)) - \lambda(b_j(\lambda) - a_j(\lambda)).
$$

Therefore $d_m(\lambda)$ can be estimated as $\hat{d}_m(\lambda) = \hat{e}_m(\lambda) - \hat{e}_1(\lambda)$ where

$$
\hat{e}_k(\lambda) = \sup_{\{C_1(\lambda),\ldots,C_k(\lambda)\} \in \mathcal{I}_k} \sum_{j=1}^{k} \left[n^{-1} \sum_{i=1}^{n} \delta(X_i; C_j(\lambda)) - \lambda(b_j(\lambda) - a_j(\lambda)) \right].
$$

The test statistic suggested by Müller and Sawitzki (1991) for testing the null hypothesis of a single mode versus m modes is then given by

$$\hat{d}_m = \sup_{\lambda \geq 0} \hat{d}_m(\lambda).$$

Methods for computing this statistic are considered in Section 5.4. The test rejects unimodality when $\hat{d}_m$ is too large. This test is equivalent to the DIP test of unimodality developed by Hartigan and Hartigan (1985).

The main difficulty with using the test statistic $\hat{d}_m$ to investigate unimodality is that critical values of the test are difficult to compute when f is unknown, even asymptotically. See Müller and Sawitzki (1991) and Polonik (1995). The usual method for computing the critical value of the test is based on simulating samples from a uniform density. This method has some asymptotic justification though empirical evidence usually shows the test is quite conservative. See Polonik (1995). Cheng and Hall (1998) develop a method for computing the critical value that provides a less conservative version of the test. Fisher and Marron (2001) consider combining the excess mass test with the test proposed by Silverman (1981, 1983).

Hall and Ooi (2004) also investigate the problem of computing confidence levels for the number of modes of an unknown smooth density. Citing the results of Donoho (1988), and using their own asymptotic investigations, they conclude that the simple approach of Efron and Tibshirani (1998) will not provide a sound method for mode investigation. In particular they show that the bandwidth required for this method would need to be larger than that is used for density estimation so that small spurious modes in the tails of the estimate are smoothed out. Hall and Ooi (2004) therefore conclude that the excess mass approach is a more suitable approach to investigating modality. Theorem 3.4 of Hall and Ooi (2004) shows that the bootstrap can be used to construct a consistent estimate of the sampling distribution of $\hat{d}_2$ and that the bootstrap percentile method can be used to construct a confidence interval for $\hat{d}_2$ that has asymptotically correct coverage accuracy. This implies that an observed confidence level for the excess mass statistic can then be computed by setting the region of interest Ψ equal to the endpoints of the bootstrap percentile method confidence interval and solving for the corresponding coverage probability using the methods of Chapter 2.

Such observed confidence levels will not provide confidence levels on the number of modes of an unknown smooth density. They can, however, be used to investigate the shapes of the density in terms of the excess mass using methods similar to those presented in Hall and Ooi (2004). As an example, suppose we compute that the observed confidence level that the excess mass is within a region $\Psi = [a, b]$ is equal to α. Using the data sharpening algorithms presented in the appendices of Hall and Ooi (2004), smooth estimates of f based on kernel density estimation can be computed under the constraint that the excess mass of the estimate is equal to a grid of values between a and b. These estimates would indicate the estimated shapes of the corresponding densities that are contained within that region of excess mass. Examples using this method are presented in Section 5.3. For further details on data sharpening algorithms see Section 5.4, Choi and Hall (1999), Braun and Hall (2004), Hall and Kang (2005), and Hall and Minnottee (2002).

5.2.3 An Empirical Study

This section considers an empirical study of computing observed confidence levels on the number of modes of an unknown density using two methods based on the ideas of Efron and Tibshirani (1998) and the method described in Section 5.2.2 based on the results of Donoho (1988). The densities under consideration are normal mixtures studied in Marron and Wand (1992), and are plotted in Figure 5.4. Two of these mixtures are unimodal, three are bimodal, and one is trimodal. Samples of size $n = 25$, 50, 100 and 200 where simulated from each of these densities and the observed confidence level that the number of modes is greater than or equal to 2 was computed using all three methods. Two or more modes corresponds to the region Ψ_2 using the notation of Section 5.2.2. Bandwidth selection for the method of Efron and Tibshirani (1998) was based on the two-stage plug-in method described in Section 5.2.1. The original method proposed by Efron and Tibshirani (1998) uses a single bandwidth estimate from the original data and for each density estimated from the resamples. In addition to this method, the use of a new bandwidth estimate computed on each resample was also studied. These two estimated observed confidence levels will be denoted as $\hat{\alpha}^*_{\text{et}}(\Psi_2)$ and $\hat{\alpha}^*_{\text{perc}}(\Psi_2)$, respectively. Both methods used $b = 1000$ resamples to estimate the confidence level. For the method based on the results of Donoho (1988), the Kolmogorov distance between the empirical distribution function of the sample and the nearest unimodal density must be computed. To approximate this distance, the distance between the empirical distribution function and the distribution function corresponding to the kernel density estimate with the smallest bandwidth that results in a unimodal density is used. The observed confidence level estimated using this method will be denoted as $\hat{\alpha}^*_{\text{don}}(\Psi_2)$. For further information on computing both of these estimates, see Section 5.4.

The results of the study are presented in Table 5.1. For the unimodal distributions both $\hat{\alpha}^*_{\text{et}}$ and $\hat{\alpha}^*_{\text{perc}}$ would appear to overstate the amount of confidence, even for large sample sizes, while $\hat{\alpha}^*_{\text{don}}$ tends to do a better job of not putting to much confidence into more than one mode. Note that choosing a new bandwidth for each resample only causes this problem to be worse. For the skewed unimodal distribution, however, the average observed confidence levels appear to be stable or increasing with the sample size. This is probably due to spurious modes appearing in the long tail of the skewed distribution. The measure $\hat{\alpha}^*_{\text{don}}$ may also have the additional issues that the confidence statement in Donoho (1988) provides only a lower bound for the confidence level of the interval and the approximate nature of the method for finding the

Figure 5.4 *Normal mixtures used in the simulation study: (a) Standard Normal, (b) Skewed Unimodal, (c) Symmetric Bimodal, (d) Separated Biomodal, (e) Asymmetric Bimodal, and (f) Trimodal.*

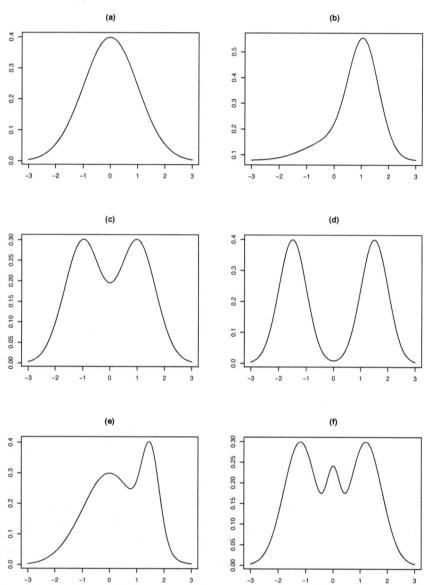

closest unimodal density to the original one from the data. All three methods are able to detect the bimodal distributions. All of the methods are less able to detect the trimodal distribution, almost certainly due to the fact that the small third mode between the two larger modes makes samples from this distribution appear to be unimodal at times.

5.3 Density Estimation Examples

5.3.1 A Problem from Astronomy

Consider the data on the percentage of silica in a sample of chondrite meteorites introduced in Section 1.3.6. It is clear from Figure 1.8 that the kernel density estimate based on this data exhibits two clear modes, though the dip between the modes is not substantial. The existence of at least two modes has been the conclusion of several researchers including Good and Gaskins (1980), and indirectly Hall and Ooi (2004). Observed confidence levels for the conclusion that there are at least two modes in this density based on the methods described in Section 5.2.2 are given in Table 5.2. Both of the methods based on Efron and Tibshirani (1998) show strong support for the contention that there is at least two modes in the underlying density for the data, while the method based on the confidence interval of Donoho (1998) shows much less support for this contention. Most analyses of this data have shown a clear preference for two or more modes. Recall that the method based on Donoho (1998) may understate the observed confidence level in this case due to the conservative nature of the corresponding confidence bound.

The second approach to investigating the modality of this data is based on the excess mass approach. The excess mass statistic d_2 is always between 0 and 1 and therefore the bootstrap method of Hall and Ooi (2004) can be used to find observed confidence levels that the excess mass is in the regions $\Theta_1 = [0, 0.10)$, $\Theta_2 = [0.10, 0.20)$, $\Theta_3 = [0.20, 0.30)$, $\Theta_4 = [0.30, 0.40)$, $\Theta_5 = [0.40, 0.50)$, and $\Theta_6 = [0.50, 1]$. The observed confidence levels based on the bootstrap percentile method with $b = 1000$ resamples are given in Table 5.3. To get an idea of what type of shapes the density estimate has for various values of excess mass, the data sharpening method, based on a simulated annealing algorithm was used to produce density estimates whose excess mass is constrained to the endpoints of the regions with the largest observed confidence levels. These density estimates are given in Figure 5.5. The value of excess mass with the largest observed confidence levels appear between plots (a) and (c). It appears that the estimates of the densities in this range of excess mass are bimodal, though the mode is quite weak for the lower end of the range. For the range of the excess mass with the largest observed confidence level, $\Psi_3 = [0.20, 0.30)$, whose bounds correspond to plots (b) and (c), the second mode is more pronounced. Note that plot (d) hardly resembles the original plot of the data. This arises from the fact that there is little chance that the excess mass for

Table 5.1 *The results of the simulation study on counting the number of modes of an unknown smooth density. The reported results are the average observed confidence levels, in percent, for the region Ψ_2, which includes all densities with two or more modes. The observed confidence levels $\hat{\alpha}^*_{et}(\Psi_2)$, $\hat{\alpha}^*_{perc}(\Psi_2)$, and $\hat{\alpha}^*_{don}(\Psi_2)$ correspond to the methods of Efron and Tibshirani (1998) using a single bandwidth and a bandwidth for each resample, and the method based on the results of Donoho (1998).*

Distribution	Sample Size	Method $\hat{\alpha}^*_{et}(\Psi_2)$	$\hat{\alpha}^*_{perc}(\Psi_2)$	$\hat{\alpha}^*_{don}(\Psi_2)$
Standard Normal	25	45.51	66.38	7.86
	50	40.83	61.67	8.49
	100	35.43	55.75	5.40
	200	32.02	50.90	6.86
Skewed Unimodal	25	61.00	75.51	16.12
	50	63.89	76.63	19.28
	100	67.56	77.74	15.81
	200	72.34	80.39	23.08
Symmetric Bimodal	25	52.62	72.20	10.21
	50	70.46	78.39	18.04
	100	80.23	83.84	22.46
	200	93.57	93.38	46.00
Separated Bimodal	25	99.92	99.88	56.56
	50	100.00	100.00	80.58
	100	100.00	100.00	94.77
	200	100.00	100.00	99.74
Asymmetric Bimodal	25	53.92	74.66	12.79
	50	64.18	75.18	17.46
	100	76.51	81.60	22.92
	200	91.19	91.35	43.03
Trimodal	25	62.28	78.34	14.89
	50	71.80	79.57	20.55
	100	90.06	90.90	36.43
	200	97.92	97.57	58.06

Table 5.2 *Estimated observed confidence levels, in percent, that the density of the percentage of silica in chondrite meteorites has at least two modes.*

Method	$\hat{\alpha}^*_{et}(\Psi_2)$	$\hat{\alpha}^*_{perc}(\Psi_2)$	$\hat{\alpha}^*_{don}(\Psi_2)$
Observed Confidence Level	82.5	94.2	20.2

Table 5.3 *Observed confidence levels for the regions $\Theta_1, \ldots, \Theta_6$ based on the bootstrap percentile method with $b = 1000$ resamples for the excess d_2 computed on the chondrite meteorite data.*

Region	$\Theta_1 = [0, 0.10)$	$\Theta_2 = [0.10, 0.20)$	$\Theta_3 = [0.20, 0.30)$
$\hat{\alpha}^*_{perc}$	0.2	41.4	57.0
Region	$\Theta_4 = [0.30, 0.40)$	$\Theta_5 = [0.40, 0.50)$	$\Theta_6 = [0.50, 1]$
$\hat{\alpha}^*_{perc}$	1.4	0.0	0.0

the distribution is that large, and therefore a significant amount of perturbation of the original data was required to obtain an excess mass estimate that large.

5.3.2 A Problem from Ecology

Holling (1992) obtained data on the body mass index for 101 boreal forest birds found east of the Manitoba-Ontario border in pure conifer or mixed conifer stands. The data are given in Appendix 1 of Holling (1992). Holling suggests that there are certain biological processes operating at different scales of time and space, and that these differences can lead to discontinuities in the distributions of certain ecological variables. In particular, Holling (1992) suggests that there are eight discontinuities in the distribution of the body mass index of the 101 boreal forest birds. Manly (1996) reconsidered this data using the mode test of Silverman (1981) and concludes that the data indicate far fewer modes, and that a bimodal distribution fits the data quite well. A kernel density estimate computed on the data is given in Figure 5.6.

There are two relatively strong modes in the kernel density estimate, and a hint of a third mode in the shoulder between the two modes. Observed confidence levels for the conclusion that there are at least two modes in this density based on the methods described in Section 5.2.2 are given in Table 5.4. To compute $\hat{\alpha}^*_{don}(\Psi_2)$ the data were slightly perturbed with some random noise to avoid ties. All three methods show that there is substantial evidence that there is at least two modes in this density.

Figure 5.5 *Kernel density estimates for perturbed versions of the chondrite data with the specified values of excess mass: (a) $\hat{d}_2 = 0.10$, (b) $\hat{d}_2 = 0.20$, $\hat{d}_2 = 0.30$ and $\hat{d}_2 = 0.40$. Bandwidth selection was accomplished using a two-stage plug-in technique.*

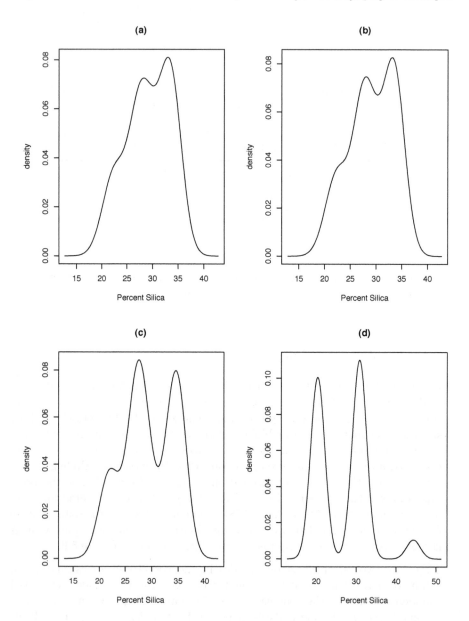

Figure 5.6 *Kernel density estimate of the log (base 10) of the body mass index of 101 boreal forest birds. The bandwidth was selected using a two-stage plug-in procedure.*

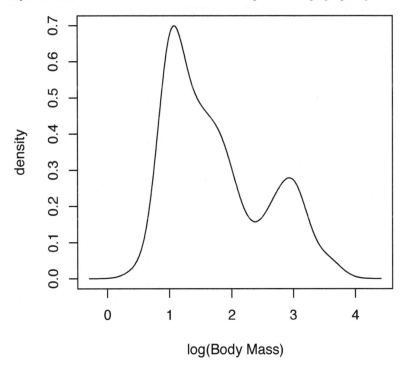

log(Body Mass)

Table 5.4 *Estimated observed confidence levels, in percent, that the density of the body mass of the boreal forest birds has at least two modes.*

Method	$\hat{\alpha}^*_{et}(\Psi_2)$	$\hat{\alpha}^*_{perc}(\Psi_2)$	$\hat{\alpha}^*_{don}(\Psi_2)$
Observed Confidence Level	99.70	99.60	95.30

The observed excess mass for this data is $\hat{d}_2 = 0.072$. The bootstrap percentile method was used to compute observed confidence levels that the excess mass d_2 is in the regions $\Theta_1 = [0, 0.05)$, $\Theta_2 = [0.05, 0.10)$, $\Theta_3 = [0.10, 0.15)$, $\Theta_4 = [0.15, 0.20)$, and $\Theta_5 = [0.20, 1.00]$. The observed confidence levels are given in Table 5.5. A plot of density estimates with excess mass constrained to the endpoints of the the regions Θ_2–Θ_4 is given in Figure 5.7. It is clear that two modes are observed in each of the plots. There is a hint of as third mode in the last plot, but the excess mass associated with this plot is well beyond the confidence range of excess mass for this data.

Figure 5.7 *Kernel density estimates for perturbed versions of the bird mass data with the specified values of excess mass: (a) $\hat{d}_2 = 0.05$, (b) $\hat{d}_2 = 0.10$, $\hat{d}_2 = 0.15$, and $\hat{d}_2 = 0.20$. Bandwidth selection was accomplished using a two-stage plug-in technique.*

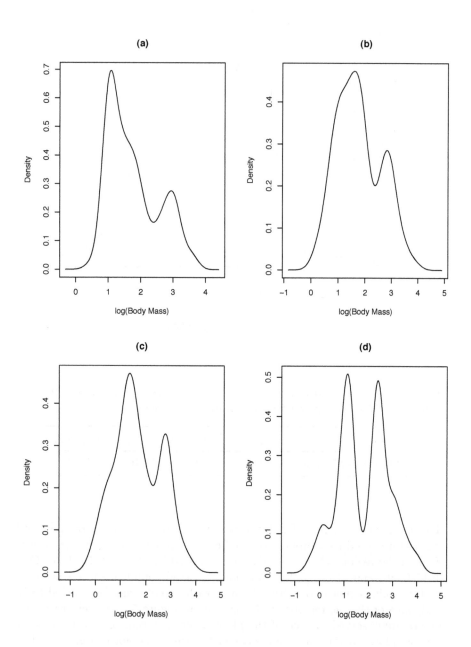

Table 5.5 *Observed confidence levels for the regions* $\Theta_1, \ldots, \Theta_5$ *based on the bootstrap percentile method with* $b = 1000$ *resamples for the excess* d_2 *computed on the bird mass data.*

Region	$\Theta_1 = [0, 0.05)$	$\Theta_2 = [0.05, 0.10)$	$\Theta_3 = [0.10, 0.15)$
$\hat{\alpha}^*_{perc}$	0.0	58.9	40.6

Region	$\Theta_4 = [0.15, 0.20)$	$\Theta_5 = [0.20, 1.00)$
$\hat{\alpha}^*_{perc}$	0.5	0.0

Table 5.6 *Estimated observed confidence levels, in percent, that the density of the body mass of the boreal forest mammals has at least two modes.*

Method	$\hat{\alpha}^*_{et}(\Psi_2)$	$\hat{\alpha}^*_{perc}(\Psi_2)$	$\hat{\alpha}^*_{don}(\Psi_2)$
Observed Confidence Level	72.9	98.6	46.59

A second set of data considered by Holling (1992) considers thirty-six species of boreal forest mammals from the same area. Holling (1992) again suggests the presence of multiple gaps in the data, whereas Manly (1996) concludes that the distribution is unimodal. A kernel density estimate computed on the data is given in Figure 5.8. There again are two strong modes in the kernel density estimate, though the low sample size may indicate considerable variability in the estimate.

Observed confidence levels for the conclusion that there are at least two modes in this density based on the methods described in Section 5.2.2 are given in Table 5.6. While the percentile method indicates that there is strong evidence that there are at least two modes in the data, the method of Efron and Tibshirani (1998) and the method based on Donoho (1988) indicate a moderate level of confidence for these hypotheses. The observed excess mass for the data is $\hat{d}_2 = 0.108$. Percentile method observed confidence levels based on $b = 1000$ resamples where computed for the excess mass d_2 begin in the regions $\Theta_1 = [0, 0.10)$, $\Theta_2 = [0.10, 0.20)$, $\Theta_3 = [0.20, 0.30)$, and $\Theta_4 = [0.30, 1.00]$. These estimated observed confidence levels are given in Table 5.7. It is clear from Table 5.7 that there is a great amount of confidence that the excess mass is in the region Θ_2. The simulated annealing program was used to create density estimates constrained to have an excess mass equal to the endpoints of this interval. A plot of these estimates is given in Figure 5.9. Both plots indicate clear bimodality.

Figure 5.8 *Kernel density estimate of the log (base 10) of the body mass index of thirty-six boreal forest mammals. The bandwidth was selected using a two-stage plug-in procedure.*

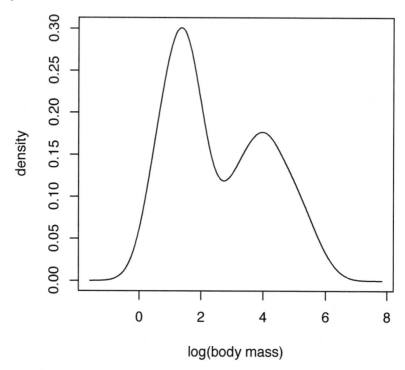

log(body mass)

Table 5.7 *Observed confidence levels for the regions $\Theta_1, \ldots, \Theta_4$ based on the bootstrap percentile method with $b = 1000$ resamples for the excess d_2 computed on the mammal mass data.*

Region	$\Theta_1 = [0, 0.10)$	$\Theta_2 = [0.10, 0.20)$	$\Theta_3 = [0.20, 0.30)$
$\hat{\alpha}^*_{perc}$	1.3	85.4	14.6
Region	$\Theta_4 = [0.30, 1.00]$		
$\hat{\alpha}^*_{perc}$	0.0		

Figure 5.9 *Kernel density estimates for perturbed versions of the mammal mass data with the specified values of excess mass: (a)* $\hat{d}_2 = 0.10$ *and (b)* $\hat{d}_2 = 0.20$. *Bandwidth selection was accomplished using a two-stage plug-in technique.*

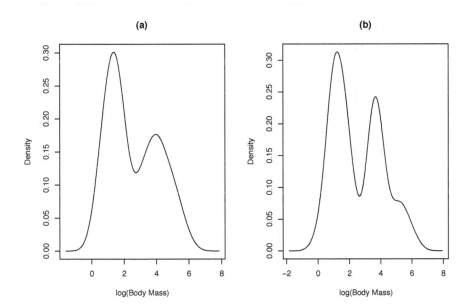

5.3.3 A Problem from Geophysics

Azzalini and Bowman (1990) considers data based on 299 eruptions of the Old Faithful geyser in Yellowstone National Park in the United States between August 1, 1985, and August 15, 1985. Figure 5.10 shows a kernel density estimate of the 299 eruption durations from this data. The data have been slightly presmoothed by a vector of independent random variates with mean 0 and standard deviation $\frac{1}{10}$ to remove ties from the data which would inhibit the calculation of $\hat{\alpha}^*_{\text{don}}$. The bandwidth for the kernel density estimate was computed using the two-stage plug-in estimate described in Section 5.2.1. There is clear evidence in the density estimate of at least two modes. In fact, the kernel density estimate shows four modes, though two of the modes are quite weak. The observed confidence levels for two or modes are given in Table 5.8. Clearly, all of the methods imply very strong evidence that there are at least two modes in this density.

5.4 Solving Density Estimation Problems Using R

The R statistical computing environment offers many functions and libraries for computing kernel density estimates. The library used to generate the plots

Figure 5.10 *Kernel density estimate of the duration of 299 eruptions of the Old Faithful geyser in Yellowstone National Park. The bandwidth was selected using a two-stage plug-in procedure.*

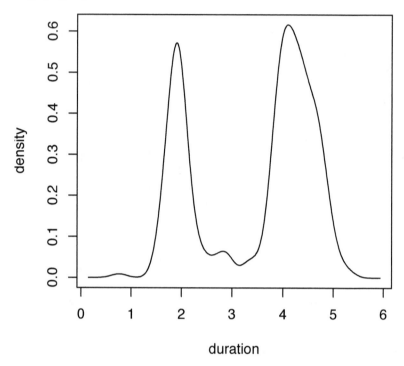

Table 5.8 *Estimated observed confidence levels, in percent, that the density of the duration of geyser eruptions has at least two modes.*

Method	$\hat{\alpha}^*_{\mathrm{et}}(\Psi_2)$	$\hat{\alpha}^*_{\mathrm{perc}}(\Psi_2)$	$\hat{\alpha}^*_{\mathrm{don}}(\Psi_2)$
Observed Confidence Level	100.00	100.00	99.99

and perform calculations in this book is the KernSmooth library that was developed in conjunction with Wand and Jones (1995). The dpik command computes the two-stage plug-in bandwidth estimate described in Secton 5.2.1 and the bkde command computes the kernel density estimate with a specified bandwidth on a grid of values. For example, to produce a plot of the kernel density estimate for an object x that contains the data, the command

```
plot(bkde(x,bandwidth=dpik(x)),type="l")
```

can be used. To count the modes in a kernel density estimate the naive function

```
count.modes <- function(kde) {
    k <- kde$y
    n <- length(k)
    mc <- 0
    for(i in 1:(n-2))
          if((k[i]<k[i+1])&&(k[i+1]>k[i+2])) mc <- mc + 1
    return(mc) }
```

was used. Note that a more sophisticated function would be required for general use. The above function may be not pick up on very wide and flat modes where the value of the kernel estimator may not change. The argument of this function is an object created by the `bkde`. Therefore, to count the number of modes of a kernel estimate on a data object x using the two-stage plug-in estimate of the bandwidth the command

```
count.modes(bkde(x,bandwidth=dpik(x)))
```

can be used. This same approach can be used to compute observed confidence levels based on the method suggested in Section 5.2.2. A function that is suitable for use with the `boot` function to compute $\hat{\alpha}_{\text{perc}}$ is given by

```
mode.boot <- function(data,i) {
    if(count.modes(bkde(data[i],bandwidth=dpik(data[i])))==1)
          return(0)
    else return(1) }
```

This particular function can be used to compute the observed confidence level that there are at least two modes in the unknown density. Simple modifications of this function make it suitable to compute observed confidence levels for other regions. The observed confidence level can then be computed using the command

```
mean(boot(data=x,statistic=mode.boot,R=1000)$t)
```

using the same data object x. The original approach used by Efron and Tibshirani (1998) used the bandwidth from the original data x for computing the kernel density estimate on each resample. This procedure is easy accomplished by modifying `mode.boot` to read in this bandwidth. A suitably modified function is

```
mode.boot.const <- function(data,i,h) {
    if(count.modes(bkde(data[i],bandwidth=h))==1) return(0)
    else return(1) }
```

and the corresponding command to calculate the observed confidence level is

```
mean(boot(data=x,statistic=mode.boot,R=1000,h=dpik(x))$t)
```

Computation of the observed confidence level $\hat{\alpha}_{\text{don}}(\Psi_k)$ requires slightly more sophisticated computations. The Kolmogorov distance and related probabilities can be computed using the ks.test function. This function performs the Kolmogorov goodness-of-fit test that either a single sample follows a specified distribution, or that two samples follow the same distribution. In the single sample case, the first argument is the observed data vector and the second argument is a function corresponding to the distribution function that data is being tested against. For example, the p-value from testing whether a data vector x follows a standard normal distribution can be obtained with the command

```
ks.test(x,pnorm)$p
```

To test against a nonstandard normal distribution, for example a normal distribution with mean equal to 7 and standard deviation equal to 10, the optional arguments for the pnorm function are appended to the end of the arguments of the ks.test function as follows

```
ks.test(x,pnorm,mean=7,sd=10)$p
```

In order to use the function ks.test to compute $\hat{\alpha}_{\text{don}}(\Psi_k)$, the function specified in the arguments must correspond to the distribution function of the closest distribution to the empirical distribution function of the data that has $k - 1$ or fewer modes. The special case of calculating $\hat{\alpha}_{\text{don}}(\Psi_2)$ will be addressed in this section. In this case the argument required is the closest unimodal distribution to the empirical distribution of the data. The approximate algorithm for finding this distribution used in Sections 5.2.3 and 5.3 was to find the kernel density estimate of the data with the smallest bandwidth that is still unimodal. The following function uses a simple search technique to find this bandwidth.

```
h.one.mode <- function(x,gridsize=100) {
    hstart <- dpik(x)
    if(count.modes(bkde(x,bandwidth=hstart))==1) {
        nm <- 1
        hlower <- hstart
        while(nm==1) {
            hlower <- hlower/2
            nm <- count.modes(bkde(x,bandwidth=hlower)) }
        for(h in seq(hlower,hstart,length.out=gridsize))
            if(count.modes(bkde(x,bandwidth=h))==1)
            return(h) }
    else {
        nm <- 2
        hupper <- hstart
        while(nm!=1) {
            hupper <- 2*hupper
```

```
        nm <- count.modes(bkde(x,bandwidth=hupper)) }
    for(h in seq(hstart,hupper,length.out=gridsize))
        if(count.modes(bkde(x,bandwidth=h))==1)
        return(h) }}
```

A function corresponding to the distribution function evaluated at t of a kernel density estimate on a sample `sample` with bandwidth h is given by

```
kdf <- function(t,sample,h) {
    s <- 0
    n <- length(sample)
    for(i in 1:n) s <- s + pnorm((t-sample[i])/h)/n
    return(s) }
```

This function assumes that the kernel function is normal. The observed confidence level $\hat{\alpha}_{\mathrm{don}}(\Psi_2)$ can then by computed using the command

```
1-ks.test(x,kdf,sample=x,h=h.one.mode(x))$p
```

To compute the excess mass statistic, the function `dip` from the R library diptest can be used, based on the fact that $\hat{d}_2$ is twice this statistic. See Müller and Sawitzki (1991) for further details. Therefore, the excess mass of a sample contained in the object x can be computed using the command `dip(x)`. A function suitable for use with the `boot` function to compute an observed confidence level based on the percentile method that the excess mass is in the range L to U is given by

```
boot.excess.mass <- function(data,i,L,U) {
    if((dip(data[i]>=L)&&(dip(data[i]<=U)) return(1)
    else return(0) }
```

Hence the percentile method observed confidence level that the excess mass of the sample contained in the object x is in the range L to U can be computed using the command

```
mean(boot(data=x,statistic=boot.excess.mass,R=1000,L=L,U=U)$t)
```

The construction of a reliable simulated annealing program is somewhat more involved and will not be addressed.

5.5 Nonparametric Regression

5.5.1 Methods for Nonparametric Regression

Nonparametric regression methods attempt to model the relationship $E(Y|x) = \theta(x)$ as a function of x where no parametric assumptions about the function $\theta(x)$ are made. The data considered in this chapter will consist of data pairs

$(x_1, Y_1), \ldots, (x_n, Y_n)$ where the values $x_1, \ldots, x_n$ are considered fixed, and not random. The value Y_i is considered random and is assumed to be observed conditional on x_i. The model generating the data is assumed to have the form $Y_i = \theta(x_i) + \epsilon_i$ for $i = 1, \ldots, n$ where $\epsilon_1, \ldots, \epsilon_n$ is a set of independent and identically distributed errors with $E(\epsilon_i) = 0$ and $V(\epsilon_i) = \sigma^2 < \infty$. Less restrictive models, that do not require homogeneity of the variances, or can be modeled using random variables for the dependent variable, can also be studied with minor modifications to the framework studied here.

Many methods have been developed for fitting nonparametric regression models. The focus of this chapter will be on local polynomial fitting methods, also commonly known as kernel regression. An overview of local polynomial fitting techniques can be found in Wand and Jones (1995). For additional techniques for performing nonparametric regression see Härdle (1990), Simonoff (1996), and Thompson and Tapia (1990). There are other types of kernel regression. Chu and Marron (1991) provides an overview of these methods along with comparisons of their practical use.

Local polynomial regression is motivated by the idea that smooth functions can be approximated locally by low order polynomials. Therefore, the essential components of local polynomial regression are the order of the polynomial that is used to locally approximate the function, and the criteria used to define the locality around the point where the function is being estimated. There is theoretical evidence that higher order polynomials will provide a fit with better asymptotic statistical performance than lower order polynomials. There is also an advantage in choosing polynomials with an odd order. The advantage of increasing the polynomial order diminishes for larger orders of the polynomial and therefore the focus of this book will not go past linear functions. Extension of the methods presented here to higher orders of polynomials is easily accomplished. Locality of the polynomial fit is accomplished through weighted least squares regression techniques where the weight of each data pair (x_i, Y_i) used in the fit is inversely proportion to the distance between the point x where the function is being fit and the value of x_i. A simple method of fitting a usual least squares estimate on points that are within a certain distance to x_i can produce estimates for $\theta(x)$ that have some nonsmooth properties, such as points where $\theta'(x)$ does not exist. A generalization of this procedure is based on using a smooth function $K(x)$ to calculate the weight of the weighted least squares fit. This function usually has the same properties as the kernel function defined in Section 5.2.1. Therefore, local polynomial regression is also referred to as *kernel regression*. As with kernel density estimation, the standard normal density will be assumed to be the kernel function in this chapter, though the methods can be adapted to other kernel functions readily. The locality of the weights of the kernel function will again be adjusted using a bandwidth h that essentially is used as a scale parameter for the kernel function.

The local polynomial estimate of $\theta(x)$ at the point x is derived by computing the weighted least squares estimate of a pth-order polynomial fit of $Y_1, \ldots, Y_n$ on $x_1, \ldots, x_n$ using the kernel function and the bandwidth to set the weights. This fit is centered at x so that the pth order polynomial fit of $\theta(x)$ will be the constant term from this fit. Therefore, let $\hat{\beta}_0, \ldots, \hat{\beta}_p$ minimize the weighted sum of squares

$$\sum_{i=1}^{n} \left[Y_i - \sum_{j=0}^{p} \beta_j (x_i - x)^j \right]^2 \frac{1}{h} K \left(\frac{x_i - x}{h} \right).$$

Then the pth-order polynomial estimate of $\theta(x)$ with bandwidth h is given by $\hat{\theta}_{p,h}(x) = \hat{\beta}_0$. Such an estimate can be written efficiently using matrix notation. Let $\mathbf{X}(x)$ be a $n \times (p+1)$ matrix of the form

$$\mathbf{X}(x) = \begin{bmatrix} 1 & x_1 - x & \cdots & (x_1 - x)^p \\ 1 & x_2 - x & \cdots & (x_2 - x)^p \\ \vdots & \vdots & \ddots & \vdots \\ 1 & x_n - x & \cdots & (x_n - x)^p \end{bmatrix},$$

and a weight matrix $\mathbf{W}(x)$ be a diagonal $n \times n$ matrix with i^{th} diagonal element equal to $K[(x_i - x)/h]/h$. The intercept of the standard weight least squares fit can then be written as

$$\hat{\theta}_{p,h}(x) = \mathbf{e}_1 [\mathbf{X}'(x)\mathbf{W}(x)\mathbf{X}(x)]^{-1}\mathbf{X}'(x)\mathbf{W}(x)\mathbf{Y}, \qquad (5.10)$$

where $\mathbf{e}_1$ is the first Euclidean basis vector of $\mathbb{R}^{p+1}$ given by $\mathbf{e}_1' = (1, 0, \ldots, 0)$.

In some cases this estimator has a closed form. When $p = 0$, or local constant fitting, the estimator can be written as

$$\hat{\theta}_{0,h}(x) = \frac{\sum_{i=1}^{n} K \left(\frac{x_i - x}{h} \right) Y_i}{\sum_{i=1}^{n} K \left(\frac{x_i - x}{h} \right)}, \qquad (5.11)$$

which is a simple local weighted average. This estimator is usually known as the *Nadaraya-Watson Estimator* and was introduced separately by Nadaraya (1964) and Watson (1964). When $p = 1$, or local linear fitting, the estimator can be written as

$$\hat{\theta}_{1,h}(x) = (nh)^{-1} \sum_{i=1}^{n} \frac{[\hat{s}_{2,h}(x) - \hat{s}_{1,h}(x)(x_i - x)] K \left(\frac{x_i - x}{h} \right) Y_i}{\hat{s}_{2,h}(x)\hat{s}_{0,h}(x) - \hat{s}_{1,h}^2(x)},$$

where

$$\hat{s}_{k,h}(x) = (nh)^{-1} \sum_{i=1}^{n} (x_i - x)^k K \left(\frac{x_i - x}{h} \right).$$

The asymptotic mean squared error of local polynomial estimators can be found, though in the general case the results can be somewhat complicated. The results for local linear estimators will be summarized below. Complete results can be found in Ruppert and Wand (1994) and in Chapter 5 of Wand and Jones (1995). Assume

1. The design is equally spaced on $[0,1]$. That is $x_i = i/n$ for $i = 1, \ldots, n$.

2. The function $\theta''(x)$ is continuous on $[0,1]$.

3. The kernel function K is symmetric about 0 with support $[-1,1]$.

4. The bandwidth h is a sequence in n such that $h \to 0$ and $nh \to \infty$ as $n \to \infty$.

5. The point x is on the interior of the interval. That is $h < x < 1-h$ for all $n \geq n_0$ for a fixed value of n_0. The interior of the interval will be denoted as $I[0,1]$.

Assumption 1 specifies a fixed equally-spaced design for the x values. Assumption 2 assures that the function is smooth enough to be approximated locally by a linear function. The third assumption is technically violated by the use of the normal kernel in this chapter, but the broad issues of the results will still demonstrate the relevant factors that affect the mean square error of the estimator. Assumption 5 will be critical in applying kernel regression methods. The behavior of kernel estimates near the edge of the range of the data is very different from the interior of the range of the independent variable. In particular the bias and variance of the estimator can be inflated when estimating $\theta(x)$ for values of x near the boundaries of the data range. For example, the Nadaraya-Watson estimator has an optimal mean square error of order $n^{-4/5}$ for estimating $\theta(x)$ when x is in the interior of the range of the data. Near the boundaries, however, the mean square error is of order $n^{-2/3}$. See Section 5.5 of Wand and Jones (1995) for further details on this problem.

Under Assumptions 1–5, the asymptotic bias of the local linear estimator of $\theta(x)$ is

$$E[\hat{\theta}_{p=1,h}(x) - \theta(x)] = \tfrac{1}{2}h^2\theta''(x)\sigma_K^2 + o(h^2) + O(n^{-1}),$$

and the asymptotic variance is

$$V[\hat{\theta}_{p=1,h}(x)] = (nh)^{-1}R(K) + o[(nh)^{-1}].$$

These results can be extended to any order of local polynomial fitting. These results generally show a preference for an odd order for the local polynomial. Further, higher orders of the polynomial have asymptotically lower mean square error. The practical gains from choosing orders greater than 3 are minimal, and a local linear or cubic fit is generally suggested. Bandwidth selection for these methods is based on minimizing a weighted integral of the asymptotic mean square error, from which an asymptotically optimal bandwidth can be derived. Typically this bandwidth depends on derivatives of the regression function $\theta(x)$. A plug-in method, similar to the one presented in Section 5.2.1, can be applied and these derivatives are estimated using derivatives of local polynomial estimates, each with their own optimal bandwidth. See Section 5.8 of Wand and Jones (1995) for further details of these methods.

An alternative method for choosing the smoothing parameter is based on a cross-validation score, as was summarized in Section 5.2.1 for choosing the smoothing parameter of a kernel density estimator. To obtain this smoothing parameter one minimizes the leave-one-out cross-validation score given by

$$\text{CV}(h) = \frac{1}{n} \sum_{i=1}^{n} [Y_i - \hat{\theta}_{p,h,(-i)}(x_i)]^2$$

where $\hat{\theta}_{p,h,(-i)}$ is the local polynomial estimate calculated when the ith data pair (x_i, Y_i) is omitted from the data.

5.5.2 Observed Confidence Levels for $\theta(x)$

The focus of this section will be on constructing observed confidence levels for $\theta(x)$ for a specified value of $x \in I[0,1]$. Problems concerned with determining the shape of the regression function are bound to encounter similar problems to those found in nonparametric density estimation. For example, one may wish to compute an observed confidence level that $\theta(x)$ is monotonic in $I[0,1]$. Unfortunately, nonmonotonic functions that are within ϵ of any monotonic function can be constructed using arbitrarily small nonmonotonic perturbations. Without specific assumptions to forbid such behavior, it would be impossible to tell the difference between the two situations on the basis of finite samples from each.

The development of confidence intervals, and the associated observed confidence levels, follows the same general framework as in Chapter 2, with a few notable differences. For simplicity, this section will deal specifically with the local constant, or Nadaraya-Watson estimator of Equation (5.11), though similar developments of this material could center around alternative methods of nonparametric regression.

The development of confidence intervals for $\theta(x)$ naturally evolves from considering the standardized and studentized distributions of $\hat{\theta}_{0,h}(x)$. That is, the distributions of the random variables $[\hat{\theta}_{0,h}(x) - \theta(x)]/\tau(x)$ and $[\hat{\theta}_{0,h}(x) - \theta(x)]/\hat{\tau}(x)$ where $\tau^2(x) = V[\hat{\theta}_{0,h}(x)|\mathbf{x}]$ is the pointwise variance of $\hat{\theta}_{0,h}(x)$ and $\hat{\tau}^2(x)$ is an estimate which will be described later. The asymptotic bias of the Nadaraya-Watson estimator is of order h^2, under the usual assumption that $x \in I[0,1]$. This bias can be significant, particularly in places where the concavity of $\theta(x)$ is large, such as regions where there are sharp peaks in $\theta(x)$. Confidence intervals centered on the estimate $\hat{\theta}_{0,h}(x)$ can therefore be unreliable. For the immediate development of the theory below, the bias problem will be ignored in the sense that the confidence intervals will be developed for $\bar{\theta}_{0,h}(x) = E[\hat{\theta}_{0,h}(x)|\mathbf{x}]$, following the development of Hall (1992b). Suggestions for correcting or reducing the bias will be discussed at the end of this section.

Define

$$G_n(x,t) = P\{[\hat{\theta}_{0,h}(x) - \bar{\theta}_{0,h}(x)]/\tau(x) \le t|\mathbf{x}\},$$

and
$$H_n(x,t) = P\{[\hat{\theta}_{0,h}(x) - \bar{\theta}_{0,h}(x)]/\hat{\tau}(x) \le t | \mathbf{x}\}.$$

The pointwise variance of $\hat{\theta}_{0,h}(x)$ is given by

$$\tau^2(x) = \frac{\sigma^2 \sum_{i=1}^n K^2\left(\frac{x-x_i}{h}\right)}{\left[\sum_{i=1}^n K\left(\frac{x-x_i}{h}\right)\right]^2}.$$

To estimate $\tau^2(x)$ the error variance σ^2 must be estimated based on the observed data. Considerable research has concentrated on this problem, and difference based estimators appear to provide simple methods that work reasonably well in practice. See Dette, Munk, and Wagner (1998), Gasser, Sroka, and Jennen-Steinmetz (1986), Müller (1988), Müller and Stadtmüller (1987, 1988), Rice (1984), and Seifert, Gasser, and Wolfe (1993). In this section an optimal difference estimator developed by Hall, Kay, and Titterington (1990) will be used to estimate σ^2. See Seifert, Gasser, and Wolfe (1993) and Dette, Munk, and Wagner (1998) before applying this type of estimator for general use, especially if periodicity may be an element of a regression problem.

To represent general difference based estimators of σ^2, let $d_{-m_1}, \ldots, d_{m_2}$ be a sequence of real numbers such that

$$\sum_{i=-m_1}^{m_2} d_i = 0,$$

and

$$\sum_{i=-m_1}^{m_2} d_i^2 = 1.$$

It will be assumed that for $i < -m_1$ or $i > m_2$ that $d_{-m_1} d_{m_2} \ne 0$ where $m_1 \ge 0$, $m_2 \ge 0$, and $m_1 + m_2 = m$. Let $Y_{[i]}$ be the element in $\mathbf{Y}$ that corresponds to the ith largest element in $\mathbf{x}$, then a general difference based estimator of σ^2 has the form

$$\hat{\sigma}^2 = \frac{1}{n-m} \sum_{i=m_1+1}^{n-m_2} \left(\sum_{j=-m_1}^{m_2} d_j Y_{[i+j]}\right)^2. \tag{5.12}$$

Therefore, $\tau^2(x)$ can be estimated with

$$\hat{\tau}^2(x) = \frac{\hat{\sigma}^2 \sum_{i=1}^n K^2\left(\frac{x-x_i}{h}\right)}{\left[\sum_{i=1}^n K\left(\frac{x-x_i}{h}\right)\right]^2}.$$

See Dette, Munk, and Wagner (1998), Hall, Kay, and Titterington (1990), and Seifert, Gasser, and Wolfe (1993) for further information on choosing m and the weights $d_{-m_1}, \ldots, d_{m_2}$. Dette, Munk, and Wagner (1998) show that the choice of m is motivated by similar ideas to choosing the bandwidth h. For functions that are fairly smooth with no periodicities, $m = 3$ with the weights $d_1 = 0.809$, $d_2 = -0.500$ and $d_3 = -0.309$ can be used. For functions that have larger variation, $m = 3$ with the weights $d_1 = -0.408$, $d_2 = 0.816$, and $d_3 = -0.408$ have been suggested.

To develop confidence intervals for $\bar{\theta}_{0,h}(x)$, let $g_\alpha(x) = G_n^{-1}(x, \alpha)$ and $h_\alpha(x) = H_n^{-1}(x, \alpha)$. The usual critical points for $\hat{\theta}_{0,h}(x)$ from Section 2.3 are

$$\hat{\theta}_{\text{ord}}(x, \alpha) = \hat{\theta}_{0,h}(x) - \tau(x) g_{1-\alpha}(x),$$

$$\hat{\theta}_{\text{stud}}(x, \alpha) = \hat{\theta}_{0,h}(x) - \hat{\tau}(x) h_{1-\alpha}(x),$$

$$\hat{\theta}_{\text{hyb}}(x, \alpha) = \hat{\theta}_{0,h}(x) - \hat{\tau}(x) g_{1-\alpha}(x),$$

and

$$\hat{\theta}_{\text{perc}}(x, \alpha) = \hat{\theta}_{0,h}(x) + \hat{\tau}(x) g_\alpha(x).$$

The corresponding confidence intervals are given as in Section 2.3. For example, a studentized $100(\omega_U - \omega_L)\%$ confidence interval for $\bar{\theta}_{0,h}(x)$ is given by

$$C_{\text{stud}}(\alpha, \boldsymbol{\omega}; \mathbf{X}, \mathbf{Y}) = [\hat{\theta}_{0,h}(x) - \hat{\tau}(x) h_{1-\omega_L}(x), \hat{\theta}_{0,h}(x) - \hat{\tau}(x) h_{1-\omega_U}(x)],$$

for a fixed value of x within the range of $\mathbf{x}$. Let $\Psi = [t_L, t_U]$ be an arbitrary interval on $\Theta = \mathbb{R}$. Then, setting $\Psi = C_{\text{stud}}(\alpha, \boldsymbol{\omega}; \mathbf{x}, \mathbf{Y})$ yields

$$\alpha_{\text{stud}}(x, \Psi) = H_n \left[x, \frac{\hat{\theta}_{0,h}(x) - t_L}{\hat{\tau}(x)} \right] - H_n \left[x, \frac{\hat{\theta}_{0,h}(x) - t_U}{\hat{\tau}(x)} \right], \qquad (5.13)$$

as the observed confidence level for Ψ. The ordinary, hybrid, and percentile observed confidence levels are similarly developed as

$$\alpha_{\text{ord}}(x, \Psi) = G_n \left[x, \frac{\hat{\theta}_{0,h}(x) - t_L}{\tau(x)} \right] - G_n \left[x, \frac{\hat{\theta}_{0,h}(x) - t_U}{\tau(x)} \right], \qquad (5.14)$$

$$\alpha_{\text{hyb}}(x, \Psi) = G_n \left[x, \frac{\hat{\theta}_{0,h}(x) - t_L}{\hat{\tau}(x)} \right] - G_n \left[x, \frac{\hat{\theta}_{0,h}(x) - t_U}{\hat{\tau}(x)} \right], \qquad (5.15)$$

and

$$\alpha_{\text{perc}}(x, \Psi) = G_n \left[x, \frac{t_U - \hat{\theta}_{0,h}(x)}{\hat{\tau}(x)} \right] - G_n \left[x, \frac{t_L - \hat{\theta}_{0,h}(x)}{\hat{\tau}(x)} \right]. \qquad (5.16)$$

Note that the ordinary observed confidence level assumes that the variance function $\tau(x)$ is known, while the remaining observed confidence levels use an estimate for $\tau(x)$. Following the usual development, if $G_n(x, t)$ and $H_n(x, t)$ are unknown then the normal approximation can be used to approximate these observed confidence levels as

$$\hat{\alpha}_{\text{stud}}(x, \Psi) = \Phi \left[\frac{\hat{\theta}_{0,h}(x) - t_L}{\hat{\tau}(x)} \right] - \Phi \left[\frac{\hat{\theta}_{0,h}(x) - t_U}{\hat{\tau}(x)} \right],$$

and

$$\hat{\alpha}_{\text{ord}}(x, \Psi) = \Phi\left[\frac{\hat{\theta}_{0,h}(x) - t_L}{\tau(x)}\right] - \Phi\left[\frac{\hat{\theta}_{0,h}(x) - t_U}{\tau(x)}\right],$$

for the cases when $\tau(x)$ is unknown, and known, respectively.

To develop bootstrap approximations for $G_n(x,t)$ and $H_n(x,t)$ let $\tilde{\epsilon}_i = Y_i - \hat{\theta}_{0,h}(x_i)$ for all i such that $x_i \in I[0,1]$. Denote this set simply by I and let $\tilde{n}$ be the number of elements in I. Let

$$\bar{\tilde{\epsilon}} = \tilde{n}^{-1} \sum_{i \in I} \tilde{\epsilon}_i$$

and define $\hat{\epsilon}_i = \tilde{\epsilon}_i - \bar{\tilde{\epsilon}}$ for $i \in I$ as the *centered residuals*, whose mean is 0. Let $\hat{F}_n$ be the empirical distribution of $\hat{\epsilon}_i$ for $i \in I$. Then the bootstrap estimate of $G_n(x,t)$ is given by

$$\hat{G}_n(x,t) = P^*\{[\hat{\theta}_{0,h}^*(x) - \bar{\theta}_{0,h}^*(x)]/\tilde{\tau}(x) \le t | \epsilon_1^*, \dots, \epsilon_n^* \sim \hat{F}_n\},$$

where $P^*(A) = P(A|\mathbf{x}, \mathbf{Y})$, $\hat{\theta}_{0,h}^*(x)$ is the local constant estimator computed on $(x_1, \hat{Y}_1^*), \dots, (x_n, \hat{Y}_n^*)$ and $\hat{Y}_i^* = \hat{\theta}_{0,h}(x) + \epsilon_i^*$ for $i = 1, \dots, n$. The function $\tilde{\tau}(x)$ is defined by

$$\tilde{\tau}^2(x) = \frac{\tilde{\sigma}^2 \sum_{i=1}^{n} K^2\left(\frac{x-x_i}{h}\right)}{\left[\sum_{i=1}^{n} K\left(\frac{x-x_i}{h}\right)\right]^2},$$

where

$$\tilde{\sigma}^2 = \tilde{n}^{-1} \sum_{i \in I} \hat{\epsilon}_i^2.$$

Finally, it follows that $\bar{\theta}_{0,h}^*(x) = \hat{\theta}_{0,h}(x)$. See Theoretical Exercise 5. The bootstrap estimate of $H_n(x,t)$ is given by

$$\hat{H}_n(x,t) = P^*\{[\hat{\theta}_{0,h}^*(x) - \bar{\theta}_{0,h}^*(x)]/\tilde{\tau}^*(x) \le t | \epsilon_1^*, \dots, \epsilon_n^* \sim \hat{F}_n\},$$

where function $\tilde{\tau}^*(x)$ is defined by

$$\tilde{\tau}^{*2}(x) = \frac{\tilde{\sigma}^{*2} \sum_{i=1}^{n} K^2\left(\frac{x-x_i}{h}\right)}{\left[\sum_{i=1}^{n} K\left(\frac{x-x_i}{h}\right)\right]^2},$$

and

$$\tilde{\sigma}^{*2} = \frac{1}{n-m} \sum_{i=m_1+1}^{n-m_2} \left(\sum_{j=-m_1}^{m_2} d_j \epsilon_{i+j}^*\right)^2.$$

See Section 4.5.3 of Hall (1992a). Correspondingly, the bootstrap estimates of the observed confidence levels given in Equations (5.13)–(5.16) are given by

$$\hat{\alpha}_{\text{stud}}^*(x, \Psi) = \hat{H}_n\left[x, \frac{\hat{\theta}_{0,h}(x) - t_L}{\hat{\tau}(x)}\right] - \hat{H}_n\left[x, \frac{\hat{\theta}_{0,h}(x) - t_U}{\hat{\tau}(x)}\right],$$

$$\hat{\alpha}_{\text{ord}}^*(x, \Psi) = \hat{G}_n\left[x, \frac{\hat{\theta}_{0,h}(x) - t_L}{\tau(x)}\right] - \hat{G}_n\left[x, \frac{\hat{\theta}_{0,h}(x) - t_U}{\tau(x)}\right],$$

$$\hat{\alpha}_{\text{hyb}}^{*}(x, \Psi) = \hat{G}_n \left[x, \frac{\hat{\theta}_{0,h}(x) - t_L}{\hat{\tau}(x)} \right] - \hat{G}_n \left[x, \frac{\hat{\theta}_{0,h}(x) - t_U}{\hat{\tau}(x)} \right],$$

and

$$\hat{\alpha}_{\text{perc}}^{*}(x, \Psi) = \hat{G}_n \left[x, \frac{t_U - \hat{\theta}_{0,h}(x)}{\hat{\tau}(x)} \right] - \hat{G}_n \left[x, \frac{t_L - \hat{\theta}_{0,h}(x)}{\hat{\tau}(x)} \right].$$

The asymptotic analysis of these methods again follows from Edgeworth expansion theory. As before, the parameter $\bar{\theta}_{0,h}(x)$ will be assumed to be the parameter of interest. It follows from Section 4.5.2. of Hall (1992a) that

$$\begin{aligned} G_n(x,t) &= \Phi(t) + (nh)^{-1/2}\gamma p_1(t)\phi(t) + (nh)^{-1}[\kappa p_2(t) + \gamma^2 p_3(t)]\phi(t) \\ &\quad + O[n^{-1} + (nh)^{-3/2}], \end{aligned}$$

and

$$\begin{aligned} H_n(x,t) &= \Phi(t) + (nh)^{-1/2}\gamma p_1(t)\phi(t) + (nh)^{-1}[\kappa p_2(t) + \gamma^2 p_3(t)]\phi(t) \\ &\quad + n^{-1/2}h^{1/2}\gamma p_4(t)\phi(t) + O[n^{-1} + (nh)^{-3/2}]. \end{aligned}$$

The polynomials $p_1(t), \ldots, p_4(t)$ are functions of x, h, n, K and $\mathbf{x}$, but are not functions of any unknown parameters. This presentation will only use various general properties of these functions so that their specific form need not be known. For specific definitions of these polynomials see Section 4.5.2 of Hall (1992a). The coefficients γ and κ are defined as $\gamma = E[(\epsilon_i/\sigma)^3]$ and $\kappa = E[(\epsilon_i/\sigma)^4] - 3$.

The bootstrap estimates of $G_n(x,t)$ and $H_n(x,t)$ also have Edgeworth expansions of the form

$$\begin{aligned} \hat{G}_n(x,t) &= \Phi(t) + (nh)^{-1/2}\hat{\gamma} p_1(t)\phi(t) + (nh)^{-1}[\hat{\kappa} p_2(t) + \hat{\gamma}^2 p_3(t)]\phi(t) \\ &\quad + O_p[n^{-1} + (nh)^{-3/2}]. \end{aligned}$$

and

$$\begin{aligned} \hat{H}_n(x,t) &= \Phi(t) + (nh)^{-1/2}\hat{\gamma} p_1(t)\phi(t) + (nh)^{-1}[\hat{\kappa} p_2(t) + \hat{\gamma}^2 p_3(t)]\phi(t) \\ &\quad + n^{-1/2}h^{1/2}\hat{\gamma} p_4(t)\phi(t) + O_p[n^{-1} + (nh)^{-3/2}]. \end{aligned}$$

The polynomials in the bootstrap expansions are exactly the same as in the conventional expansions. The only change between the bootstrap and conventional expansions is that γ and κ are replaced be their sample versions, which generally have the property $\hat{\gamma} = \gamma + O_p(n^{-1/2})$ and $\hat{\kappa} = \kappa + O_p(n^{-1/2})$. Therefore it follows that $\hat{G}_n(x,t) = G_n(x,t) + O_p(n^{-1}h^{-1/2})$ and $\hat{H}_n(x,t) = H_n(x,t) + O_p(n^{-1}h^{-1/2})$. See Sections 4.5.2 and 4.5.3 of Hall (1992a) for more in-depth discussion about the rates of convergence of the error terms of these expansions. Finally, the corresponding bootstrap estimates of the percentiles of $G_n(x,t)$ and $H_n(x,t)$ are given by $\hat{g}_\alpha(x) = \hat{G}_n^{-1}(x,\alpha)$ and $\hat{h}_\alpha(x) = \hat{H}_n^{-1}(x,\alpha)$.

The asymptotic analysis of the observed confidence levels defined above follows the same general arguments used in Chapter 2 with some notable differences.

Take $C_{\text{stud}}(\alpha, \boldsymbol{\omega}; \mathbf{x}, \mathbf{Y})$ as the standard confidence interval for the case when $\tau(x)$ is unknown and note of course that α_{stud} is accurate. Similarly α_{ord} is accurate for the case when $\tau(x)$ is known and the ordinary confidence interval is used as the standard interval.

To analyze the hybrid observed confidence level note that

$$\alpha_{\text{hyb}}[C_{\text{stud}}(\alpha, \boldsymbol{\omega}; \mathbf{x}, \mathbf{Y})] = G_n[x, h_{1-\omega_L}(x)] - G_n[x, h_{1-\omega_U}(x)].$$

Assume that h is of larger order that $n^{-1/3}$, then the Edgeworth expansion for $H_n(x, t)$ can be inverted to yield the Cornish-Fisher expansion

$$h_{1-\omega_L} = z_{1-\omega_L} - (nh)^{-1/2}\gamma p_1(z_{1-\omega_L}) + O[(nh)^{-1} + n^{-1/2}h^{1/2}].$$

Substituting this expansion into the Edgeworth expansion for $H_n(x, t)$ and using appropriate Taylor expansions yields

$$G_n(x, h_{1-\omega_L}) = 1 - \omega_L + O[(nh)^{-1} + n^{-1/2}h^{1/2}],$$

with a similar result for $G_n(x, h_{1-\omega_L})$. It then follows that

$$\alpha_{\text{hyb}}[C_{\text{stud}}(\alpha, \boldsymbol{\omega}; \mathbf{x}, \mathbf{Y})] = \alpha + O[(nh)^{-1} + n^{-1/2}h^{1/2}].$$

The optimal bandwidth h for the local constant estimator is of order $O(n^{-1/5})$ so that in this case the hybrid method has accuracy of order $O(n^{-3/5})$. For the percentile method it follows that

$$\alpha_{\text{perc}}[C_{\text{stud}}(\alpha, \boldsymbol{\omega}; \mathbf{x}, \mathbf{Y})] = G_n[x, -h_{1-\omega_U}(x)] - G_n[x, -h_{1-\omega_L}(x)],$$

where it follows from the Cornish-Fisher expansion of $h_{1-\omega_U}$ that

$$G_n(x, -h_{1-\omega_U}(x)) = \omega_U + 2(nh)^{-1/2}p_1(z_{1-\omega_U})\phi(z_{1-\omega_U}) + O[(nh)^{-1} + n^{-1/2}h^{1/2}],$$

with a similar result for $G_n(x, -h_{1-\omega_L})$. It then follows that

$$\begin{aligned}\alpha_{\text{perc}}[C_{\text{stud}}(\alpha, \boldsymbol{\omega}; \mathbf{x}, \mathbf{Y})] &= \alpha + 2(nh)^{-1/2}p_1(z_{1-\omega_U})\phi(z_{1-\omega_U}) \\ &\quad - 2(nh)^{-1/2}p_1(z_{1-\omega_L})\phi(z_{1-\omega_L}) \\ &\quad + O[(nh)^{-1} + n^{-1/2}h^{1/2}] \\ &= \alpha + O[(nh)^{-1/2}].\end{aligned}$$

Therefore, when $h = O(n^{-1/5})$, it follows that

$$\alpha_{\text{perc}}[C_{\text{stud}}(\alpha, \boldsymbol{\omega}; \mathbf{x}, \mathbf{Y})] = \alpha + O(n^{-2/5}),$$

which is less accurate than the hybrid method. The analysis of the normal approximation is similar. Note that

$$\hat{\alpha}_{\text{stud}}[C_{\text{stud}}(\alpha, \boldsymbol{\omega}; \mathbf{x}, \mathbf{Y})] = \Phi[h_{1-\omega_L}(x)] - \Phi[h_{1-\omega_U}(x)].$$

Again, the Cornish-Fisher expansion of $h_{1-\omega_L}(x)$ yields

$$\Phi(h_{1-\omega_L}) = 1 - \omega_L + (nh)^{-1/2}p_1(z_{1-\omega_L})\phi(z_{1-\omega_L}) + O[(nh)^{-1} + n^{-1/2}h^{1/2}],$$

so that

$$\hat{\alpha}_{\text{stud}}[C_{\text{stud}}(\alpha, \boldsymbol{\omega}; \mathbf{x}, \mathbf{Y})] = \alpha + O[(nh)^{-1/2} + n^{-1/2}h^{1/2}],$$

which matches the accuracy of the percentile method.

While the studentized method is accurate, the bootstrap version is not due to the bootstrap approximation of $H_n(x, t)$. In particular,

$$
\begin{aligned}
\hat{\alpha}^*_{\text{stud}}[C_{\text{stud}}(\alpha, \boldsymbol{\omega}; \mathbf{x}, \mathbf{Y})] &= \hat{H}_n[x, h_{1-\omega_L}(x)] - \hat{H}_n[x, h_{1-\omega_U}(x)] \\
&= H_n[x, h_{1-\omega_L}(x)] - H_n[x, h_{1-\omega_U}(x)] \\
&\quad + O_p(n^{-1}h^{-1/2}) \\
&= \alpha + O_p(n^{-1}h^{-1/2})
\end{aligned}
$$

where the accuracy of the bootstrap approximation which was described earlier is implemented. This indicates that when $h = O(n^{-1/5})$ it follows that $\hat{\alpha}^*_{\text{stud}}[C_{\text{stud}}(\alpha, \boldsymbol{\omega}; \mathbf{x}, \mathbf{Y})] = \alpha + O(n^{-9/10})$, an improvement over the normal approximation and the nonbootstrap versions of the percentile and hybrid methods. In fact, the accuracy of these methods stay the same as the non-bootstrap case. This is due to the fact that the bootstrap approximation of $G_n(x, t)$ is more accurate than the corresponding observed confidence levels. For example, for the hybrid method

$$
\begin{aligned}
\hat{\alpha}^*_{\text{hyb}}[C_{\text{stud}}(\alpha, \mathbf{X}, \mathbf{Y})] &= \hat{G}_n[x, h_{1-\omega_L}(x)] - \hat{G}_n[x, h_{1-\omega_U}(x)] \\
&= G_n[x, h_{1-\omega_L}(x)] - G_n[x, h_{1-\omega_U}(x)] \\
&\quad + O_p(n^{-1}h^{-1/2}) \\
&= \alpha + O_p[n^{1/2}h^{-1/2} + n^{-1}h^{-1/2} + (nh)^{-1}],
\end{aligned}
$$

where if $h = O(n^{-1/5})$ it follows that the largest order in the error term is $O(n^{-3/5})$, matching the result from earlier. A similar argument applies to the percentile method.

As indicated earlier, the results for the observed confidence levels given in this section are technically for $\bar{\theta}_{0,h}(x)$ and not the true parameter $\theta(x)$ due to the significant bias that can be induced by the smoothing method. There are two basic approaches to account for this bias. To motivate these approaches first consider the asymptotic bias of the local constant estimator which for an equally spaced design is given by

$$
\bar{\theta}_{0,h}(x) - \theta(x) = h^2 \theta''(x)\sigma_K^2/2 + o_p(h^2).
$$

It is clear that the asymptotic bias is heavily influenced by both the bandwidth h and the concavity of $\theta(x)$ at the point x. One approach to account for the bias is to explicitly estimate it based on the data, which essentially requires an estimate of $\theta''(x)$. Local polynomial estimators can be used for this purpose, though the estimates require estimation of an additional smoothing parameter. See Section 5.7 of Wand and Jones (1995) for further details. A second approach chooses the bandwidth to be much smaller than is usually implemented for the estimation of $\theta(x)$. For example, when $h = O(n^{-1/3})$ it follows that $\bar{\theta}_{0,h}(x) - \theta(x) = O_p(n^{-2/3})$, but if $h = O(n^{-3/4})$ then $\bar{\theta}_{0,h}(x) - \theta(x) = O_p(n^{-3/2})$. The particular form of the bandwidth is difficult to specify in this case. See Hall (1992b) for a more detailed discussion of these issues. A third approach is based on using infinite-order kernel functions. See McMurry and Politis (2004). As these methods are beyond the scope of this book, the bias problem will not be dealt with explicitly in this book. In the simulations and applications that follow, observed confidence levels will only be computed when it is apparent that the concavity of the regression function appears to be near 0, where the asymptotic bias is negligible.

5.5.3 An Empirical Study

This section presents the results of a small empirical study designed to investigate how well the asymptotic results of the previous section translate to finite sample size problems. Due to the availability of software for bandwidth selection, local linear regression will be the focus of this investigation. For the study 100 sets of data were simulated from the regression model $\theta(x) = 1 - e^{-5x}$ where x was a set of n evenly spaced observations on $[0, 1]$. This function is plotted in Figure 5.11. The error vector ϵ was generated as a set of n independent and identically distributed random variables from a normal distribution with mean 0 and variance $\sigma^2 = 1/25$. The observed confidence levels α_{stud}, $\hat{\alpha}_{stud}$, $\hat{\alpha}_{stud}^*$, $\hat{\alpha}_{hyb}^*$ and $\hat{\alpha}_{perc}^*$ where computed that $\theta(x)$ is in the region $\Psi = [0.8, 0.9]$ when $x = 0.40$. Bandwidth selection for the local linear estimate was based on the direct plug-in methodology described by Ruppert, Sheather, and Wand (1995). The variance is estimated using the difference estimator with $m_1 = m_2 = 1$ and the weights $d_{-1} = 0.809$, $d_0 = -0.500$ and $d_1 = -0.309$. Therefore, the estimator has the form

$$\hat{\sigma}^2 = \frac{1}{n-2} \sum_{i=2}^{n-1} [(0.809)Y_{[i-1]} + (-0.500)Y_{[i]} + (-0.309)Y_{[i+1]}]^2.$$

Samples of size $n = 25$, 50, 100, and 200 were used in the simulation. The average observed confidence levels for each method are reported in Table 5.9. The studentized observed confidence levels were estimated by simulated observations from the true model. The bootstrap methods used residuals for values of x in the range $(0.10, 0.90)$ to avoid boundary effects.

The results of the simulations show no clear winning method for the simulated problem, though none of the methods perform particularly poorly from an absolute error standpoint. The studentized method does show its superior performance as the sample size increases, indicating that perhaps larger samples are required before the asymptotic superiority of this method is apparent. It should be noted that the estimate of the true studentized method (α_{stud}) does not include a bias-correction so that it is not possible to quantify the true effect of the smoothing bias in this problem.

Figure 5.11 *The regression function used in the empirical study.*

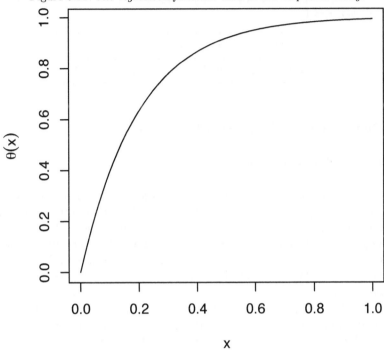

Table 5.9 *Results of the empirical study of observed confidence levels using local linear regression. The reported values are the average observed confidence level over 100 replications that $\theta(x) \in \Psi = [0.80, 0.90]$ when $x = 0.40$. Bandwidth selection for the local linear estimate is based on a direct plug-in methodology. The value closest to the true estimated value given by α_{stud} is italicized.*

n	α_{stud}	$\hat{\alpha}_{\text{stud}}$	$\hat{\alpha}^*_{\text{stud}}$	$\hat{\alpha}^*_{\text{hyb}}$	$\hat{\alpha}^*_{\text{perc}}$
25	37.13	35.71	36.68	*37.42*	41.10
50	48.92	47.09	42.29	45.61	*49.20*
100	56.97	61.39	*57.50*	55.82	63.20
200	68.86	75.29	*67.64*	66.69	78.70

Table 5.10 *Observed confidence levels, in percent, that $\hat{\theta}(x)$ is in the region $\Psi_1 =$ [75, 80] when $x = 4$.*

Region	$\hat{\alpha}_{\text{stud}}$	$\hat{\alpha}^*_{\text{stud}}$	$\hat{\alpha}^*_{\text{hyb}}$	$\hat{\alpha}^*_{\text{perc}}$
$\Psi = [75, 80]$	97.7	96.5	95.6	98.6

5.6 Nonparametric Regression Examples

5.6.1 A Problem from Geophysics

Consider the duration and waiting time data for the Old Faithful geyser presented in Section 1.3.7. The data are plotted in Figure 1.9. Based on this data it is of interest to determine an observed confidence level that $\theta(x)$ is in the region $\Psi = [75, 80]$ when $x = 4$. The methods described in Section 5.5.2 where implemented to compute this observed confidence level. The estimated optimal bandwidth using the direct plug-in method described by Ruppert, Sheather, and Wand (1995), is given by $\hat{h}_{\text{pi}} = 0.2036$, which yields an estimate of $\hat{\theta}(4) = 76.53$. The estimated standard error of $\hat{\theta}(4)$ is $\hat{\tau}(4) = 0.7621$. The bootstrap methods used $b = 1000$ resamples each. To avoid the effects of boundary bias, residuals where only used from for durations that where between 2 and 5 minutes. The results of these calculations are given in Table 5.10. One can observed from the table that there is a great degree of confidence that $\theta(x) \in [75, 80]$ when $x = 4$, or that there is a great deal of confidence that the mean waiting time to the next eruption is between 75 and 80 minutes when the duration of the previous eruption was 4 minutes.

5.6.2 A Problem from Chemistry

Radiocarbon dating is a method that uses the half life of the naturally occurring unstable isotope carbon-14 (^{14}C) to determine the age of materials that are rich in carbon. As the relationship is not perfect, calibration of the method based on items of known ages is important. Pearson and Qua (1993) report data on high precision measurements of radiocarbon on Irish oak, which can be used to construct a calibration curve for radiocarbon dating. The age indicated by radiocarbon data versus the actual calendar age for samples of Irish oaks that have an actual calendar age between 5000 and 6000 years are plotted in Figure 5.12. The data are also available in the software library associated with Bowman and Azzalini (1997), where a different subset of the data is analyzed. A similar set of calibration data is also reported by Stuiver, Reimer and Braziunas (1998). A local linear kernel regression fit is included in Figure 5.12. The optimal bandwidth was estimated using the direct plug-in algorithm

Table 5.11 *Observed confidence levels, in percent, that $\hat{\theta}(x)$ is in the region $\Theta_1 =$ [4750, 4850] when $x = 5500$.*

Region	$\hat{\alpha}_{stud}$	$\hat{\alpha}^*_{stud}$	$\hat{\alpha}^*_{hyb}$	$\hat{\alpha}^*_{perc}$
$\Theta_1 = [4750, 4850]$	92.4	98.6	99.7	91.3

of Ruppert, Sheather, and Wand (1995), which for this data produces a band-width estimate of $\hat{h} = 37.56$. As Bowman and Azzalini (1997) report, the regression curve exhibits nonlinear fluctuations that result from the natural production of radiocarbon. Given the curve exhibited in the figure, it may be of interest to hypothesize about the location of the regression function based on the observed data. For example, suppose an object has an actual age of 5500 years. How much confidence is there that the mean radiocarbon age is between 4750 and 4850 years?

The methods described in Section 5.5.2 where implemented to compute the observed confidence level that $\theta(5500) \in \Psi = [4750, 4850]$ The estimated optimal bandwidth using the direct plug-in method described by Ruppert, Sheather and Wand (1995), is given by $\hat{h}_{pi} = 37.56$, which yields an estimate of $\hat{\theta}(5500) = 4773.50$, which is an element of Ψ. The estimated standard error of $\hat{\theta}(5500)$ is $\hat{\tau}(5500) = 16.40$. The bootstrap estimates are based on 1000 re-samples. To avoid the effects of boundary bias, only residuals from actual ages between 5100 and 5900 years were used for resampling. The results of the cal-culations are given in Table 5.11. For the region Ψ there is general agreement between the measures that there is considerable observed confidence that the mean radiocarbon dated age is in the range 4750 to 4850 years when the actual age is 5500 years. The normal approximation and percentile method assigns slightly less confidence to this region than the remaining methods.

5.7 Solving Nonparametric Regression Problems Using R

Several R libraries are capable of computing local polynomial estimates. For simplicity the KernSmooth library used in Section 5.4 will also be used for local linear fitting. Two functions are important to local linear fitting in this library. The dpill function computes a direct plug-in estimate of the optimal bandwidth based on the algorithm of Ruppert, Sheather, and Wand (1995). The usage is quite simple. To store the estimate of the optimal bandwidth for a local polynomial estimator of the object Y over the grid of values in x in the object h, the command

```
h <- dpill(x,Y)
```

Figure 5.12 *Radiocarbon dated age versus actual age for samples of Irish oaks between 5000 and 6000 years old. A local linear fit is indicated by the solid line.*

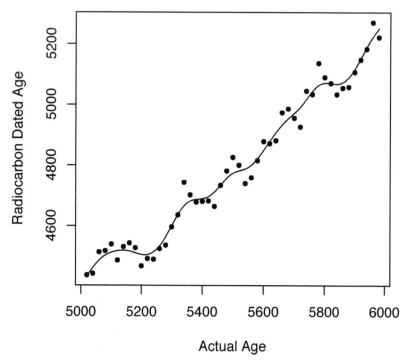

is used. The corresponding local linear estimate can then be stored in the object `ll.fit` using the command

```
ll.fit <- locpoly(x,Y,bandwidth=h)
```

The object `ll.fit` contains the fitted values in `ll.fit$y` and the corresponding grid of x values in `ll.fit$x`. The grid of x values used for the fit can be adjusted using the `range.x` and `gridsize` options in the `locpoly` function. These options are discussed below.

The difference estimator of σ^2 described in Equation (5.12) is easily implemented in R. For the specific form of this estimator used in Sections 5.5.3 and 5.6 the function

```
npr.diff.est <- function(Y) {
    w <- c(0.809,-0.500,-0.309)
    n <- length(Y)
    out <- 0
    for(i in 2:(n-1)) out <- out + (sum(w*Y[(i-1):(i+1)]))^2
    return(out/(n-2.0)) }
```

can be used. Note that this function assumes that the **Y** vector has been sorted with respect to **x**. The variance function estimate $\hat{\tau}(x)$ can then be computed for the local linear estimator using the function

```
npr.tau.hat <- function(xx,x,Y,h) {
    n <- length(x)
    s2 <- npr.diff.est(Y)
    bottom <- n*n*(npr.s(xx,x,h,2)*npr.s(xx,x,h,0)-
        npr.s(xx,x,h,1)*npr.s(xx,x,h,1))^2
    top <- 0
    for(i in 1:n) top <- top + ((npr.s(xx,x,h,2)-
        npr.s(xx,x,h,1)*(x[i]-xx))*dnorm((x[i]-xx)/h)/h)^2
    return(s2*top/bottom) }
```

where **xx** is the value of x that the function is being evaluated at.

Implementing the bootstrap is more difficult for local linear regression as only residuals for x values away from the boundary are used. Therefore, the **boot** function is not used here, and the residuals are resampled using the **sample** function. This can cause many potential technical problems, especially when using the **locpoly** function. In particular, one must assure that the fitted values for the desired x values are available in the fitted object. This can be controlled to some extent using the **range.x** and **gridsize** options in the **locpoly** function. The function given below was used to compute the bootstrap estimate of $H_n(x,t)$ for the empirical study of Section 5.5.3.

```
npr.student.boot <- function(xx,x,Y,b) {
    n <- length(x)
    h <- dpill(x,Y)
    ll.fit <- locpoly(x,Y,bandwidth=h,range.x=c(0,1),gridsize=n)
    ll.fit.default <- locpoly(x,Y,bandwidth=h)
    that <- ll.fit.default$y[ll.fit.default$x==xx]
    xI <- x[(x>0.10)&(x<0.90)]
    YI <- Y[(x>0.10)&(x<0.90)]
    nt <- length(xI)
    et <- matrix(0,nt,1)
    for(i in 1:nt) et[i] <- YI[i]-ll.fit$y[ll.fit$x==xI[i]]
    ehat <- et - mean(et)
    T.star <- matrix(0,b,1)
    for(i in 1:b) {
        ehat.star <- sample(ehat,n,replace=T)
        Y.star <- ll.fit$y + ehat.star
        h.star <- dpill(x,Y.star)
        ll.fit.star <- locpoly(x,Y.star,bandwidth=h.star)
        that.star <- ll.fit.star$y[ll.fit.star$x==xx]
        tau.star <- sqrt(npr.tau.hat(xx,x,Y.star,h.star))
        T.star[i] <- (that.star-that)/tau.star }
```

```
    return(T.star) }
```

This function returns b simulated values of $[\hat{\theta}^*_{1,h^*}(x) - \hat{\theta}_{1,h}(x)]/\hat{\tau}^*(x)$ where again xx is the value of x that the functions are being evaluated at. For a given value of t, $\hat{H}_n(x,t)$ can then be estimated using the command

```
sum(T.star<t)/b
```

Note that two separate local linear fits are used in this function. The first object 11.fit is computed so that the values in the object x have fitted values. This is possible because the values in x are a regularly spaced grid. This allows the residuals to be calculated. The second object 11.fit.default is computed so that the value stored in the object xx has a fitted value. Note further that the interior range for the x values has been hard coded into this function. Obviously this code depends on a certain values for the x and xx objects, and the function would need to be modified before it could be used in a more general setting. An alternative approach involves interpolating between the fitted values using a function such as the following

```
npr.interpolate <- function(xx,11.fit) {
    x <- 11.fit$x
    y <- 11.fit$y
    n <- length(x)
    if(xx<x[1]) return(y[1])
    if(xx>x[n]) return(y[n])
    for(i in 2:n) if(xx<x[i]) return(y[i-1]+(y[i]-y[i-1])
        *(xx-x[i-1])/(x[i]-x[i-1])) }
```

though the efficiency of the algorithm would almost certainly suffer. A version of the npr.student.boot function that uses the interpolation function is

```
npr.student.boot.interpolate <- function(xx,x,Y,b,LB,UB) {
    n <- length(x)
    h <- dpill(x,Y)
    11.fit <- locpoly(x,Y,bandwidth=h)
    that <- npr.interpolate(xx,11.fit)
    xI <- x[(x>LB)&(x<UB)]
    YI <- Y[(x>LB)&(x<UB)]
    nt <- length(xI)
    et <- matrix(0,nt,1)
    for(i in 1:nt)
        et[i] <- YI[i]-npr.interpolate(xI[i],11.fit)
    ehat <- et - mean(et)
    T.star <- matrix(0,b,1)
    for(i in 1:b) {
        ehat.star <- sample(ehat,n,replace=T)
        Y.star <- matrix(0,n,1)
```

```
      for(j in 1:n) Y.star[j] <-
            npr.interpolate(x[j],ll.fit) + ehat.star[j]
      h.star <- dpill(x,Y.star)
      ll.fit.star <- locpoly(x,Y.star,bandwidth=h.star)
      that.star <- npr.interpolate(xx,ll.fit.star)
      tau.star <-
            sqrt(npr.tau.hat.linear(xx,x,Y.star,h.star))
      T.star[i] <- (that.star-that)/tau.star }
   return(T.star) }
```

Note that this function has added the arguments LB and UB, which specify the range over which the residuals used by the bootstrap calculated. The use of the locpoly function is usually preferable to writing a new function for the calculation of the local linear estimator because the locpoly function takes advantage of the fast computation techniques based on binning. See Appendix D of Wand and Jones (1995). To compute hybrid and percentile observed confidence levels, a function that returns b simulated values of $[\hat{\theta}^*_{1,h^*}(x) - \hat{\theta}_{1,h}(x)]/\hat{\tau}(x)$ can be obtained by making small changes to the functions shown above.

5.8 Exercises

5.8.1 Theoretical Exercises

1. The symmetric bimodal distribution used in the empirical study of Section 5.2.3 has density

$$f(t) = \frac{3}{4}\phi\left(\frac{3(t+1)}{2}\right) + \frac{3}{4}\phi\left(\frac{3(t-1)}{2}\right).$$

Suppose λ is an arbitrary positive value. Write out and expression for the excess mass $d_2(\lambda)$ in terms of λ. Using R, or some other software package, plot $d_2(\lambda)$ versus λ and approximate the value of d_2.

2. Derive Equation (5.10). That is, using standard estimation theory based on weighted least squares, show that

$$\hat{\theta}_{p,h}(x) = e_1[\mathbf{X}'(x)\mathbf{W}(x)\mathbf{X}(x)]^{-1}\mathbf{X}'(x)\mathbf{W}(x)\mathbf{Y}$$

is the constant term $\hat{\beta}_0$ where $\hat{\beta}_0, \ldots, \hat{\beta}_p$ minimizes the weighted sum of squares

$$\sum_{i=1}^{n}\left[Y_i - \sum_{j=0}^{p}\beta_j(x_i - x)^j\right]^2 \frac{1}{h}K\left(\frac{x_i - x}{h}\right).$$

3. Using the form of the local polynomial estimator $\hat{\theta}_{p,h}$ given in Equation

(5.10), show that the local constant estimator can be written as

$$\hat{\theta}_{0,h}(x) = \frac{\sum_{i=1}^{n} K\left(\frac{x_i - x}{h}\right) Y_i}{\sum_{i=1}^{n} K\left(\frac{x_i - x}{h}\right)}.$$

4. Using the form of the local polynomial estimator $\hat{\theta}_{p,h}$ given in Equation (5.10), show that the local linear estimator can be written as

$$\hat{\theta}_{1,h}(x) = (nh)^{-1} \sum_{i=1}^{n} \frac{[\hat{s}_{2,h}(x) - \hat{s}_{1,h}(x)(x_i - x)]K\left(\frac{x_i - x}{h}\right) Y_i}{\hat{s}_{2,h}(x)\hat{s}_{0,h}(x) - \hat{s}_{1,h}^2(x)},$$

where

$$\hat{s}_{k,h}(x) = (nh)^{-1} \sum_{i=1}^{n} (x_i - x)^k K\left(\frac{x_i - x}{h}\right).$$

5. Consider the local constant estimator $\hat{\theta}_{0,h}$ given by

$$\hat{\theta}_{0,h}(x) = \frac{\sum_{i=1}^{n} K\left(\frac{x_i - x}{h}\right) Y_i}{\sum_{i=1}^{n} K\left(\frac{x_i - x}{h}\right)},$$

where $Y_i = \theta(x_i) + \epsilon_i$ for $i = 1, \ldots, n$ and $\epsilon_1, \ldots, \epsilon_n$ is a set of independent and identically distributed random variables with mean 0 and variance σ^2.

(a) Show that

$$\hat{\theta}_{0,h}(x) - \bar{\theta}_{0,h}(x) = \frac{\sum_{i=1}^{n} K\left(\frac{x - x_i}{h}\right) \epsilon_i}{\sum_{i=1}^{n} K\left(\frac{x - x_i}{h}\right)}.$$

(b) Use the result in Part (a) to show that

$$\frac{\hat{\theta}_{0,h}(x) - \bar{\theta}_{0,h}(x)}{\tau(x)} = \frac{1}{L(x)\sigma} \sum_{i=1}^{n} \epsilon_i K\left(\frac{x - x_i}{h}\right),$$

and

$$\frac{\hat{\theta}_{0,h}(x) - \bar{\theta}_{0,h}(x)}{\hat{\tau}(x)} = \frac{1}{L(x)\hat{\sigma}} \sum_{i=1}^{n} \epsilon_i K\left(\frac{x - x_i}{h}\right),$$

where

$$L(x) = \left[\sum_{i=1}^{n} K^2\left(\frac{x - x_i}{h}\right)\right]^{1/2}.$$

See Equations (4.105) and (4.106) of Hall (1992a).

6. Demonstrate that there is a bimodal density F such that $\|\Phi - F\| < \epsilon$ for every $\epsilon > 0$.

5.8.2 Applied Exercises

1. A educational researcher in mathematics is interested in the effect of optional tutorial sessions offered for students in a first-year calculus course.

She suspects that there may be two distinct populations of final exam grades in the class corresponding to those students who attended the session and those who did not. Unfortunately, attendance was not taken at the sessions and therefore there is no data on which students attended and which did not. She decides that she will look at a kernel density estimate for the scores on the final exams. If there is an indication of two sub-populations, that is two modes, in the density then she will pursue the matter further with a more formal investigation the next semester. The scores on the final exam for the fifty students in her class are given in the table below.

76	76	100	81	88	81	71	93	88	58
74	98	77	91	86	68	83	86	85	87
79	83	27	72	59	60	40	65	46	69
81	58	37	67	67	41	76	37	72	36
67	58	72	41	69	57	80	56	71	59

(a) Compute a kernel density estimate based on the data using the two-stage plug-in bandwidth described in this chapter. How many modes are apparent in the density?

(b) Compute observed confidence levels using the three methods described in Section 5.2.2 for the condition that the density has at least two modes. What general conclusions would you make about the contention that there may be two subpopulations in the data?

(c) Compute observed confidence levels for the excess mass d_2 using regions of the form $\Theta_1 = [0, 0.10)$, $\Theta_2 = [0.10, 0.20)$, and $\Theta_3 = [0.20, 0.30)$. What regions have the greatest observed confidence? *Optional*: Using a simulated annealing algorithm, construct density estimates whose excess mass is constrained to the endpoint values of the region with the highest observed confidence. What do these plots indicate about the possible number of modes of the density?

2. An biologist spends a sunny Sunday afternoon collecting sixty pine needles in a forest near her home. That night she carefully measures each of the pine needles in millimeters. These observations are given in the table below. The biologist is interested in whether there is more than one population of pine needles in the sample.

41	41	35	31	43	36	31	35	27	44
26	34	32	39	34	38	30	36	37	36
39	33	41	43	34	34	29	38	76	71
61	55	62	57	63	49	53	65	56	66
67	54	65	70	75	59	67	62	47	63
58	74	67	50	58	40	43	49	55	66

Table 5.12 *Electricity bill and average monthly high temperature (°F) for a single home over 36 months.*

Month	1	2	3	4	5	6
Temperature	30.8	35.7	52.5	67.7	79.8	78.6
Electric Bill	96.71	90.28	74.69	70.13	81.45	101.94
Month	7	8	9	10	11	12
Temperature	86.9	77.8	80.6	68.2	51.0	38.5
Electric Bill	121.45	107.52	99.63	76.85	73.71	91.51
Month	13	14	15	16	17	18
Temperature	34.7	34.9	56.4	60.3	73.8	78.7
Electric Bill	96.73	89.59	77.52	67.61	78.73	97.95
Month	19	20	21	22	23	24
Temperature	92.0	82.4	84.1	65.9	52.9	42.8
Electric Bill	121.97	109.66	101.70	76.52	74.55	88.80
Month	25	26	27	28	29	30
Temperature	41.2	36.3	54.8	64.7	74.1	88.1
Electric Bill	102.70	91.26	73.88	71.40	79.48	101.60
Month	31	32	33	34	35	36
Temperature	88.5	81.1	74.1	67.6	63.3	34.1
Electric Bill	120.77	109.08	102.89	75.20	76.64	89.13

(a) Compute a kernel density estimate based on the data using the two-stage plug-in bandwidth described in this chapter. How many modes are apparent in the density?

(b) Compute observed confidence levels using the three methods described in Section 5.2.2 for the condition that the density has at least two modes. What general conclusions would you make about the contention that there may be two subpopulations in the data?

(c) Compute observed confidence levels for the excess mass d_2 using regions of the form $\Theta_1 = [0, 0.10)$, $\Theta_2 = [0.10, 0.20)$, and $\Theta_3 = [0.20, 0.30)$. What regions have the greatest observed confidence? *Optional*: Using a simulated annealing algorithm, construct density estimates whose excess mass is constrained to the endpoint values of the region with the highest observed confidence. What do these plots indicate about the possible number of modes of the density?

3. The data in Table 5.12 gives the results of a study on the mean monthly high temperature (°F) of a house with electric heating and air conditioning and the corresponding monthly electric bill. A total of 36 months were observed.

Table 5.13 *Corn yield and rainfall for twenty-five counties.*

County	1	2	3	4	5
Rainfall	17.8	16.4	14.3	14.4	12.8
Yield	140.6	144.3	133.9	134.2	117.4
County	6	7	8	9	10
Rainfall	15.6	14.1	17.4	12.6	11.2
Yield	145.8	141.4	147.9	117.9	122.0
County	11	12	13	14	15
Rainfall	13.2	21.5	15.2	14.2	16.2
Yield	127.1	145.2	142.7	122.8	136.5
County	16	17	18	19	20
Rainfall	13.6	13.0	8.2	9.9	13.4
Yield	132.8	130.5	102.6	113.1	135.6
County	21	22	23	24	25
Rainfall	13.3	20.3	14.2	12.4	16.5
Yield	134.6	148.9	136.2	114.0	145.7

(a) Fit a local linear regression line to the data and comment on the result. Use the direct plug-in estimate of Ruppert, Sheather, and Wand (1995) to estimate the asymptotically optimal bandwidth. Comment on the trend observed in the local linear fit and use the fitted model to estimate the mean electric bill when the average high temperature in a month is 80 degrees.

(b) Compute observed confidence levels that the mean electric bill is in the regions $\Theta_1 = [70, 80)$, $\Theta_2 = [80, 90)$, $\Theta_3 = [90, 100)$, $\Theta_4 = [100, 110)$, and $\Theta_5 = [110, 120)$, when the average high temperature for the month is 80 degrees. Use the normal approximation and the bootstrap estimates of the studentized, hybrid and percentile methods. For the bootstrap methods use $b = 1000$ resamples. The standard error of $\hat{\theta}(x)$ can be estimated using the method outlined in Section 5.5.3.

4. The amount and timing of rainfall are important variables in predicting corn yield. An early example of a study of the effect of rainfall on corn yield is given by Hodges (1931). Suppose that twenty-five counties in the mid-western United States were randomly selected during the previous growing season (May through August) and the growing season precipitation and corn yield, measured in bushels per acre, were reported. The data are given in Table 5.13.

(a) Fit a local linear regression line to the data and comment on the result. Use the direct plug-in estimate of Ruppert, Sheather, and Wand (1995) to estimate the asymptotically optimal bandwidth. Comment on the trend observed in the local linear fit and use the fitted model to estimate the mean yield when the rainfall is 14 inches.

(b) Compute observed confidence levels that the mean corn yield is in the regions $\Theta_1 = (129, 131)$, $\Theta_2 = (128, 132)$, $\Theta_3 = (127, 133)$, $\Theta_4 = (126, 134)$, and $\Theta_5 = (125, 135)$ when the rainfall is 14 inches. Use the normal approximation and the bootstrap estimates of the studentized, hybrid, and percentile methods. For the bootstrap methods use $b = 1000$ resamples. The standard error of $\hat{\theta}(x)$ can be estimated using the method outlined in Section 5.5.3.

CHAPTER 6

Further Applications

6.1 Classical Nonparametric Methods

Classical nonparametric methods, which are usually based on ranking or counting, can be used to compute distribution-free confidence regions for many problems where no specific assumptions about the form of the population of interest can be made. These methods can therefore be used to compute observed confidence levels for the associated parameters as well. Classical nonparametric methods usually focus on slightly different parameters than their parametric counterparts. For example, the median of a distribution is usually the focus instead of the mean. The problems also usually use simplifying assumptions, such as symmetry, to translate these problems into more familiar parameters. For example, in the case of symmetry, the mean and the median will coincide. This section will briefly discuss the use of these methods for the particular type of problem where the possible shift in a symmetric distribution is of interest. Several references provide book length treatments of nonparametric statistics. For more applied presentations, see Conover (1980), Gibbons and Chakraborti (2003), Hollander and Wolfe (1999), and Lehmann (2006). For theoretical presentations, see Hàjek (1969), Hàjek and Šidák (1967), Hettmansperger (1984), and Randles and Wolfe (1979).

The two sample location problem is usually studied in nonparametric statistics with the location-shift model. In this model it is assumed that $X_1, \ldots, X_m$ is a set of independent and identically distributed random variables from a distribution F and that $Y_1, \ldots, Y_n$ is a set of independent and identically distributed random variables from a distribution G where $G(t) = F(t - \theta)$ for every $t \in \mathbb{R}$. The two samples are assumed to be mutually independent. The parameter θ is called the *shift parameter*. The null hypothesis that $\theta = 0$ is usually tested using the Wilcoxon two-sample rank statistic. This statistic is computed by combining and ranking the two samples. The test statistic then corresponds to the sum of the ranks associated with $Y_1, \ldots, Y_n$. Under the null hypothesis the distribution of the test statistic is distribution free over the set of continuous distributions. This allows one to construct a distribution-free test of the null hypothesis that $\theta = 0$. The test was originally introduced by Wilcoxon (1945). An equivalent test was also proposed by Mann and Whitney (1947).

Inversion of the Wilcoxon rank sum test yields a distribution-free confidence interval for θ. Let w_ξ be the ξth percentile of the distribution of the Wilcoxon rank sum test statistic under the null hypothesis that $\theta = 0$. Tables of this distribution can be found in most applied nonparametric statistics textbooks, though one should be very careful as there is not a standard definition of this test statistic. See, for example, Table A.6 in Hollander and Wolfe (1999). To calculate a distribution-free confidence interval for θ compute all mn differences of the form $Y_i - X_j$ for $i = 1, \ldots, n$ and $j = 1, \ldots, m$. Denote the ordered version of these differences as $U_{(1)}, \ldots, U_{(mn)}$. A $100(1 - \omega_L - \omega_U)\%$ confidence interval for θ is then given by $(U_{(C_{\omega_L})}, U_{(mn+1-C_{\omega_U})})$ where

$$C_\xi = \frac{n(2m + n + 1)}{2} + 1 - w_{1-\xi}.$$

Let Θ be the parameter space of θ, which will generally be $\mathbb{R}$, and let $\Psi = (t_L, t_U)$ be any interval subset of Θ. The observed confidence level of Ψ based on the Wilcoxon rank sum confidence interval is computed by setting $t_L = U_{(C_{\omega_L})}$ and $t_U = U_{(mn+1-C_{\omega_U})}$ and solving for ω_L and ω_U, where it can be noted that

$$mn + 1 - C_{\omega_U} = w_{1-\omega_u} - \frac{n(n + 1)}{2}.$$

Suppose that $t_L = U_{(C_{\omega_L})}$ implies that $C_{\omega_L} = C_L$ and $t_U = U_{(mn+1-C_{\omega_U})}$ implies that $mn + 1 - C_{\omega_U} = C_U$. Then $w_{1-\omega_L} = n(2m + n + 1)/2 + 1 - C_L$ and $w_{1-\omega_U} = C_U - n(n + 1)/2$, from which the values of ω_L and ω_U, and the corresponding observed confidence levels can be obtained.

As an example consider the data of Thomas and Simmons (1969) analyzed by Hollander and Wolfe (1999). The data consist of observations on the level of histamine (μg) per gram of sputum for nine allergic smokers and thirteen non-allergic smokers. The data are given in Table 6.1. Suppose it is of interest to compute observed confidence levels that the histamine levels for allergics is at least 25, 50, and 100 μg/g more than that of nonallergics. That is, assuming the two samples fit within the framework described above, compute observed confidence levels that θ, the shift parameter between the two populations, is in the regions $\Theta_1 = (-25, 0)$, $\Theta_2 = (-50, 0)$, and $\Theta_3 = (-100, 0)$ where the sample of allergic subjects is assumed to be the values $X_1, \ldots, X_m$. There are $mn = 117$ pairwise differences between the two groups of observations. For brevity these differences will not be listed here, however it follows that $U_{(27)} \simeq -100$, $U_{(62)} \simeq -50$, $U_{(86)} \simeq 25$, and $U_{(107)} \simeq 0$. Typically, as in this problem, one will not be able to find differences that are exactly equal to the endpoints of the interval regions of interest. In this example the closest difference was used.

To compute the observed confidence level for Θ_1 it is noted that $C_L = 86$ from which it follows that $w_{1-\omega_L} = 123$ which implies that $\omega_L = 0.9587$. Similarly $C_U = 107$ from which it follows that $w_{1-\omega_L} = 198$ which implies that $\omega_L = 0.0002$. This implies that the observed confidence level for $\Theta_1 = 1 - 0.9587 - 0.0002 = 0.0411$. Similar calculations can be used to show that the

Table 6.1 *Observations on the level of histamine (μg) per gram of sputum for 9 allergic smokers (Group 1) and 13 non-allergic smokers (Group 2). The subject observations have been sorted within each group. The data originate in Thomas and Simmons (1969).*

Subject	Group	Histamine Level
1	1	31.0
2	1	39.6
3	1	64.7
4	1	65.9
5	1	67.6
6	1	100.0
7	1	102.4
8	1	1112.0
9	1	1651.0
10	2	4.7
11	2	5.2
12	2	6.6
13	2	18.9
14	2	27.3
15	2	29.1
16	2	32.4
17	2	34.3
18	2	35.4
19	2	41.7
20	2	45.5
21	2	48.0
22	2	48.1

observed confidence levels for Θ_2 and Θ_3 are 0.4478 and 0.9872, respectively. Therefore, it is clear that there is little confidence that the shift parameter is within 25 units below the origin, but there is substantial confidence that the shift parameter is within 100 below the origin.

An approximate approach can be developed by using the normal approximation given in Equation (4.41) of Hollander and Wolfe (1999). This approximation states that

$$C_\xi = mn/2 - z_{1-\xi}[mn(m+n+1)/12]^{1/2}.$$

Setting C_L equal to this approximation yields an approximation for ω_L given by

$$\omega_L \simeq 1 - \Phi[(C_L - mn/2)[mn(m+n+1)/12]^{-1/2}].$$

Similarly, setting $C_U = mn + 1 - C_{\omega_L}$ yields an approximation for ω_U given

by
$$\omega_U \simeq 1 - \Phi[(C_U - 1 - mn/2)[mn(m + n + 1)/12]^{-1/2}].$$
Applying these approximations to the example data yields approximate observed confidence levels equal to 0.0169, 5916 and 0.9661 for Θ_1, Θ_2, and Θ_3, respectively. The approximation will be more reliable when both sample sizes are larger.

6.2 Generalized Linear Models

Generalized linear models provide an extension of the linear model framework to log-linear, probit, and other models. This generalization was first described by Nelder and Wedderburn (1972). A book-length treatment of this topic is given by McCullagh and Nelder (1989). This section will briefly discuss the application of observed confidence levels to these types of models.

In the classical linear model there is an observed random vector $\mathbf{Y}$ and a matrix of known constants $\mathbf{X}$ such that $E(\mathbf{Y}) = \mathbf{X}\boldsymbol{\theta}$ where $\boldsymbol{\theta}$ is a vector of unknown parameters. The random vector $\mathbf{Y}$ is called the *random component* of the model and $\mathbf{X}\boldsymbol{\theta}$ is called the *systematic component*. The function that relates the random component to the systematic component is called the *link function*. In the classical linear model the link function is the identity function. It is further typical in linear models to assume that $\mathbf{Y}$ is a multivariate normal random vector, though this was not necessarily the case in Chapter 4.

The generalized linear model extends the framework of the classical linear model in two fundamental ways. First, $\mathbf{Y}$ is no longer assumed to follow a multivariate normal distribution, rather each component of $\mathbf{Y}$ is assumed to follow a distribution that is a member of the exponential family. Such distributions have the general form

$$f(y) = \exp\{[y\beta - b(\beta)]/a(\phi) + c(y, \phi)\},$$

where a, b and c are functions and β and ϕ are parameters. The parameter β is usually called the *canonical parameter* and ϕ is usually called the *dispersion parameter*.

The second extension of the generalized linear model over the linear model is in the link function. The link function can be a function other than the identity, and its specification is an important device in defining the type of model being studied. The linear model uses the identity link which insures a linear relationship between $E(\mathbf{Y})$ and the parameters in the systematic component. There are models where $\mathbf{Y}$ consists of non-negative counts where such a link function may be unrealistic. In this case the log function is often used as the link function resulting in a log-linear model, usually associated with the Poisson distribution for the components of $\mathbf{Y}$. Similarly a logit link and a binomial distribution can be used to model components of $\mathbf{Y}$ that are counts with a maximum value m. A special case, which focuses on binary data, can use any number of possible link functions including the logit, inverse normal, complementary log-log, or the log-log link functions along with the binary distribution. For other applications and link functions, see McCullagh and Nelder (1989).

As an example consider the leukemia survival data of Feigl and Zelen (1965). The data consist of the survival time in weeks of thirty-three subjects who are from two groups. The log (base 10) of the white blood cell count of each subject was also observed. The data are given in Table 6.2. Davison and Hinkley (1997) fit this data using a logarithmic link function with a model of the form

$$log(\mu_{ij}) = \theta_i + \theta_3 x_{ij}$$

where $\theta' = (\theta_1, \theta_2, \theta_3)$, μ_{ij} is the mean and x_{ij} is the log (base 10) of the white blood cell count of the jth subject in group i. Hence the specified model assumes that the effect of the white blood cell count is the same for both groups, but that each group has its own intercept. Let Y_{ij} be the survival time of the jth subject in group i, then under this model Y_{ij}/μ_{ij} has an exponential distribution with mean equal to 1. Davison and Hinkley (1997) conclude that this model appears to fit the data adequately. The likelihood based parameter estimates for the fit are $\hat{\theta}_1 = 6.829$, $\hat{\theta}_2 = 5.811$, and $\hat{\theta}_3 = -0.700$ with an estimated covariance matrix for θ given by

$$\hat{V}(\hat{\theta}) = \begin{bmatrix} 1.746 & 1.716 & -0.411 \\ 1.716 & 1.819 & -0.419 \\ -0.411 & -0.419 & 0.100 \end{bmatrix}.$$

See Equation (7.4) of Davison and Hinkley (1997). Given these results, how much confidence is there that θ_1 exceeds θ_2?

There are many approaches to computing confidence intervals for the parameters of such a model. For simplicity this section will consider a simple approach based on the normal approximation. There are certainly methods that are more more appropriate for this model. See Section 7.2 of Davison and Hinkley (1997) for several alternative methods. The normal approximation states that $\hat{V}^{-1/2}(\hat{\theta})(\hat{\theta} - \theta)$ approximately follows a $N_3(0, I)$ distribution. Following the development of Section 3.2, it follows that $\hat{C}_{stud}(\alpha; X, Y) = \hat{\theta} - \hat{V}^{1/2}(\hat{\theta})N_\alpha$ is a $100\alpha\%$ confidence region for θ. The corresponding observed confidence level for an arbitrary region $\Psi \subset \Theta$, where Θ is the parameter space of θ is given by

$$\hat{\alpha}_{stud}(\Psi) = \int_{\hat{V}^{-1/2}(\hat{\theta})(\hat{\theta}-\Psi)} \phi_p(t)dt.$$

For the region $\Psi = \theta : \theta_1 > \theta_2$ this observed confidence level is $\hat{\alpha}_{stud}(\Psi) = 99.75$ in percent, indicating a great deal of confidence that θ_1 exceeds θ_2.

Table 6.2 *Leukemia data from Feigl and Zelen (1965). The white blood cell count is the log (base 10) of the actual white blood cell count and the survival time is in weeks.*

Subject	Group	White Blood Cell Count	Survival Time
1	1	3.36	65
2	1	2.88	156
3	1	3.63	100
4	1	3.41	134
5	1	3.78	16
6	1	4.02	108
7	1	4.00	121
8	1	4.23	4
9	1	3.73	39
10	1	3.85	143
11	1	3.97	56
12	1	4.51	26
13	1	4.54	22
14	1	5.00	1
15	1	5.00	1
16	1	4.72	5
17	1	5.00	65
18	2	3.64	56
19	2	3.48	65
20	2	3.60	17
21	2	3.18	7
22	2	3.95	16
23	2	3.72	22
24	2	4.00	3
25	2	4.28	4
26	2	4.43	2
27	2	4.45	3
28	2	4.49	8
29	2	4.41	4
30	2	4.32	3
31	2	4.90	30
32	2	5.00	4
33	2	5.00	43

6.3 Multivariate Analysis

Many problems in mutlivariate analysis can be set within the framework of the problem of regions, and therefore are open to solution through observed confidence levels. Several problems have already been addressed. For example in Chapter 3, order-restricted and other types of inference of mean vectors is considered in several examples and exercise problems. This section will briefly focus on the application of observed confidence levels in the classical problem of principal component analysis.

Let $\mathbf{X}_1, \ldots, \mathbf{X}_n$ be a set of independent and identically distributed d dimensional random vectors from a distribution F with mean vector $\boldsymbol{\mu}$ and covariance matrix $\boldsymbol{\Sigma}$, where the elements of $\boldsymbol{\Sigma}$ are assumed to be finite. Let $\mathbf{X} = [\mathbf{X}_1, \ldots, \mathbf{X}_n]$ be the $d \times n$ matrix whose columns correspond to the random vectors in the sample and $\hat{\boldsymbol{\Sigma}}$ be the usual sample covariance matrix. The first *principal component* of $\mathbf{X}$ corresponds to the linear function $\boldsymbol{p}_1' \mathbf{X}$ that maximizes the sample variance $\boldsymbol{p}_1' \hat{\boldsymbol{\Sigma}} \boldsymbol{p}_1$ subject to the constraint that $\boldsymbol{p}_1'$ is normalized so that $\boldsymbol{p}_1' \boldsymbol{p}_1 = 1$. Simple optimization theory can be used to show that the vector $\boldsymbol{p}_1$ is the eigenvector that corresponds to the largest eigenvalue of $\hat{\boldsymbol{\Sigma}}$. See, for example, Section 8.2 of Morrison (1990). The remaining orthogonal eigenvectors define the remaining $d - 1$ principal components based on the decreasing values of the corresponding eigenvalues. That is, the kth principal component is computed using the eigenvector $\mathbf{p}_k$ that corresponds to the kth largest eigenvalue under the constraints that $\boldsymbol{p}_k' \boldsymbol{p}_k = 1$ and $\boldsymbol{p}_k' \boldsymbol{p}_j = 0$ for all $k \neq j$. The eigenvalues are interpretable as the corresponding sample variance of the linear function $\mathbf{p}_k' \mathbf{X}$. Therefore, the first few principal components are usually of interest as they often show the directions in which most of the variability of the data is concentrated. Such results are useful in effectively reducing the dimension of the data, and the corresponding complexity of any statistical analysis of the data. In the most useful cases, the coefficients of the eigenvectors are also interpretable in terms of the components of the data vectors. Note that principal component analysis can also be performed on the sample correlation matrix if a standardized approach is desired.

As an example consider the data on Swiss banknotes from Flury and Riedwyl (1988). The data consist of six measurements of 100 genuine Swiss banknotes: the length of the bill (X_1), the width of the bill along the left edge (X_2), the width of the bill along the right edge (X_3), the width of the margin (to the image) at the bottom (X_4), the width of the margin (to the image) at the top (X_5), and the length along the diagonal of the image (X_6). The sample covariance matrix of the data is given in Table 6.3.

The eigenvalues of the sample covariance matrix in Table 6.3 are equal to 0.6890, 0.3593, 0.1856, 0.0872, 0.0802 and 0.0420. These indicate that the first four principal components account for over 90% of the variability in the

Table 6.3 *Sample covariance matrix of the Swiss banknote data from Flury and Riedwyl (1988).*

	X_1	X_2	X_3	X_4	X_5	X_6
X_1	0.1502	0.0580	0.0573	0.0571	0.0145	0.0055
X_2	0.0580	0.1326	0.0859	0.0567	0.0491	-0.0431
X_3	0.0573	0.0859	0.1263	0.0582	0.0306	-0.0238
X_4	0.0571	0.0567	0.0582	0.4132	-0.2635	-0.0002
X_5	0.0145	0.0491	0.0306	-0.2635	0.4212	-0.0753
X_6	0.0055	-0.0431	-0.0238	-0.0002	-0.0753	0.1998

data. The eigenvectors corresponding to these eigenvalues are

$$
\mathbf{p}_1 = \begin{bmatrix} 0.061 \\ 0.013 \\ 0.037 \\ 0.697 \\ -0.706 \\ 0.106 \end{bmatrix}, \quad
\mathbf{p}_2 = \begin{bmatrix} -0.378 \\ -0.507 \\ -0.454 \\ -0.358 \\ -0.365 \\ 0.364 \end{bmatrix}, \quad
\mathbf{p}_3 = \begin{bmatrix} -0.472 \\ -0.101 \\ -0.196 \\ 0.108 \\ -0.074 \\ -0.844 \end{bmatrix},
$$

and

$$
\mathbf{p}_4 = \begin{bmatrix} 0.786 \\ -0.244 \\ -0.281 \\ -0.242 \\ -0.243 \\ -0.354 \end{bmatrix}.
$$

The first principal component is interpretable as the difference between the top and bottom margins to the image, an indication of the vertical position of the banknote image on the bill. The remaining eigenvectors have less convincing interpretations. This analysis suggests several problems that can be studied in terms of observed confidence levels. For example, how much confidence is there that four principal components are required to account for 90% of the variability in the data? Also, it appears that the top and bottom margins are important in accounting for this variability. How much confidence is there that these two measurements have the largest absolute weights in the first principal component?

Standard methods for computing standard errors based on the multivariate normal distribution are available for principal components. However, given the complex nature of the problems suggested above, a simple bootstrap analysis of the problem based on the percentile method will be presented. Bootstrap analysis of the principal component problem is not new. See, for example, Diaconis and Efron (1983), Jackson (1993, 1995), Mehlman, Shepherd, and

Table 6.4 *Observed confidence levels, in percent, that k principal components are required to account for 90% of the variability in the data. The calculations are based on 10,000 bootstrap resamples.*

k	1	2	3	4	5	6
$\hat{\alpha}_{\text{perc}}(k)$	0.00	0.00	0.91	97.47	1.62	0.00

Table 6.5 *Observed confidence levels, in percent, for all possible pairs of measurements having the largest two absolute weights in the first principal component. Only pairs with nonzero observed confidence levels are shown. The calculations are based on 10,000 bootstrap resamples.*

Measurements	Observed Confidence Level
length, bottom margin	0.03
left edge width, bottom margin	0.08
right edge width, bottom margin	0.06
right edge width, top margin	0.01
bottom margin, top margin	99.74
top margin, diagonal	0.08

Kelt (1995), Milan and Whittaker (1995), and Stauffer, Garton, and Steinhorst (1985). For the first problem, 10,000 resamples of size 100 drawn from the original observations from the banknotes were generated and a principal component analysis was performed on each. The number of eigenvalues from the sample covariance of each resample needed to account for 90% of the variability in the data was computed. The proportion of resamples that required k principal components, where $k = 1, \ldots, 6$, was then calculated as the observed confidence level. The results of these calculations are given in Table 6.4. It is clear from this analysis that there is a large degree of confidence that four principal components will account for at least 90% of the variability in the banknote data. The same general methodology was used to compute the amount of confidence that the top and bottom margins have the largest weights in the first principal component. For a more complete analysis, the observed confidence level was computed for each possible pair of measurements. The results of these calculations are given in Table 6.5. It is clear that there is substantial confidence that the bottom and top margin measurements will generally be the measurements with the largest absolute weights in the first principal component.

6.4 Survival Analysis

Problems in survival analysis and reliability are concerned with estimating and comparing the survival functions of random variables that represent lifetimes or failure times. These problems often have the added difficulty that some of the observed data is right censored, so for some observations it is only known that their failure time exceeds a certain amount, but the actual failure time is not observed. This section briefly describes the potential application of observed confidence levels to these problems and suggests areas of further research.

Let $X_1, \ldots, X_n$ be a set of n independent and identically distributed failure times from a continuous distribution F. Let $C_1, \ldots, C_n$ be a set of n independent and identically distributed censoring times from a continuous distribution G. The censoring time C_i is associated with the failure time X_i so that due to the censoring the actual observed observations are $Y_1, \ldots, Y_n$ where $Y_i = \min\{X_i, C_i\}$ for $i = 1, \ldots, n$. In order to keep track of which observed values are censored, let $o_i = \delta(Y_i; X_i)$ for $i = 1, \ldots, n$. Of interest in these types of problems is the estimation of the survival function $S(x) = 1 - F(x) = P(X_i > x)$, which is the probability of survival past time x. In the nonparametric setting the most common estimate of $S(x)$ that accounts for the censoring is given by the *product limit* estimator of Kaplan and Meier (1958). To compute this estimator, let $x_{(1)} < x_{(2)} < \cdots < x_{(k)}$ be the ordered distinct uncensored failure times. Define

$$n_i = \sum_{j=1}^{n} \delta[Y_j; (-\infty, x_{(i)})],$$

and

$$d_i = \sum_{j=1}^{n} \delta(Y_j; x_{(i)})(1 - o_j).$$

Then the product limit estimator of $S(x)$ is given by

$$\hat{S}(x) = \prod_{x_{(i)} \leq x} \left(1 - \frac{d_i}{n_i}\right).$$

Note that when there are no censored values in the data the product limit estimator is equal to one minus the empirical distribution function of the failure times. In this case the estimator is known as the empirical survival function.

Observed confidence levels for regions of the survival probability defined at a specified time x can be computed using any of the numerous pointwise confidence intervals for $S(x)$ based on the product limit estimator. For example, Hollander and Wolfe (1999) given an asymptotic confidence interval of the form

$$C(\alpha, \omega; \mathbf{X}) = [\hat{S}_n(x) - z_{1-\omega_L} \hat{v}^{1/2}(x), \hat{S}_n(x) - z_{1-\omega_U} \hat{v}^{1/2}(x)]$$

where $\alpha = \omega_U - \omega_L$ and

$$\hat{v}(x) = \hat{S}_n^2(x) \sum_{x_{(i)} \le x} \frac{d_i}{n_i(n_i - d_i)}.$$

Let $\Psi(x) = (t_L(x), t_U(x)) \subset [0,1]$, then setting $C(\alpha, \boldsymbol{\omega}; \mathbf{X}) = \Phi(x)$ yields an asymptotic observed confidence level for $\Psi(x)$ as

$$\alpha[\Psi(x)] = \Phi \left[\frac{\hat{S}_n(x) - t_L(x)}{\hat{v}^{1/2}(x)} \right] - \Phi \left[\frac{\hat{S}_n(x) - t_U(x)}{\hat{v}^{1/2}(x)} \right]. \qquad (6.1)$$

This type of observed confidence level can be used to assess the level of confidence that the survival probability is within a specified region at time x. Many other approximate confidence intervals for $S(x)$ have also been proposed. See Section 11.6 of Hollander and Wolfe (1999) for a summary of some of these procedures.

More interesting applications come from comparing two survival curves. In biomedical applications this usually involves comparing survival curves of a treatment and control groups. See, for example, Heller and Venkatraman (1996). In such cases it is often on interest to asses whether one curve is strictly dominated by the other after time 0 (where both curves equal 1) or whether the curves cross at some point. Crossing survival curves would indicate that one group had better survival probabilities for early time periods, and worse survival probabilities for later time periods. The time at which such a crossing occurs is also of interest. For these types of applications the extension of confidence limits for $S(x)$, or even confidence bands for $S(x)$ over the entire range of x, is not straightforward to computing observed confidence levels that two survival curves cross. One can visualize this problem as an extension of the multiple parameter problem of ordering the components of two mean vectors, but the associated theory may be more complicated. For the sake of exposition in this section the bootstrap percentile method approach will be implemented. Further research may be required to obtain more satisfactory results.

The bootstrap percentile method approach to determining an observed confidence level for two survival curves crossing is implemented by computing the proportion of time that one estimated survival curve crosses another given independent resampling from each set of survival data. The simplest approach to resampling from the censored survival data is to resample the pairs (Y_i, o_i) from the original data and compute the product limit estimator on the resulting resampled data, with the attached censoring information. For further ideas on implementing the bootstrap with censored data, see Akritas (1986), Barber and Jennison (1999), Efron (1981b), Efron and Tibshirani (1986), Ernst and Hutson (2003), Heller and Venkatraman (1996), James (1997), Reid (1981), and Strawderman and Wells (1997).

As an example consider the research of Muenchow (1986) who describes a

Figure 6.1 *Product limit estimates for the waiting times for flying insects visiting female (solid line) and male (dotted line) flowers. The crosses on each plot indicate the location of censored observations.*

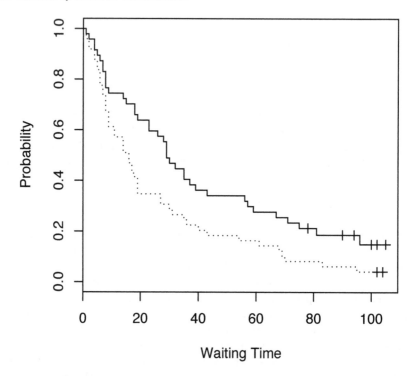

Waiting Time

study to determine whether male and female flowers are equally attractive to flying insects. The data consist of waiting times for forty-nine male flower pairs and forty-seven female flower pairs. Each flower pair was observed until one of the flowers was visited by a flying insect. Two of the waiting times for the male flower pairs, and eight of the waiting times for the female flower pairs are right censored in that no insect was observed during the study time. The product limit estimates for each type of flower pair are plotted in Figure 6.1.

The first observed confidence levels computed for this problem concerns a waiting time of 60 minutes. The parameter space of $S(x)$ was divided into regions of the form $\Theta_i = [0, i/10)$ for $i = 1, \ldots, 10$. The approximate observed confidence level was computed that $S(x)$ is in each of these regions for each gender of flower. The results of these calculations are given in Table 6.6. It is clear that the most confident interval for the male flower pairs is $\Theta_2 = [0.10, 0.20)$ while for the female flowers the most confident interval is $\Theta_3 = [0.20, 0.30)$. A more complicated problem concerns whether the two survival curves cross or not. The percentile method was used to compute observed

Table 6.6 *Observed confidence levels in percent for the survival probability at a waiting time of 60 minutes for each gender of flower.*

Gender	Θ_1	Θ_2	Θ_3	Θ_4	Θ_5	Θ_6	$\Theta_7-\Theta_{10}$
Male	11.43	64.12	23.87	0.48	0.00	0.00	0.00
Female	0.34	11.66	52.01	33.06	2.89	0.03	0.00

confidence levels that the two survival curves cross at least one time in the range of the data and that they cross at least once after 10 minutes. Note that both curves are equal to 1 for very small time values, so that a crossing is defined as crossing after initially being equal. The bootstrap percentile method based on 1000 resamples estimates the observed confidence level of any crossing of the two survival curves as 52.5%, while the observed confidence level for at least one crossing after 10 minutes is 21.8%.

6.5 Exercises

1. A test of space visualization was given to two groups of students at a local university. The first group, which consisted of twelve subjects, all had majors within the business school. The second group, which consisted of seventeen subjects, all had majors within the school of engineering. The test consisted of timing how long it took, in minutes, for the student to assemble a set of oddly shaped blocks into a cube. The data are given in the table below. Assume that the shift model described in Section 6.1 is true for these populations.

Business School Majors				
17.66	10.67	4.20	13.04	0.53
7.56	12.65	5.75	31.85	8.18
4.36	9.89			

Engineering School Majors				
2.55	4.55	1.30	6.70	9.44
1.26	1.54	8.18	1.55	1.59
1.08	1.62	5.09	1.98	2.84
1.21	2.99			

(a) Use the Wilcoxon rank sum test to test the null hypothesis that the shift parameter is zero versus the alternative that the engineering students take less time. Use a significance level of $\alpha = 0.05$ to run the test.

(b) Using the method described in Section 6.1, compute a 95% confidence interval for the shift parameter.

(c) Compute observed confidence levels that the shift parameter θ is in the regions $\Theta_1 = (-1, 1)$ and $\Theta_2 = (-2, 2)$. Use the exact method and then repeat the analysis using the normal approximation. How do the two methods compare? How do the observed confidence levels compare to the conclusions reached using the first two parts of this exercise?

2. Two types of charcoal are compared in terms of the length of time (in minutes) that the stack temperature in a indirect smoking system stays within a specified optimal range. The two types of charcoal being compared are wood chunk charcoal, which is made from whole chunks of wood, and compressed dust charcoal, which is manufactured from compressed pieces of sawdust. Ten samples of each type of charcoal are tested. The observed data are given in the table below. Assume that the shift model described in Section 6.1 is true for these populations.

Chunk Charcoal									
75.6	75.2	71.4	71.4	74.9	72.1	64.4	73.0	78.6	73.7

Compressed Dust Charcoal									
66.4	68.8	67.2	68.4	65.1	72.6	69.2	72.1	66.7	74.1

(a) Use the Wilcoxon rank sum test to test the null hypothesis that the shift parameter is zero versus the alternative that there is a difference between the two types of charcoal. Use a significance level of $\alpha = 0.01$ to run the test.

(b) Using the method described in Section 6.1, compute a 99% confidence interval for the shift parameter.

(c) Compute observed confidence levels that the shift parameter θ is in the regions $\Theta_1 = (-1, 1)$ and $\Theta_2 = (-5, 5)$. Use the exact method and then repeat the analysis using the normal approximation. How do the two methods compare? How do the observed confidence levels compare to the conclusions reached using the first two parts of this exercise?

3. A small grocery store is experimenting with the use of coupons to bring in new customers. In a study of coupon effectiveness, coupons are put in three different local newspapers (labeled A, B, and C) on three different days: Friday (F), Saturday (Sa) and Sunday (Su). The coupons have codes written on them that will allow the store to distinguish which paper the coupon appeared in and on what day. The number of coupons redeemed at the store were counted over the next week. The data are given in the table below.

Newspaper	Day	Number Redeemed
A	F	15
	Sa	22
	Su	55
B	F	44
	Sa	67
	Su	74
C	F	15
	Sa	34
	Su	64

(a) Fit a log-linear model for the data. Fit both factors, including an interaction. Does it appear that both factors and the interaction are significant?

(b) Ignoring the day effect and the interaction, compute observed confidence levels on the six possible relative rankings of the newspapers by their effectiveness for bringing in customers based on the estimated factor effects. Use the normal approximation. If the interaction is significant, does this affect the interpretation of the outcome of these results?

(c) Ignoring the paper effect and the interaction, compute observed confidence levels on the six possible relative rankings of the days by their effectiveness for bringing in customers based on the estimated factor effects. Use the normal approximation. If the interaction is significant, does this affect the interpretation of the outcome of these results?

4. A study of allergy medication is designed to compare a placebo versus a standard treatment and a new treatment. In the study seven subjects where given the treatments in a random order. It is known that each treatment no longer has any effect after 24 hours, so that each subject was given the three treatments over a three-day period. Each subject kept track of the number of times that they sneezed each day. The data are given in the table below.

Subject	Placebo	Standard Treatment	New Treatment
1	12	11	4
2	16	8	2
3	10	13	7
4	23	9	6
5	16	7	2
6	25	19	7
7	10	6	0

(a) Fit a log-linear model to this data. Assume that the treatment effects are the same for each subject, but that each subject has a different baseline. That is, fit an intercept for each subject. Does the data appear to support this model?

(b) Estimate the effect of each treatment. Using the normal approximation, compute observed confidence levels for each possible ordering of the treatment effects. What orderings have the largest observed confidence levels? Interpret these results in terms of the study.

5. A manufacturer is interested in examining the structure of the variability in aluminum rods that are being produced at her plant. Four critical measurements (in cm) are taken on each rod: top end diameter (X_1), bottom end diameter (X_2), length (X_3), and middle circumference (X_4). In a study of this variability thirty such rods are sampled and measured. The data are given in the table below.

X_1	X_2	X_3	X_4	X_1	X_2	X_3	X_4
1.014	1.049	49.966	3.079	1.104	1.082	50.118	2.883
1.059	1.075	50.008	3.247	0.966	1.006	49.939	3.109
0.939	0.918	49.872	3.031	0.927	0.960	50.032	2.956
1.065	1.064	50.029	3.089	0.986	1.036	50.087	3.176
0.950	0.985	49.970	3.110	0.989	0.996	50.066	2.965
1.024	1.017	49.991	3.104	1.038	1.038	49.734	3.149
1.001	1.022	49.902	3.194	1.054	1.016	50.176	3.195
0.950	0.979	49.942	3.167	0.903	0.915	50.039	3.111
1.014	0.997	49.965	3.061	0.996	0.979	49.907	3.272
1.012	0.979	49.865	3.191	0.960	0.987	49.960	3.163
0.949	0.924	49.945	2.657	1.088	1.104	50.124	2.983
0.957	0.977	50.010	3.023	1.044	1.033	49.855	3.019
1.042	1.017	49.928	2.938	1.026	1.021	50.090	3.275
1.062	1.086	50.074	3.217	1.044	1.060	49.883	3.252
0.991	0.969	50.093	3.011	0.947	0.935	49.973	3.057

(a) Compute the sample covariance matrix and the sample correlation matrix for this data. What measurements appear to be correlated? Does the strength and direction of the correlations have intuitive appeal in terms of the measurements?

(b) Compute the eigenvectors and their associated eigenvalues for the sample covariance matrix of this data. Interpret these calculations in terms of a principal component analysis.

(c) Using the bootstrap percentile method and 10,000 resamples, compute observed confidence levels on the number of principal components required to account for 90% of the variability in the data.

(d) Using the bootstrap percentile method and 10,000 resamples, compute observed confidence levels that each of the four measurements has the largest absolute weight in the first principal component.

(e) Using the bootstrap percentile method and 10,000 resamples, compute observed confidence levels that each possible pair of the four measurements have the largest two absolute weights in the first principal component.

6. The data given Table 6.7 originate from Jolicoeur and Mosimann (1960) who studied shape and size variation of painted turtles. The data consist of three measurements on the carapaces of twenty-four female and twenty-four male turtles. The measurements are: length (X_1), width (X_2) and height (X_3). Note that Johnson and Wichern (1998) suggest that a logarithmic transformation be applied to the measurements.

(a) Compute the sample covariance matrix and the sample correlation matrix for this data, for each gender separately. What measurements appear to be correlated? Does the strength and direction of the correlations have intuitive appeal in terms of the measurements? Do the data seem to be consistent between the genders?

(b) Compute the eigenvectors and their associated eigenvalues for the sample covariance matrix of this data, for each gender separately. Interpret these calculations in terms of a principal component analysis. Do the results appear to be consistent between the genders?

(c) Using the bootstrap percentile method and 10,000 resamples, compute observed confidence levels on the number of principal components required to account for 90% of the variability in the data for each gender separately.

(d) Using the bootstrap percentile method and 10,000 resamples, compute observed confidence levels that each of the three measurements has the largest absolute weight in the first principal component for each gender separately.

(e) Treating the female and male samples as being independent of one another, compute an observed confidence level based on the bootstrap percentile method that the two genders have the same measurement with the largest absolute weight in the first principal component. Use 10,000 resamples.

7. A local church is interested in knowing whether a new member class will encourage new members to the congregation to stay with the church for a longer period of time. Over the five-year study period, 100 new members were tracked. Fifty of these new members were randomly selected to take part in the new member class. The remaining fifty new members did not attend the class, but were given a free coffee mug with picture of the church on it. The number of weeks that each member attended the church until they stopped attending for each group is given in Table 6.8. Data that are

Table 6.7 *Data on the size and shape of painted turtles from Jolicoeur and Mosimann (1960).*

Female			Male		
X_1	X_2	X_3	X_1	X_2	X_3
98	81	38	93	74	37
103	84	38	94	78	35
103	86	42	96	80	35
105	86	42	101	84	39
109	88	44	102	85	38
123	92	50	103	81	37
123	95	46	104	83	39
133	99	51	106	83	39
133	102	51	107	82	38
133	102	51	112	89	40
134	100	48	113	88	40
136	102	49	114	86	40
138	98	51	116	90	43
138	99	51	117	90	41
141	105	53	117	91	41
147	108	57	119	93	41
149	107	55	120	89	40
153	107	56	120	93	44
155	115	63	121	95	42
155	117	60	125	93	45
158	115	62	127	96	45
159	118	63	128	95	45
162	124	61	131	95	46
177	132	67	135	106	47

in boldface indicate censored observations. That is, the participants in the study where still attending church when the study was completed.

(a) Compute product-limit estimates for the survival curve for each of the groups and plot them on a common set of axes. From these estimates, does it appear that new members will attend longer if they attend the new member class? Consider the assumptions associated with the product-limit estimator. Do these assumptions appear to be true for this data?

(b) Using the approximate confidence interval for the survival curve, compute observed confidence levels that each of the survival curves are in the regions $\Theta_i = [(i-1)/10, i/10)$ for $i = 1, \ldots, 10$ at one year from when the person started attending.

Table 6.8 *Church attendance data.*

\multicolumn{10}{c}{Treatment Group: Attended New Member Class}									

Treatment Group: Attended New Member Class

169	1521	**190**	74	630	1332	4	532	943	24
943	24	**366**	638	**1016**	267	515	63	56	59
331	390	5	247	117	1	588	77	182	481
713	**512**	296	78	101	489	804	378	480	28
466	267	25	714	111	376	221	138	**625**	175

Control Group: No New Member Class

258	192	173	678	537	550	487	199	327	104
525	163	90	138	420	350	699	**184**	235	166
97	209	188	417	104	645	67	74	286	46
730	376	**1664**	435	345	43	137	**152**	485	24
24	**51**	462	1251	**33**	170	67	43	27	**270**

(c) Using the percentile method, compute an observed confidence level that the two survival curves never cross, though they may be equal to some points. Compute observed confidence levels that the survival curve for the treatment group is always greater than or equal to that of the control group. Compute an additional observed confidence level that the survival curve of the control group is always greater to or equal than that of the treatment group.

8. Three types of cooling devices are being considered to cool a new type of microprocessor on laptop computers. One device is a dedicated fan module, the second device is circulates a special cooling liquid around the processor and the third device is based on a new super efficient heat sink material. The manufacturer of the laptop computers is interested in which cooling system will allow the microprocessor to function the longest in adverse conditions. Twenty prototype notebook computers were equipped with each type of cooling system and where tested in a controlled environment of 50 degrees Celsius. The number of days until each laptop failed due to a microprocessor problem was then recorded. Any laptop that failed due to another reason was repaired and put back into service without changing the microprocessor. The observed results of the experiment are given in Table 6.9. The study had a set length of 100 days, so that any laptops still functioning at the end of the study have censored values, indicated by boldface.

(a) Compute product-limit estimates for the survival curve for each of the groups and plot them on a common set of axes. Does it appear that one of the methods is better than the others as far as providing longer lifetimes for the microprocessors? Consider the assumptions associated

Table 6.9 *Microprocessor failure data.*

Dedicated Fan Module				
99.9	4.1	26.2	15.4	108.6
10.8	42.8	**100.0**	**100.0**	44.6
36.8	59.6	**100.0**	63.0	12.2
18.9	52.3	35.5	38.2	47.2

Liquid Cooling System				
34.0	12.9	**100.0**	17.7	**100.0**
54.4	65.2	39.3	**100.0**	14.5
100.0	97.5	20.2	**100.0**	81.6
36.6	**100.0**	83.9	23.8	**100.0**

Heat Sink				
87.8	18.4	26.2	77.7	13.5
109.5	50.2	47.5	64.0	**100.0**
100.0	27.3	56.7	96.7	35.2
56.0	73.9	**100.0**	**100.0**	76.6

with the product-limit estimator. Do these assumptions appear to be true for this data?

(b) Using the approximate confidence interval for the survival curve, compute observed confidence levels that each of the survival curves are in the regions $\Theta_i = [(i-1)/10, i/10)$ for $i = 1, \ldots, 10$ at 50 days.

(c) Using the bootstrap percentile method, compute observed confidence levels that each of the survival curves is strictly above the other two for the range 50 to 100 days.

CHAPTER 7

Connections and Comparisons

7.1 Introduction

This chapter compares the method of computing observed confidence levels as a solution to multiple testing problems to standard solutions based on statistical hypothesis testing, multiple comparison techniques, and Bayesian posterior probabilities. The frequentists confidence levels proposed by Efron and Tibshirani (1998), referred to in this chapter as *attained confidence levels* are also studied. Section 7.2 considers the relationship between observed confidence levels and a single statistical hypothesis test. In particular this section explores the level at which points inside and outside a specified region would be rejected by a hypothesis test based on the observed confidence level. Sequences of multiple tests are considered in Section 7.3, where multiple comparison techniques are used to make the comparisons and control the overall error rate. Section 7.4 considers attained confidence levels, and demonstrates that certain paradoxes observed with these levels do not occur for observed confidence levels. Section 7.5 briefly considers Bayesian confidence levels which are based on the posterior probabilities. An example from Section 7.4 is used to highlight the differences between observed confidence levels and Bayesian confidence levels.

7.2 Statistical Hypothesis Testing

A more in-depth understanding of the use and interpretation of observed confidence levels can be obtained by exploring the connections the method has with statistical hypothesis testing. Let $\mathbf{X}$ be sample vector from a probability model with parameter $\boldsymbol{\theta}$ and sample space Θ. Let $\Theta_0 \subset \Theta$ be an arbitrary region of the sample space, Then a level ξ statistical test of the hypothesis $H_0 : \boldsymbol{\theta} \in \Theta_0$ consists of a real function $U(\mathbf{X}, \Theta_0)$, called a *test statistic*, and two regions $R(\xi, \Theta_0, U)$ and $A(\xi, \Theta_0, U)$. The region $R(\xi, \Theta_0, U)$ is called the *rejection region* of the test and is usually selected so that

$$\sup_{\boldsymbol{\theta} \in \Theta_0} P_{\boldsymbol{\theta}}[U(\mathbf{X}, \Theta_0) \in R(\xi, \Theta_0, U)] = \xi, \qquad (7.1)$$

where $\xi \in (0, 1)$ is called the *significance level* of the test. The region $A(\xi, \Theta_0, U)$ is usually constructed as $A(\xi, \Theta_0, U) = \mathbb{R} \setminus R(\xi, \Theta_0, U)$, and is called the *acceptance region* of the test. In practice, if $U(\mathbf{X}, \Theta_0) \in R(\xi, \Theta_0, U)$, then the conclusion is made that the hypothesis $H_0 : \boldsymbol{\theta} \in \Theta_0$ is false. From Equation (7.1) it is clear that ξ is the maximum probability that such a conclusion is made in error. Note that it is possible that the acceptance and rejection regions may also depend on the sample $\mathbf{X}$, though it is not indicated by the notation.

The connection between statistical tests of hypotheses and observed confidence levels is obtained by considering the close relationship between confidence regions of tests of hypotheses. This relationship states that for every confidence region $C(\alpha, \omega; \mathbf{X})$ for the $\boldsymbol{\theta}$ based on the sample $\mathbf{X}$, there exists a statistical hypothesis test of $H_0 : \boldsymbol{\theta} = \boldsymbol{\theta}_0$ such that $U(\mathbf{X}, \boldsymbol{\theta}_0) \in A(\xi, \boldsymbol{\theta}_0, U)$ for every $\boldsymbol{\theta}_0 \in \mathbf{C}(\alpha, \omega; \mathbf{X})$ where $\xi = 1 - \alpha$. Correspondingly, for every $\boldsymbol{\theta}_0 \notin \mathbf{C}(\alpha, \omega; \mathbf{X})$, $U(\mathbf{X}, \boldsymbol{\theta}_0) \in R(\xi, \boldsymbol{\theta}_0, U)$. See Theorem 9.22 of Casella and Berger (2002) for complete details on this result.

Now consider an arbitrary region $\Psi \subset \Theta$, and suppose that $C(\alpha, \omega; \mathbf{X}) = \Psi$, conditional on the observed sample $\mathbf{X}$, so that the observed confidence level of Ψ is α. The connection between confidence regions and tests of hypothesis described above then implies that there exists a statistical hypothesis test of $H_0 : \boldsymbol{\theta} = \boldsymbol{\theta}_0$ such that $U(\mathbf{X}, \boldsymbol{\theta}_0) \in A(\xi, \boldsymbol{\theta}_0, U)$ for every $\boldsymbol{\theta}_0 \in \Psi$, where $\xi = 1 - \alpha$. That is, if the observed confidence level of Ψ is α, then the hypothesis test associated with the confidence region used to compute the observed confidence level would fail to reject the hypothesis $H_0 : \boldsymbol{\theta}_0 \in \Psi$ for all $\boldsymbol{\theta}_0 \in \Psi$, at the $\xi = 1 - \alpha$ significance level. Correspondingly, the same test would reject all points outside Ψ at the same significance level.

Suppose that the observed confidence level of Ψ is near 1. That implies that points outside of Ψ are rejected at a very low (near 0) significance level. Equation (7.1) then implies that the points outside of Ψ would be rejected for a very low error rate. That is, there is substantial evidence that $\boldsymbol{\theta}$ is not outside Ψ as these points can be rejected with little chance of error. Therefore, an observed confidence level near 1 indicates that there is strong evidence that $\boldsymbol{\theta} \in \Psi$. Similarly, if the observed confidence level is near 0, there are points outside Ψ which would be rejected, but only when the error rate ξ is near 1. It is therefore difficult to conclude that $\boldsymbol{\theta} \in \Psi$, and hence an observed confidence level near 0 indicates that there is little evidence that $\boldsymbol{\theta} \in \Psi$. In a comparative sense, if the observed confidence level of a region $\Theta_i \subset \Theta$ is larger than the observed confidence level for another region $\Theta_j \subset \Theta$, where $i \neq j$, the one can conclude that points outside Θ_i can be rejected at a lower significance level that points outside Θ_j. That is, there is less chance of erroneously rejecting points outside Θ_i than Θ_j. This would indicate that $\boldsymbol{\theta} \in \Theta_i$ is a preferable model to $\boldsymbol{\theta} \in \Theta_j$.

Some care must be taken when interpreting observed confidence levels in terms of statistical tests. In particular, the results of this section do not guarantee what properties the statistical test associated with a corresponding observed

Figure 7.1 *An example of a set of regions of* Θ *that may lead to an unreasonable test. The value of* $\hat{\theta}$ *is indicated by the "+" symbol.*

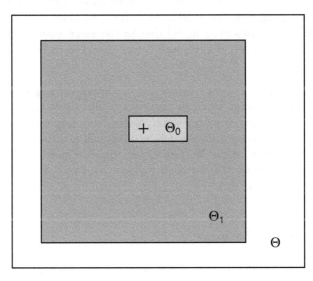

confidence level have. For example, most reasonable tests would fail to reject $H_0 : \boldsymbol{\theta} \in \Theta_0$ except for very large significance levels when $\hat{\boldsymbol{\theta}} \in \Theta_0$, particularly when $\hat{\boldsymbol{\theta}}$ has reasonable properties, such as being an unbiased estimator of $\boldsymbol{\theta}$. The test associated with some observed confidence levels may not have this property. A typical example is constructed as shown in Figure 7.1. Because Θ_1 covers a large portion of the parameter space, it would not be out of the question for the observed confidence level of Θ_1 to be quite large even though $\hat{\theta}$ is not an element of Θ_1. But if the observed confidence level of Θ_1 is near 1, then there exists a statistical test that would reject points outside Θ_1 at a very small significance level, including the test of $H_0 : \boldsymbol{\theta} = \boldsymbol{\theta}_0$ where $\boldsymbol{\theta}_0 = \hat{\boldsymbol{\theta}}$. Therefore, the value of $\hat{\boldsymbol{\theta}}$ would be rejected at a very small significance level. In general these types of problems are much less likely to occur when $\hat{\boldsymbol{\theta}}$ is a reasonable estimator of $\boldsymbol{\theta}$, say one that has small bias and is consistent, and the sample size is moderately large.

When the sample size is small such problems may arise even when the estimator $\hat{\boldsymbol{\theta}}$ has reasonable properties. It is important to note, however, that this is not a deficiency of the method of observed confidence levels, but is a deficiency of attempting to interpret the observed confidence levels in terms of statistical hypothesis tests. As an example consider once again the situation represented in Figure 7.1 and suppose that while $\hat{\boldsymbol{\theta}}$ is asymptotically unbiased, the bias may be substantial for smaller sample sizes. In this case it may be reasonable to conclude that $\boldsymbol{\theta} \in \Theta_1$ even though $\hat{\boldsymbol{\theta}}$ is observed in Θ_0. This is because it is more likely that $\boldsymbol{\theta} \in \Theta_1$ and $\hat{\boldsymbol{\theta}}$ was observed in Θ_0 due to the substantial bias rather than the situation where $\theta \in \Theta_0$ and $\hat{\theta}$ was observed in Θ_0 *despite* the bias. A similar situation can occur even when $\hat{\theta}$ is unbiased, but has a large standard error. In this case it may be again reasonable to conclude that $\theta \in \Theta_1$ even though $\hat{\theta}$ is observed in Θ_0. This is because it is more likely that $\theta \in \Theta_1$ and $\hat{\theta}$ was observed in Θ_0 due to the large variability of $\hat{\theta}$, rather than the situation where $\theta \in \Theta_0$ and $\hat{\theta}$ was observed in Θ_0 *despite* the large variability in $\hat{\theta}$.

An extreme example occurs when $\hat{\theta}$ has a continuous sampling distribution and Θ_0 is a single point in Θ, or more generally where Θ_0 is any region of Θ that has probability 0 with respect to the sampling distribution of $\hat{\theta}$. In such a case the observed confidence level of Θ_0 will invariably be 0, regardless of the observed data vector $\mathbf{X}$. This occurs because most confidence regions have a confidence level of 0 for such regions. An example of this problem is considered in Section 4.6.1.

7.3 Multiple Comparisons

Multiple comparison techniques have been the standard approach to solving multiple testing problems within the frequentist framework. In essence, these methods attempt to solve the problem of regions through the use of repeated statistical hypothesis tests. The overall error rate, which is the probability of committing at least one Type I error, of a sequence of hypothesis tests will generally exceed the error rate specified for each test. Therefore methods for controlling the error rate are usually used in conjunction with the sequence of tests. There have been many techniques developed to both make such comparisons and control the overall error rate, each with their own particular assumptions and properties. This section will only look at the types of conclusions made for two very general techniques without getting into specific methods. There are certainly other cromulent techniques which may perform well for specific models and assumptions. This section is not an exhaustive comparison, but rather a glimpse of the type of comparisons that can be made. As the results are rather general in nature, no specific mechanism for controlling the overall error rate of the multiple comparison methods will be assumed.

For general accounts of multiple testing procedures see Hochberg and Tamhane (1987), Miller (1981), and Westfall and Young (1993). Multiple testing procedures have generated significant controversy over which tests should be performed and how the overall error rate should be controlled. See, for example, Section 1.5 of Westfall and Young (1993) and Saville (1990). This controversy often extends outside the field of statistics to those fields whose data analysis rely heavily on multiple comparison procedures. For example Curran-Everett (2000) states that approximately 40% of articles published by the American Physiological Society cite a multiple comparison procedure. It is the conclusion of Curran-Everett (2000) that many of the common procedures are of limited practical value. Games (1971) and Hancock and Klockars (1996) consider the use of multiple comparison techniques in educational research. O'Keefe (2003) suggests that adjusting errors rates should be abandoned in the context of human communication research. Miwa (1997) considers controversies in the application of multiple comparisons in agronomic research.

Consider the usual framework where the regions of interest are given by the sequence $\{\Theta_i\}_{i=1}^k$. To simplify the problem k will be assumed to be finite. Many multiple comparison techniques require this assumption. The inference in this section will be assumed to be based on a sample $\mathbf{X}_1, \ldots, \mathbf{X}_n$ with associated point estimate $\hat{\boldsymbol{\theta}}$. The estimator $\hat{\boldsymbol{\theta}}$ will be considered to be a reasonable estimator of $\boldsymbol{\theta}$ in that it will be assumed to be consistent, so that if $\boldsymbol{\theta} \in \Theta_i$ for some $i \in \{1, \ldots, k\}$,

$$\lim_{n \to \infty} P(\hat{\boldsymbol{\theta}} \in \Theta_i) = 1.$$

It will further be assumed that a confidence region for $\boldsymbol{\theta}$, denoted generically in this section by $C(\alpha; \mathbf{X})$, is asymptotically accurate. That is

$$\lim_{n \to \infty} P[\boldsymbol{\theta} \in C(\alpha; \mathbf{X})] = \alpha.$$

Secondly, it will be assumed that the consistency of $\hat{\boldsymbol{\theta}}$ insures that the confidence region $C(\alpha; \mathbf{X})$ shrinks to a point mass at $\boldsymbol{\theta}$ as $n \to \infty$ when the nominal level of the confidence region α is held constant.

No specific method of computing observed confidence levels will be considered in this section, except that the observed confidence levels are based on confidence regions that have some general properties that follow from the assumptions stated above. Let $\alpha(\Theta_i)$ be a generic method for computing an observed confidence level for the region Θ_i. It will be assumed that α is an asymptotically accurate method for computing observed confidence levels in that

$$\lim_{n \to \infty} \alpha[C(\alpha; \mathbf{X})] = \alpha,$$

though no specific order of accuracy will be assumed in this section. It will also be assumed that α is consistent in that if $\boldsymbol{\theta}$ is not on the boundary of Θ_i,

$$\lim_{n \to \infty} \alpha(\Theta_i) = \begin{cases} 1 & \text{if } \boldsymbol{\theta} \in \Theta_i, \\ 0 & \text{if } \boldsymbol{\theta} \notin \Theta_i. \end{cases} \tag{7.2}$$

The hypothesis test used in the multiple comparison procedure will also be assumed to have reasonable statistical properties. Let $T_{ij}(\mathbf{X})$ be the test statistic that is used for testing the null hypothesis $H_0 : \boldsymbol{\theta} \in \Theta_i$ versus the alternative hypothesis $H_1 : \boldsymbol{\theta} \in \Theta_j$. Assume that this test has rejection region $R_{ij}(\alpha)$ where α is the significance level for the test. The test will be considered to be asymptotically accurate in that

$$\lim_{n \to \infty} P[T_{ij}(\mathbf{X}) \in R_{ij}(\alpha) | \boldsymbol{\theta} \in \Theta_i] \leq \alpha,$$

with asymptotic power function

$$\lim_{n\to\infty} P[T_{ij}(\mathbf{X}) \in R_{ij}(\alpha)|\boldsymbol{\theta} \in \Theta_j] = 1,$$

for each fixed value of α. A second approach uses a test of the null hypothesis $H_0 : \boldsymbol{\theta} \in \Theta_i$ versus the alternative $H_1 : \boldsymbol{\theta} \notin \Theta_i$. Let $T_i(\mathbf{X})$ be the test statistic that is used for testing these hypotheses with rejection region $R_i(\alpha)$ where α is again the significance level for the test. As with the previous test is will be assumed that

$$\lim_{n\to\infty} P[T_i(\mathbf{X}) \in R_i(\alpha)|\boldsymbol{\theta} \in \Theta_i] \le \alpha,$$

and

$$\lim_{n\to\infty} P[T_i(\mathbf{X}) \in R_i(\alpha)|\boldsymbol{\theta} \notin \Theta_i] = 1.$$

Both tests are assumed to be unbiased , that is

$$P[T_{ij}(\mathbf{X}) \in R_{ij}(\alpha)|\boldsymbol{\theta} \in \Theta_i] \le P[T_{ij}(\mathbf{X}) \in R_{ij}(\alpha)|\boldsymbol{\theta} \in \Theta_j],$$

and

$$P[T_i(\mathbf{X}) \in R_i(\alpha)|\boldsymbol{\theta} \in \Theta_i] \le P[T_i(\mathbf{X}) \in R_i(\alpha)|\boldsymbol{\theta} \notin \Theta_i].$$

Finally, it will be assumed that the condition $\hat{\boldsymbol{\theta}} \in \Theta_i$ never constitutes substantial evidence against the null hypothesis. That is, the probability of rejecting the null hypothesis is 0 when $\hat{\boldsymbol{\theta}} \in \Theta_i$ for both of the tests discussed above.

The first approach to multiple testing considered in this section chooses a single region Θ_i as a global null hypothesis and considers all pairwise tests of the null hypothesis $H_0 : \boldsymbol{\theta} \in \Theta_i$ versus the alternative hypothesis $H_1 : \boldsymbol{\theta} \in \Theta_j$ using the test statistic $T_{ij}(\mathbf{X})$ with rejection region $R_{ij}[\xi(\alpha)]$, where $\xi : [0,1] \to [0,1]$ is an unspecified adjustment function that is designed to control the overall Type I error rate. These hypotheses are tested for each $j \in \{1, \dots, k\}$ such that $j \ne i$, where i is fixed before the data is observed.

There are a few key situations to consider when comparing the possible outcomes of this multiple comparison procedure with the computation of observed confidence levels. For the first case consider the global null hypothesis $H_0 : \boldsymbol{\theta} \in \Theta_i$ to be true, and consider the situation where the observed point estimate of $\boldsymbol{\theta}$ is also in the null hypothesis region. That is $\boldsymbol{\theta} \in \Theta_i$ and $\hat{\boldsymbol{\theta}} \in \Theta_i$. In this situation the stated assumptions imply that the test of $H_0 : \boldsymbol{\theta} \in \Theta_i$ versus $H_1 : \boldsymbol{\theta} \in \Theta_j$ will fail to reject the null hypothesis for all $k - 1$ tests. The result from the multiple comparison technique would then be the weak conclusion that there is not substantial evidence that $\boldsymbol{\theta}$ is within any other region. In the same case it is very likely that $\alpha(\Theta_i)$ would have the largest observed confidence level of all the regions, though this would not necessarily be the case. For example, Polansky (2003b) presents an example based on the observed confidence levels for the mean presented in Section 2.2.1 where the largest observed confidence level is not always in the same region as $\hat{\boldsymbol{\theta}}$. The assumption of Equation (7.2), however, indicates that this is likely to only happen for small sample sizes.

Consider again the case where $\boldsymbol{\theta} \in \Theta_i$ but $\hat{\boldsymbol{\theta}} \in \Theta_l$ for some $l \in \{1, \ldots, k\}$ such that $l \neq i$. The consistency of the estimator of $\boldsymbol{\theta}$ implies that $\hat{\boldsymbol{\theta}}$ will typically be in a region close to Θ_i except for small samples. In this situation the results of the multiple comparison technique may vary greatly depending on the structure of the regions, the location of $\hat{\boldsymbol{\theta}}$ within Θ_l and relative to the boundary of Θ_i, and other factors. In the best case it is possible that none of the hypotheses $H_0 : \boldsymbol{\theta} \in \Theta_i$ would be rejected in favor of $H_1 : \boldsymbol{\theta} \in \Theta_j$ even when $j = l$. This would most likely occur when Θ_l is adjacent to Θ_i and $\hat{\boldsymbol{\theta}}$ is near the boundary separating the two regions. In this situation the weak conclusion that there is not substantial evidence that $\boldsymbol{\theta}$ is within any other region than Θ_i would be the result. For the same situation the observed confidence levels would likely indicate that there is the most confidence in Θ_l, but that there is also substantial confidence in Θ_i as well. Another likely outcome is that only the alternative hypothesis $H_1 : \boldsymbol{\theta} \in \Theta_l$ would be accepted, which would present a clear, but erroneous indication that $\boldsymbol{\theta} \in \Theta_l$. The observed confidence levels would likely follow this lead by indicating a large amount of confidence that $\boldsymbol{\theta} \in \Theta_l$.

In the worst case several alternative hypotheses of the form $H_1 : \boldsymbol{\theta} \in \Theta_j$, including $H_1 : \boldsymbol{\theta} \in \Theta_l$ are accepted. In this case there is no simple conclusion that can be made from the multiple comparison techniques about the location of $\boldsymbol{\theta}$. For example, consider two alternative hypotheses $H_1 : \boldsymbol{\theta} \in \Theta_j$ and $H_1 : \boldsymbol{\theta} \in \Theta_q$ that have been accepted. This indicates that there is substantial evidence in the data that $\boldsymbol{\theta}$ is within Θ_j or Θ_q *when compared to* Θ_i. The p-values from these tests may provide an indication as to how strong the evidence is in each case, and may indicate which region is preferred over the other, *relative to the region* Θ_i. An additional hypothesis test may be constructed to compare these two regions. But in this case one must choose which region would be in the null hypothesis. The result of this additional test may or may not indicate a preference for either region. In the case where there are several regions to contend with, the calculations and conclusions become more difficult.

When observed confidence levels are applied to this same situation it is likely that the region Θ_l would be assigned the largest observed confidence level, in which the user would again draw the erroneous conclusion that Θ_l is a viable region to contain $\boldsymbol{\theta}$. Both the multiple comparison technique and the observed confidence levels would draw this conclusion because $\hat{\boldsymbol{\theta}} \in \Theta_l$ is perhaps the best indication based on the observed data as to the location of $\boldsymbol{\theta}$. The key idea for the observed confidence levels in this case is that one can easily assess how strong the indication is and what other regions may have substantial observed confidence levels, without additional calculations.

The possible outcomes in this case can vary from the situation that the observed confidence level for Θ_l is substantially larger than the observed confidence levels for any other region, to the case where several regions, including Θ_l, have large levels of confidence. In the best case the observed confidence level for the region Θ_i would be comparable to these observed confidence levels so that the region Θ_i would still be considered a viable candidate for the region that contains $\boldsymbol{\theta}$. However, this is not a likely outcome for regions that have been accepted over the null region Θ_i. The advantage of using observed confidence levels in this case is that is easy to compare the relative observed confidence levels of two regions without any further calculation. Further, these comparisons do not require the assignment of the null hypothesis to one of the regions.

Now consider the case where the null hypothesis is false, in that $\boldsymbol{\theta} \in \Theta_l$ for some $l \neq i$. The conclusions elicited from the multiple comparison techniques as well as the observed confidence levels would essentially be the same as those described above based on the location of $\hat{\boldsymbol{\theta}}$ relative to the null hypothesis, only the relative truth of the conclusions will be different. For example, if $H_1 : \boldsymbol{\theta} \in \Theta_l$ is the only accepted alternative hypothesis then the conclusion from the multiple comparison techniques will be the strong conclusion that $\boldsymbol{\theta} \in \Theta_l$. This is the most favorable conclusion for the multiple comparison technique in that a strong conclusion is made by rejecting a null hypothesis, and the conclusion is correct. As described above, the observed confidence level for Θ_l in this case would likely be the largest due to the acceptance of the corresponding alternative hypothesis.

The problem is now considered from the asymptotic viewpoint. If $\boldsymbol{\theta} \in \Theta_i$, the global null hypothesis region, then the consistency of $\hat{\boldsymbol{\theta}}$ guarantees that in large samples that it is very likely that $\hat{\boldsymbol{\theta}} \in \Theta_i$. Therefore, for large samples it is very likely that the global null hypothesis will not be rejected for any of the tests. Hence, from an asymptotic viewpoint the method will provide the correct conclusion, although it again is a weak conclusion obtained by failing to reject null hypotheses. If $\boldsymbol{\theta} \in \Theta_l$ where $l \neq i$, then it is very likely that $\hat{\boldsymbol{\theta}} \in \Theta_l$ and that the alternative hypothesis $H_1 : \boldsymbol{\theta} \in \Theta_l$ is accepted. This would result in the strong conclusion that $\boldsymbol{\theta} \in \Theta_l$, though it should be noted that even in the asymptotic framework it is possible that other alternative hypotheses will be accepted as well. In either case the observed confidence level for the region that contains $\boldsymbol{\theta}$ will be near 1 for large samples.

Note that the region Θ_i is afforded a special status in this type of multiple comparison analysis, as it is the global null hypothesis. Therefore, Θ_i can be thought of as the default region in the tests. A test will only reject the null hypothesis $H_0 : \boldsymbol{\theta} \in \Theta_i$ if there is significant evidence that the null the hypothesis is not true. It is this structure that results in the rejection of all of the null hypotheses to be a weak conclusion that $\boldsymbol{\theta} \in \Theta_i$. When observed confidence levels are applied to the same problem no region is given special status over any of the other regions, because no null hypothesis needs to be defined.

One way to avoid setting a single region to be the global null hypothesis is to consider a set of tests that consider each region, in turn, to be the null hypothesis. In this approach one would test the null hypothesis $H_0 : \boldsymbol{\theta} \in \Theta_i$ versus the alternative hypothesis $H_1 : \boldsymbol{\theta} \notin \Theta_i$ for $i = 1, \ldots, k$. There are essentially two basic situations to consider for this approach. First consider the case where $\boldsymbol{\theta} \in \Theta_i$ and $\hat{\boldsymbol{\theta}} \in \Theta_i$. Given the assumptions of this section the null hypothesis $H_0 : \boldsymbol{\theta} \in \Theta_i$ would not be rejected in favor of the alternative hypothesis $H_1 : \boldsymbol{\theta} \notin \Theta_i$. In the best case the remaining $k - 1$ null hypotheses of the form $H_0 : \boldsymbol{\theta} \in \Theta_j$ where $j \neq i$ would all be rejected. It is also very likely in this case that the observed confidence level for Θ_i would be large compared to the observed confidence levels for the remaining regions. Therefore, both methods would provide a strong indication that $\boldsymbol{\theta} \in \Theta_i$, which is the correct conclusion.

Another likely outcome is that $H_0 : \boldsymbol{\theta} \in \Theta_i$ would not be rejected along with several other regions. The additional regions that are not rejected would most likely be near, or perhaps border, the region Θ_i. The result is much weaker in this case, essentially providing several possibilities for the region that contains $\boldsymbol{\theta}$, including the correct one. Observed confidence levels would likely provide a similar conclusion in this case, by assigning relatively large levels of confidence to the regions that are not rejected. It is likely that Θ_i would have the largest observed confidence level, though as stated earlier, this need not always be the case. As in the previous discussion, it is clear that while observed confidence levels would provide similar results to the multiple comparison method, the observed confidence levels would allow one to easily compare the regions of interest without additional calculations.

The second case to consider has $\boldsymbol{\theta} \in \Theta_i$ but $\hat{\boldsymbol{\theta}} \in \Theta_j$ for some $j \neq i$. The worst possible outcome in this case occurs when $H_0 : \boldsymbol{\theta} \in \Theta_j$ is not rejected, as would be indicated by the assumptions of this section, while the remaining $k - 1$ null hypotheses would be rejected. As indicated above, this result would provide a clear, but erroneous indication that $\boldsymbol{\theta} \in \Theta_j$. As with the previous testing method, the observed confidence levels would most likely indicate this solution as well by assigning Θ_j to have the largest observed confidence level, relative to the remaining regions. It is also possible that several regions would avoid rejection, including perhaps Θ_i. In this case these regions would likely have comparatively large observed confidence levels compared to the remaining regions. The possibility that the correct conclusion that $\boldsymbol{\theta} \in \Theta_i$ is virtually not possible in this case with either method. However, if the region Θ_i avoids rejection, in which case it would likely have a relatively large observed confidence level, then the region Θ_i could not be ruled out as a possibility to be the region that contains Θ_i.

From an asymptotic viewpoint both methods would provide strong evidence of the correct conclusion. For large samples the null hypothesis $H_0 : \boldsymbol{\theta} \in \Theta_i$ would not be rejected, and all of the remaining $k - 1$ hypotheses of the form $H_0 : \boldsymbol{\theta} \in \Theta_j$ for $j \neq i$ would be rejected under the assumption that $\boldsymbol{\theta}$ is in the interior of Θ_i. The corresponding observed confidence levels would also proved the correct conclusion as $\alpha(\Theta_i) \to 1$ as $n \to \infty$.

Figure 7.2 *Observed confidence levels for the the regions* Θ_0 *(solid),* Θ_1 *(dashed) and* Θ_2 *(dotted) from the example.*

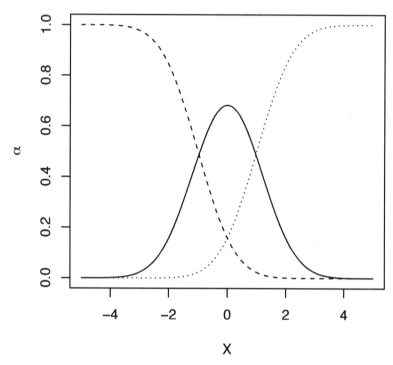

As an example consider the simple problem where a single value X is observed from a normal distribution with mean θ and unit variance. Suppose the regions of interest are of the form $\Theta_0 = [-1, 1]$, $\Theta_1 = (-\infty, -1)$ and $\Theta_2 = (1, \infty)$. Simple normal theory yields a $100(\omega_U - \omega_L)\%$ confidence interval for θ to be

$$C(\alpha, \boldsymbol{\omega}, X) = [\hat{\theta} - z_{1-\omega_L}, \hat{\theta} - z_{1-\omega_U}],$$

where $\hat{\theta} = X$. Therefore, the observed confidence levels for Θ_0, Θ_1 and Θ_2 are given by

$$\alpha(\Theta_0) = \Phi(\hat{\theta} + 1) - \Phi(\hat{\theta} - 1),$$
$$\alpha(\Theta_1) = 1 - \Phi(\hat{\theta} + 1),$$

and

$$\alpha(\Theta_2) = \Phi(\hat{\theta} - 1),$$

respectively. A plot of these observed confidence levels is given in Figure 7.2.

From Figure 7.2 it is clear that Θ_0 will have the dominant observed confidence level when $\hat{\theta}$ is in the interval $[-1, 1]$, and as $\hat{\theta}$ moves away from 0 in either direction the observed confidence level of either Θ_1 or Θ_2 becomes dominant. Note that due to the small sample size there will never be an overwhelming

Table 7.1 *Observed confidence levels in percent and the result of the two tests from the multiple comparison technique using the global null hypothesis* $H_0 : \theta \in \Theta_0$.

| $\hat{\theta}$ | Observed Confidence Levels | | | Reject $H_0 : \theta \in \Theta_0$ | |
	$\alpha(\Theta_0)$	$\alpha(\Theta_1)$	$\alpha(\Theta_2)$	versus $H_1 : \theta \in \Theta_1$	versus $H_1 : \theta \in \Theta_2$
0	68.27	15.87	15.87	No	No
$\frac{1}{2}$	62.47	6.68	30.85	No	No
1	47.73	2.28	49.99	No	No
2	15.73	0.13	84.14	No	No
3	2.28	0.00	97.72	No	Yes

amount of confidence for Θ_0, with the observed confidence level for Θ_0 being maximized when $\hat{\theta} = 0$ for which $\alpha(\Theta_0) = 0.6827$.

To consider multiple comparison techniques for this example consider first testing the global null hypothesis $H_0 : \theta \in \Theta_0$ versus the two possible alternative hypotheses $H_1 : \theta \in \Theta_1$ and $H_1 : \theta \in \Theta_2$. A level α test of $H_0 : \theta \in \Theta_0$ versus $H_1 : \theta \in \Theta_1$ consists of rejecting $H_0 : \theta \in \Theta_0$ when $\hat{\theta} < -1 + z_\alpha$, and a level α test of $H_0 : \theta \in \Theta_0$ versus $H_1 : \theta \in \Theta_2$ consists of rejecting $H_0 : \theta \in \Theta_0$ when $\hat{\theta} > 1 + z_{1-\alpha}$. A simple Bonferroni type technique will be used to control the Type I error rate. Therefore, using a 5% significance level for the overall error rate of the procedure implies that $H_0 : \theta \in \Theta_0$ will be rejected in favor of $H_1 : \theta \in \Theta_1$ when $\hat{\theta} < -2.96$ and $H_0 : \theta \in \Theta_0$ will be rejected in favor of $H_1 : \theta \in \Theta_2$ when $\hat{\theta} > 2.96$. Hence at this significance level the null hypothesis $H_0 : \theta \in \Theta_0$ would not be rejected for much of the range of $\hat{\theta}$ shown in Figure 7.2. In this same range the observed confidence levels would move from favoring Θ_0, to showing virtually the same preference for Θ_0 and Θ_1 or Θ_0 and Θ_2 and finally showing a preference for Θ_1 or Θ_2.

Some example calculations are given in Table 7.1. As the problem is symmetric about the origin, Table 7.1 only considers non-negative values of $\hat{\theta}$. Note that for virtually all of the table entries the multiple comparison technique only provides the weak conclusion that $\theta \in \Theta_0$. In this same range of values the observed confidence levels show a preference for the region Θ_0 when $\hat{\theta}$ is near 0, and equal preference for Θ_0 and Θ_2 when $\hat{\theta}$ is near 1, and a strong preference for Θ_2 as $\hat{\theta}$ becomes larger. The multiple comparison technique does not give a conclusion other than $\theta \in \Theta_0$ until $\hat{\theta}$ moves past 2.96.

The second multiple comparison technique allows each of the regions to be considered in the null hypothesis. Hence, this technique considers three tests: $H_0 : \theta \in \Theta_0$ versus $H_1 : \theta \notin \Theta_0$, $H_0 : \theta \in \Theta_1$ versus $H_1 : \theta \notin \Theta_1$, and $H_0 : \theta \in \Theta_2$ versus $H_1 : \theta \notin \Theta_2$. The first test rejects $H_0 : \theta \in \Theta_0$ when $\hat{\theta} < -1 + z_{\alpha/2}$ or $\hat{\theta} > 1 + z_{1-\alpha/2}$. The second test rejects $H_0 : \theta \in \Theta_1$ when

Table 7.2 *Observed confidence levels in percent and the result of the three tests from the multiple comparison technique using the each region for the null hypotheses.*

| | Observed Confidence Levels | | | Reject | | |
$\hat{\theta}$	$\alpha(\Theta_0)$	$\alpha(\Theta_1)$	$\alpha(\Theta_2)$	$H_0 : \theta \in \Theta_0$	$H_0 : \theta \in \Theta_1$	$H_0 : \theta \in \Theta_2$
0	68.27	15.87	15.87	No	No	No
$\frac{1}{2}$	62.47	6.68	30.85	No	No	No
1	47.73	2.28	49.99	No	No	No
2	15.73	0.13	84.14	No	Yes	No
3	2.28	0.00	97.72	No	Yes	No
4	0.13	0.00	99.87	Yes	Yes	No

$\hat{\theta} > -1 + z_{1-\alpha}$ and the third test rejects $H_0 : \theta \in \Theta_2$ when $\hat{\theta} > 1 - z_{1-\alpha}$. As before, the simple Bonferroni method will be used to control the Type I error rate so that $H_0 : \theta \in \Theta_0$ will be rejected if $\hat{\theta} < -3.39$ or $\hat{\theta} > 3.39$, $H_1 : \theta \in \Theta_1$ will be rejected if $\hat{\theta} > 1.12$ and $H_0 : \theta \in \Theta_2$ will be rejected if $\hat{\theta} < -1.12$.

Some example calculations are given in Table 7.2. Note that when $\hat{\theta}$ is near 0, none of the null hypotheses are rejected which essentially provides no credible information about θ. In this same range the observed confidence levels go from having a preference for Θ_0, in which case Θ_0 has a little more than four times the observed confidence of Θ_1 or Θ_2, to approximately equal preference for Θ_0 and Θ_2, but little confidence in Θ_1. Clearly the observed confidence levels are more informative when $\hat{\theta}$ is in this range. As $\hat{\theta}$ moves above this range, the multiple comparison technique first rejects Θ_1 as a viable possibility, and finally when $\hat{\theta}$ nears 3.4, both Θ_0 and Θ_1 are rejected as possible regions that contain θ.

7.4 Attained Confidence Levels

The method for solving the problem of regions used by Efron and Tibshirani (1998) is based on one-sided p-values, or *attained confidence levels*. This section will compare observed confidence levels to this method through examples similar to those exhibited in Efron and Tibshirani (1998).

The first example described here is based on Example 1 of Efron and Tibshirani (1998). To make the problem easier to visualize the two-dimensional case will be considered here. The four-dimensional example in Efron and Tibshirani (1998) proceeds in a similar manner. Suppose $\mathbf{Y} \sim \mathbf{N}_2(\boldsymbol{\theta}, \mathbf{I})$, a bivariate normal distribution with mean vector $\boldsymbol{\theta}$ and identity covariance matrix. The parameter of interest is $\boldsymbol{\theta}$ with $\Theta = \mathbb{R}^2$ and $\mathbf{Y}' = (\sqrt{49/2}, \sqrt{49/2}) \simeq (4.95, 4.95)$ is observed. Suppose there are two regions defined as $\Theta_0 = \{\boldsymbol{\theta} : \|\boldsymbol{\theta}\| > 5\}$

Figure 7.3 *The cylindrical shell example with two regions. The observed value of* **Y** *is indicated by the "+" symbol.*

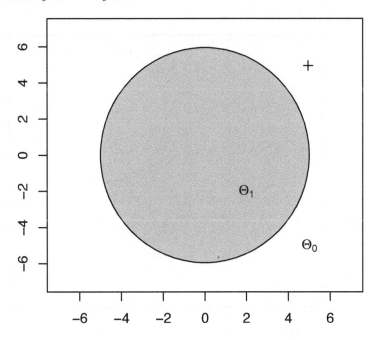

and $\Theta_1 = \{\boldsymbol{\theta} : \|\boldsymbol{\theta}\| \leq 5\}$. Figure 7.3 presents a plot of this situation. The attained confidence level of Efron and Tibshirani (1998) is computed using a one-sided p-value where Θ_1 is the preselected null hypothesis region. That is, the attained confidence level of Θ_0 is given by

$$\alpha_{\mathrm{acl}}(\Theta_0) = P(Q_1 \geq 7^2) = 0.9723,$$

where $Q_1 \sim \chi_2^2(25)$, a noncentral χ^2 distribution with 2 degrees of freedom and non-centrality parameter equal to 25. Efron and Tibshirani (1998) also compute a first-order approximation to α_{acl} that is a parametric bootstrap version of the bootstrap percentile method observed confidence level defined in Equation 3.11. This attained confidence level is computed as

$$\tilde{\alpha}_{\mathrm{acl}}(\Theta_0) = P(Q_2 \geq 5^2) = 0.9814,$$

where $Q_2 \sim \chi_2^2(49)$.

The observed confidence level for Θ_0 can be computed using the ordinary observed confidence level given in Equation (3.6) where $n = 1$, $\boldsymbol{\Sigma} = \mathbf{I}$ and $g_n(\mathbf{t}) = \phi_2(\mathbf{t})$. Therefore

$$\alpha_{\mathrm{ord}}(\Theta_0) = \int_{\hat{\boldsymbol{\theta}}-\Theta_0} \phi_2(\mathbf{t})d\mathbf{t} = 0.9815. \qquad (7.3)$$

Figure 7.4 *Region of integration for computing the observed confidence level in Equation (7.3). For comparison, the dotted region contains a standard bivariate normal random variable with a probability of 0.99.*

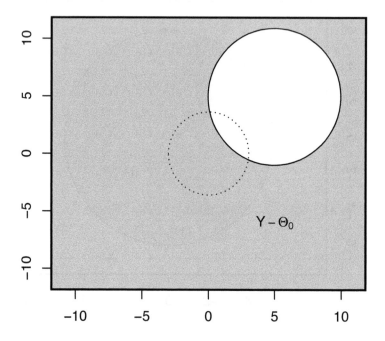

The region of integration in Equation (7.3) is plotted in Figure 7.4. The observed confidence level is close to the first-order approximation of the attained confidence level in this example.

A second example considered in Efron and Tibshirani (1998) again considers observing $\mathbf{Y} \sim N_2(\boldsymbol{\theta}, I)$ with $\Theta_0 = \{\boldsymbol{\theta} : 5 < \|\boldsymbol{\theta}\| < 9.5\}$, $\Theta_1 = \{\boldsymbol{\theta} : \|\boldsymbol{\theta}\| \leq 5\}$ and $\Theta_2 = \{\boldsymbol{\theta} : \|\boldsymbol{\theta}\| \geq 9.5\}$. As with the previous example, the two-dimensional case is considered here to allow for for easier visualization. Suppose once again that $\mathbf{Y}' = (\sqrt{49/2}, \sqrt{49/2}) \simeq (4.95, 4.95)$ is observed. Figure 7.5 presents a plot of this situation. Note that the distance between the observed value of $\mathbf{Y}$ and the nearest point not in Θ_0 is 2. Therefore, the attained confidence level of Θ_0 is computed by minimizing the probability that the distance between $\mathbf{Y}$ and the nearest vector $\boldsymbol{\theta}$ not in Θ_0 exceeds 2, over all $\boldsymbol{\theta} \in \Theta_0$. This probability is equivalent to

$$\alpha_{\text{acl}}(\Theta_0) = 1 - P(7^2 \leq Q_1 \leq (7.5)^2) = 0.9800.$$

This results in a paradoxical situation in that the region Θ_0 is smaller than it was in the previous example, but the attained confidence level has increased. The observed confidence level of Θ_0 is computed in an identical fashion as in

Figure 7.5 *The cylindrical shell example with three regions. The observed value of* **Y** *is indicated by the "+" symbol.*

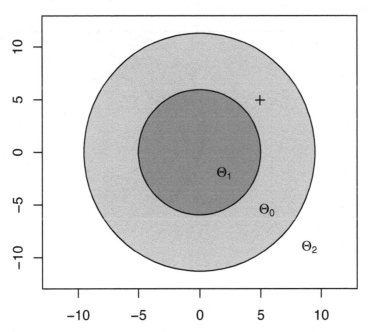

the previous example. That is

$$\alpha_{\mathrm{ord}}(\Theta_0) = \int_{\hat{\boldsymbol{\theta}} - \Theta_0} \phi_2(\mathbf{t}) d\mathbf{t} = 0.9741, \tag{7.4}$$

where the region of integration is plotted in Figure 7.6. Note the the observed confidence levels do not suffer from this paradox as the confidence level for Θ_0 decreases as the size of Θ_0 is decreased.

The third example is considered in Remark F of Efron and Tibshirani (1998) and again considers a spherical shell example with $\mathbf{Y} \sim \mathbf{N}_2(\boldsymbol{\theta}, \mathbf{I})$ and two regions $\Theta_0 = \{\boldsymbol{\theta} : \|\boldsymbol{\theta}\| \leq 1.5\}$ and $\Theta_1 = \{\boldsymbol{\theta} : \|\boldsymbol{\theta}\| > 1.5\}$. The observed value of $\mathbf{Y}$ is $(\sqrt{2}/4, \sqrt{2}/4)'$. See Figure 7.7 for a plot of this situation. The attained confidence level computed by Efron and Tibshirani (1998) is $\alpha_{\mathrm{acl}}(\Theta_0) = 0.959$ with $\tilde{\alpha}_{\mathrm{acl}}(\Theta_0) = 0.631$. The observed confidence level of Θ_0 is computed as

$$\alpha_{\mathrm{ord}}(\Theta_0) = \int_{\hat{\boldsymbol{\theta}} - \Theta_0} \phi_2(\mathbf{t}) d\mathbf{t} = 0.630. \tag{7.5}$$

See Figure 7.8 for a plot of the region of integration. Efron and Tibshirani (1998) argue that $\alpha_{\mathrm{acl}}(\Theta_0)$ is too large and $\tilde{\alpha}_{\mathrm{acl}}(\Theta_0)$ provides a more reasonable measure of confidence for this situation. The observed confidence level of Θ_0 closely matches this more reasonable measure.

Figure 7.6 *Region of integration for computing the observed confidence level in Equation (7.4). For comparison, the dotted region contains a standard bivariate normal random variable with a probability of 0.99.*

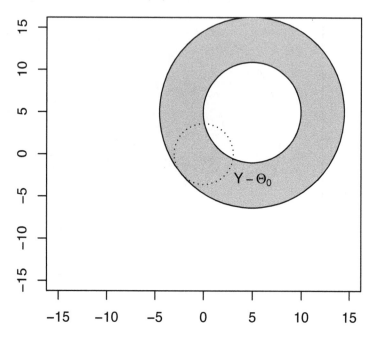

7.5 Bayesian Confidence Levels

The Bayesian solution to the problem of regions has, in principle, a relatively straightforward solution. Let $\pi(\boldsymbol{\theta})$ be the *prior distribution* of $\boldsymbol{\theta}$ on Θ. This distribution describes the random behavior of the parameter vector $\boldsymbol{\theta}$, and usually reflects the knowledge of $\boldsymbol{\theta}$ known by the experimenter, or an expert on the subject matter of the experiment. Let $\mathbf{X}' = (X_1, \dots, X_d)$ be an observed random vector from a d-dimensional parent distribution $f(\mathbf{x}|\boldsymbol{\theta})$. Note that this distribution is taken to be conditional on $\boldsymbol{\theta}$. The *posterior distribution* of the parameter vector $\boldsymbol{\theta}$, given the observed sample $\mathbf{x}$, is

$$\pi(\boldsymbol{\theta}|\mathbf{x}) = \frac{f(\mathbf{x}|\boldsymbol{\theta})\pi(\boldsymbol{\theta})}{\int_\Theta f(\mathbf{x}|\boldsymbol{\theta})\pi(\boldsymbol{\theta})d\boldsymbol{\theta}}.$$

Suppose that Ψ is a specific region of interest of Θ. Then the posterior probability that $\boldsymbol{\theta} \in \Psi$, which will be called the *Bayesian confidence level*, is then given by the posterior probability that $\boldsymbol{\theta} \in \Psi$. That is

$$\alpha_{\mathrm{bcl}}(\Psi, \pi) = \int_\Psi \pi(\boldsymbol{\theta}|\mathbf{x})d\boldsymbol{\theta} = \frac{\int_\Psi f(\mathbf{x}|\boldsymbol{\theta})\pi(\boldsymbol{\theta})d\boldsymbol{\theta}}{\int_\Theta f(\mathbf{x}|\boldsymbol{\theta})\pi(\boldsymbol{\theta})d\boldsymbol{\theta}}.$$

Figure 7.7 *The second cylindrical shell example with two regions. The observed value of* **Y** *is indicated by the "+" symbol.*

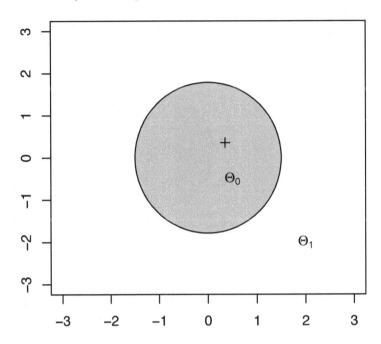

where the notation emphasizes the dependence of the measure on the prior distribution π. Methods based on observed confidence levels or attained confidence levels do not depend on a prior distribution for the unknown parameter $\boldsymbol{\theta}$. This is an advantage if no prior information is available, but may be perceived as a disadvantage if such information is available. In the case where no such information is available one can use a flat, or noninformative, prior to create an objective Bayes procedure. The simplest noninformative framework is based on the uniform, or flat, prior $\pi_f(\boldsymbol{\theta}) = c$ for some constant $c \in \mathbb{R}$ for all $\boldsymbol{\theta} \in \Theta$. In the strictest since this prior may not be a true density function when Θ is not bounded. As is demonstrated below, this does not affect the computation of the posterior probability. The Bayesian confidence level corresponding to the flat prior for a region $\Psi \in \Theta$ is given by

$$\bar{\alpha}_{\mathrm{bcl}}(\Psi) = \alpha_{\mathrm{bcl}}(\Psi, \pi) = \frac{\int_{\Psi} f(\mathbf{x}|\boldsymbol{\theta})d\boldsymbol{\theta}}{\int_{\Theta} f(\mathbf{x}|\boldsymbol{\theta})d\boldsymbol{\theta}}.$$

Efron and Tibshirani (1998) conclude that a flat prior may not provide confidence values that are consistent with p-values and attained confidence levels. As an alternative they suggest using the prior developed by Welch and Peers (1963) that is specifically designed to provide a closer match to attained confidence levels. This form of this prior varies depending on the structure of the

Figure 7.8 *Region of integration for computing the observed confidence level in Equation (7.5). For comparison, the dotted region contains a standard bivariate normal random variable with a probability of 0.99.*

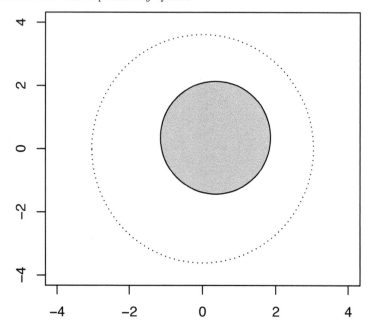

problem, but will be generally denoted by π_w with the corresponding Bayesian confidence level for a region $\Psi \in \Theta$ denoted as $\tilde{\pi}_{\mathrm{bcl}}(\Psi) = \pi_{\mathrm{bcl}}(\Psi, \pi_w)$.

A modest comparison of the Bayesian confidence levels discussed here may be accomplished by revisiting one of the examples considered in Section 7.4. In the first example $\mathbf{Y} \sim \mathbf{N}_2(\boldsymbol{\theta}, \mathbf{I})$ is observed and the regions of interest are $\Theta_0 = \{\boldsymbol{\theta} : \|\boldsymbol{\theta}\| > 5\}$ and $\Theta_1 = \{\boldsymbol{\theta} : \|\boldsymbol{\theta}\| \le 5\}$. The value $\mathbf{Y}' = (\sqrt{49/2}, \sqrt{49/2}) \simeq (4.95, 4.95)$ is observed. See Figure 7.3 for a visual representation of this problem. The first Bayesian confidence level considered here is based on the conjugate family for the multivariate normal model. The conjugate family for the multivariate normal model with known covariance matrix is again normal. For this problem suppose the prior distribution on $\boldsymbol{\theta}$ is a $\mathbf{N}_2(\boldsymbol{\lambda}, \boldsymbol{\Lambda})$ distribution where the parameters $\boldsymbol{\lambda}$ and $\boldsymbol{\Lambda}$ are known. The posterior distribution of $\boldsymbol{\theta}$ is also normal with mean vector

$$\boldsymbol{\lambda}_p = (\boldsymbol{\Lambda}^{-1} + \mathbf{I})^{-1}(\boldsymbol{\Lambda}^{-1}\boldsymbol{\lambda} + \mathbf{Y})$$

and covariance matrix $\boldsymbol{\Lambda}_p = (\boldsymbol{\Lambda}^{-1} + \mathbf{I})^{-1}$. See Section 3.6 of Gelman et. al. (1995). Therefore, for the region Θ_0, the Bayesian confidence level is given by

$$\alpha_{\mathrm{bcl}}[\Theta_0, \pi = N_2(\boldsymbol{\lambda}, \boldsymbol{\Lambda})] = \int_{\Theta_0} \phi(\boldsymbol{\theta}; \boldsymbol{\lambda}_p, \boldsymbol{\Lambda}_p)d\boldsymbol{\theta} = \int_{\Lambda_p^{-1/2}(\Theta_0 - \boldsymbol{\lambda}_p)} \phi(\boldsymbol{\theta})d\boldsymbol{\theta}$$

where $\phi_2(\boldsymbol{\theta}; \boldsymbol{\lambda}_p, \boldsymbol{\Lambda}_p)$ represents a multivariate normal density with mean vector $\boldsymbol{\lambda}_p$ and covariance matrix $\boldsymbol{\Lambda}_p$. Obviously, this measure will depend on the choices of the prior distribution parameters $\boldsymbol{\lambda}$ and $\boldsymbol{\Lambda}$. For example, if we take $\boldsymbol{\lambda} = \mathbf{0}$ and $\boldsymbol{\Lambda} = \mathbf{I}$ it follows that $\boldsymbol{\lambda}_p = \frac{1}{2}\mathbf{Y} = (2.475, 2.475)'$ and $\boldsymbol{\Lambda}_p = \frac{1}{2}\mathbf{I}$. Therefore

$$\alpha_{\text{bcl}}[\Theta_0, \pi = N_2(\mathbf{0}, \mathbf{I})] = \int_{\sqrt{2}(\Theta_0 - \frac{1}{2}\mathbf{Y})} \phi(\boldsymbol{\theta}) d\boldsymbol{\theta} \simeq 0.0034, \qquad (7.6)$$

which is well below the level of confidence indicated by the observed confidence level and the attained confidence level. The region of integration is shown in Figure 7.9. However, if we take a somewhat empirical Bayesian approach and use $\boldsymbol{\lambda} = \mathbf{Y} = (4.95, 4.95)'$ and $\boldsymbol{\Lambda} = \mathbf{I}$ it follows that $\boldsymbol{\lambda}_p = \mathbf{Y} = (4.95, 4.95)'$ and $\boldsymbol{\Lambda}_p = \frac{1}{2}\mathbf{I}$. Therefore

$$\alpha_{\text{bcl}}[\Theta_0, \pi = N_2(\mathbf{Y}, \mathbf{I})] = \int_{\sqrt{2}(\Theta_0 - \mathbf{Y})} \phi(\boldsymbol{\theta}) d\boldsymbol{\theta} \simeq 0.5016. \qquad (7.7)$$

The region of integration is shown in Figure 7.10. These two calculations demonstrate the strong dependence of the Bayesian confidence level on the prior distribution.

An objective Bayesian procedure uses one of the non-informative prior distributions studied earlier. In the case of the flat prior the measure is

$$\bar{\alpha}_{\text{bcl}}(\Theta_0) = \frac{\int_{\Theta_0} \phi_2(\mathbf{Y}; \boldsymbol{\theta}, \mathbf{I}) d\boldsymbol{\theta}}{\int_{\Theta} \phi_2(\mathbf{Y}; \boldsymbol{\theta}, \mathbf{I}) d\boldsymbol{\theta}}.$$

Note that, taken as a function of $\boldsymbol{\theta}$, $\phi(\mathbf{x}; \boldsymbol{\theta}, \mathbf{I}) = \phi(\boldsymbol{\theta}; \mathbf{x}, \mathbf{I})$. In this case $\Theta = \mathbb{R}^2$ and therefore the denominator is 1. Therefore

$$\bar{\alpha}_{\text{bcl}}(\Theta_0) = \int_{\Theta_0} \phi(\mathbf{Y}; \boldsymbol{\theta}, \mathbf{I}) d\boldsymbol{\theta} = \int_{\Theta_0 - \mathbf{Y}} \phi(\boldsymbol{\theta}) d\boldsymbol{\theta}.$$

As Efron and Tibshirani (1998) point out, this measure is then equal to the first-order bootstrap confidence value $\tilde{\alpha}_{\text{acl}}(\Theta_0) = 0.9814$. Tibshirani (1989) argues that the prior distribution developed by Welch and Peers (1963) for this example is a uniform distribution in polar coordinates in $\mathbb{R}^2$, that is $\pi_w(\boldsymbol{\theta}) = ||\boldsymbol{\theta}||^{-1}$ so that the corresponding Bayesian confidence level is

$$\tilde{\alpha}_{\text{bcl}}(\Theta_0) = \frac{\int_{\Theta_0} \phi_2(\mathbf{x}; \boldsymbol{\theta}, \mathbf{I})||\boldsymbol{\theta}||^{-1} d\boldsymbol{\theta}}{\int_{\Theta} \phi_2(\mathbf{x}; \boldsymbol{\theta}, \mathbf{I})||\boldsymbol{\theta}||^{-1} d\boldsymbol{\theta}} = \frac{\int_{\Theta_0} \phi_2(\boldsymbol{\theta}; \mathbf{x}, \mathbf{I})||\boldsymbol{\theta}||^{-1} d\boldsymbol{\theta}}{\int_{\Theta} \phi_2(\boldsymbol{\theta}; \mathbf{x}, \mathbf{I})||\boldsymbol{\theta}||^{-1} d\boldsymbol{\theta}} \simeq 0.9027.$$

7.6 Exercises

1. Suppose a region Ψ has an observed confidence level equal 0.97. What can be concluded about the results of a hypothesis test of $H_0 : \boldsymbol{\theta} = \boldsymbol{\theta}_0$ versus $H_1 : \boldsymbol{\theta} \neq \boldsymbol{\theta}_0$ for $\boldsymbol{\theta}_0 \in \Psi$ and $\boldsymbol{\theta}_0 \notin \Psi$? Assume that the hypothesis test was created by inverting the same confidence region that was used to derive the observed confidence level.

Figure 7.9 *Region of integration for computing the Bayesian confidence level in Equation (7.6). For comparison, the dotted region contains a standard bivariate normal random variable with a probability of 0.99.*

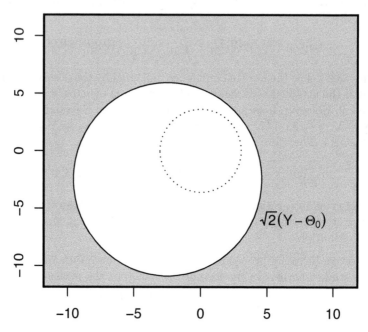

2. Suppose $X_1, \ldots, X_n$ is a set of independent and identically distributed random variables from a normal distribution with mean θ and unit variance. Using the usual normal confidence interval for θ demonstrate the following properties using specific interval regions of the form $\Theta_0 = [L_0, U_0]$ and $\Theta_1 = [L_1, U_1]$ that are subsets of the parameter space $\Theta = \mathbb{R}$.

   (a) A smaller region, that is one that has a shorter length, can have a larger observed confidence level than a larger region.

   (b) A region that contains $\hat{\theta}$ need not always have the largest observed confidence level.

   (c) There exist two regions such that $\alpha(\Theta_0) + \alpha(\Theta_1) = 1$ but $\Theta_0 \cup \Theta_1 \neq \mathbb{R}$.

3. Consider Applied Exercise 2 from Chapter 4 where three automobile engine software control systems are compared using a one-way analysis of variance.

   (a) Estimate the factor effects for the three software systems. What ordering, with respect to the estimated mean gasoline mileage, is observed? Can this simple analysis suggest any additional possible orderings of the means?

Figure 7.10 *Region of integration for computing the Bayesian confidence level in Equation (7.7). For comparison, the dotted region contains a standard bivariate normal random variable with a probability of 0.99.*

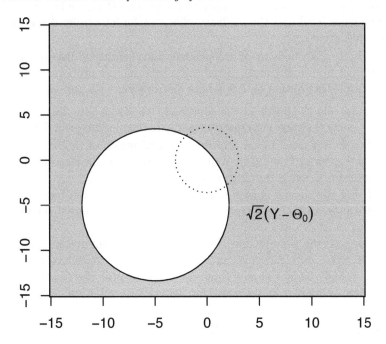

(b) Using the Bonferroni method for controlling the overall error rate of the tests, perform all possible pairwise tests of the equality of the means. What information about the mean gasoline mileage for each software system is obtained? How does these results compare to the results obtained using observed confidence levels?

(c) Repeat the analysis using Tukey's Honest Significant Difference (HSD) method instead of the Bonferroni method. What information about the mean gasoline mileage for each software system is obtained? How does these results compare to the results obtained using observed confidence levels and the Bonferroni method?

4. Let $X_1, \ldots, X_n$ be a set of independent and identically distributed random variables from a normal distribution with mean θ and unit variance. Consider two regions defined as $\Theta_0 = [-1, 1]$ and $\Theta_1 = (-\infty, -1) \cup (1, \infty)$. For a given value of $\hat{\theta} = \bar{X}$, compute the attained confidence level for Θ_0 when $\hat{\theta} \in \Theta_0$ and $\hat{\theta} \in \Theta_1$. Compare the attained confidence levels to the ordinary observed confidence level for Θ_0 under the same circumstances. Comment on any differences that are observed between the two methods.

5. Let $X_1, \ldots, X_n$ be a set of independent and identically distributed random variables from a normal distribution with mean θ and unit variance. Consider three regions defined as $\Theta_0 = [-1, 1]$, $\Theta_1 = (-2, -1) \cup (1, 2)$ and $\Theta_2 = (-\infty, -2) \cup (2, \infty)$. Suppose that $\hat{\theta} \in \Theta_1$. Describe how the attained confidence level and the ordinary observed confidence level are computed for this situation.

6. Let $X_1, \ldots, X_n$ be a set of independent and identically distributed random variables from a Bernoulli distribution with success probability θ. Suppose the prior distribution on θ is a beta distribution with parameters α and β.

   (a) Show that the posterior distribution of θ is also a beta distribution. That is, show that the beta distribution is the conjugate prior for θ.

   (b) Let $\Psi = (t_L, t_U)$ be an arbitrary interval region of the parameter space for θ given by $\Theta = (0, 1)$. Write an expression for the Bayesian confidence level for Ψ using the conjugate prior.

   (c) Using the normal approximation to the binomial distribution, write out an approximate observed confidence level for Ψ based on the sample $X_1, \ldots, X_n$.

   (d) Suppose the sample sum is 27 and $n = 50$. Compute and compare the Bayesian confidence level and the observed confidence level for the region $\Psi = [0.45, 0.55]$. Use $\alpha = 1$ and $\beta = 1$ for the prior distribution. Compute again using $\alpha = 2$ and $\beta = 2$ for the prior distribution.

7. Let $X_1, \ldots, X_n$ be a set of independent and identically distributed random variables from a normal distribution with mean θ and unit variance. Suppose the prior distribution on θ is a normal distribution with mean η and unit variance.

   (a) Show that the posterior distribution of θ is also a normal distribution. That is, show that the normal distribution is the conjugate prior for θ.

   (b) Let $\Psi = (t_L, t_U)$ be an arbitrary interval region of the parameter space for θ given by $\Theta = \mathbb{R}$. Write an expression for the Bayesian confidence level for Ψ using the conjugate prior.

   (c) Find a confidence interval for θ, and use the interval to derive an observed confidence level for Ψ based on the sample $X_1, \ldots, X_n$.

   (d) The twenty-five observations given in the table below are taken from a normal distribution with unit variance. Compute and compare the Bayesian confidence level and the observed confidence level for the region $\Psi = [-0.10, 0.10]$. Use a standard normal distribution as the prior distribution for θ.

0.92	-0.57	1.54	0.05	0.28
0.34	2.15	1.32	0.13	1.05
0.69	1.22	0.35	0.31	0.12
1.23	-0.84	-0.25	-0.48	1.66
-0.57	-1.56	-2.51	-1.30	-0.41

APPENDIX A

Review of Asymptotic Statistics

This appendix reviews some basic results, definitions, and techniques used in asymptotic statistics. Many of the results presented here are used extensively throughout the book. For more detailed treatments of asymptotic statistics and large-sample theory see Barndorff-Nielsen and Cox (1989), Ferguson (1996), Lehmann (1999), Sen and Singer (1993), and Serfling (1980).

A.1 Taylor's Theorem

One of the most basic tools used in asymptotic statistics is Taylor's Theorem. This theorem, which often allows one to approximate a smooth function with a much simpler function, is the basic tool used in the development of the Central Limit Theorem, Edgeworth and Cornish-Fisher expansions, and many other asymptotic results in statistics. The theorem is also useful in computational statistics. For example, as is demonstrated later in this section, one can approximate the standard normal distribution function with a considerable degree of accuracy without the use of numerical integration techniques. This section will provide an overview of this theorem. A more in-depth development can be found in Chapter 7 of Apostol (1967). A basic statement of the theorem is given below.

TAYLOR'S THEOREM. Assume that f is a real function with $p+1$ continuous derivatives on a neighborhood of a. Then, for every x in this neighborhood

$$f(x) = \sum_{k=0}^{p} \frac{f^{(k)}(a)}{k!}(x-a)^k + \frac{f^{(p+1)}(c)}{(p+1)!}(x-a)^{p+1}, \qquad (A.1)$$

for some $c \in [a, x]$.

The final term in Equation (A.1) is usually called the error term. The elimination of this term results in what is often called a p-term Taylor approximation, given by

$$f(x) \simeq \sum_{k=0}^{p} \frac{f^{(k)}(a)}{k!}(x-a)^k,$$

where a is generally chosen to be close to x so that the error term is not too large. In the case of asymptotic approximations it is often more convenient to use a slightly different form of Taylor's Theorem that approximates $f(x + h)$ with derivatives of $f(x)$ for h near x. Applying Taylor's Theorem to $f(x + h)$ with $a = x$ yields

$$f(x + h) = \sum_{k=0}^{p} \frac{f^{(k)}(x)}{k!} h^k + \frac{f^{(p+1)}(c)}{(p+1)!} h^{p+1}, \tag{A.2}$$

for some $c \in [x - h, x + h]$. In this case the function f is required to have $p+1$ derivatives in a neighborhood of x that contains the interval $[x - h, x + h]$. The corresponding p-term Taylor approximation has the form

$$f(x + h) \simeq \sum_{k=0}^{p} \frac{f^{(k)}(x)}{k!} h^k.$$

An analysis of the error term in Equation (A.2) reveals that as long as $f^{(p+1)}$ is bounded on $[x - h, x + h]$ then

$$\lim_{h \to 0} \frac{f^{(p+1)}(c)}{(p+1)!} h^{p+1} = 0$$

so that the approximation is more accurate for small values of h. It will be shown in Section A.4 that the approximation is also generally more accurate for larger values of p.

As an example consider approximating the exponential function near 0. A two-term Taylor approximation of e^{x+h} is given by $e^{x+h} \simeq e^x(1 + h + \frac{1}{2}h^2)$ so that

$$e^h \simeq 1 + h + \frac{1}{2}h^2. \tag{A.3}$$

Therefore, the two-term Taylor approximation in Equation (A.3) provides a simple polynomial approximation for e^h for values of h near 0. Figure A.1 presents a plot of e^h and the two-term Taylor approximation. From this plot one can clearly observe that the approximation given in Equation (A.3) is very accurate for values of h near 0. As a second example consider approximating the standard normal distribution function Φ near 0. A two-term Taylor approximation of $\Phi(x + h)$ is given by $\Phi(x + h) \simeq \Phi(x) + h\phi(x)(1 - \frac{1}{2}hx)$. Setting $x = 0$ yields a linear two-term Taylor approximation for $\Phi(h)$ given by $\Phi(h) \simeq \frac{1}{2} + h(2\pi)^{-1/2}$, which will be accurate for small values of h. A plot of $\Phi(h)$ and the two-term Taylor approximation is given in Figure A.2, which again indicates that the approximation is accurate for small values of h.

A.2 Modes of Convergence

The concepts of limit and convergence, which are easily defined for a sequence of constants, are not be uniquely defined for sequences of random variables.

Figure A.1 *The exponential function (solid line) and the two-term Taylor approximation (dashed line) of the exponential function*

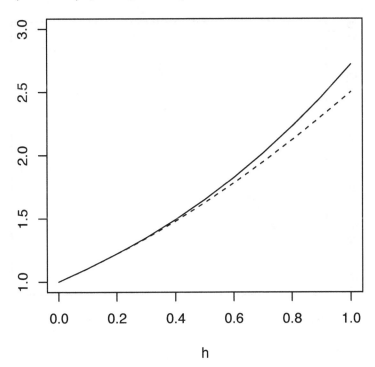

This is due to the fact that a sequence of random variables actually represents a sequence of functions mapping a sample space to the real line. Hence, the concepts of limit and convergence for random variables is essentially the same as the corresponding concepts for sequences of functions. This section will consider three basic definitions of limit and convergence for a sequence of random variables.

Let $\{X_n\}_{i=n}^{\infty}$ be a sequence of random variables and let X be another random variable. Then X_n *converges in probability* to X as $n \to \infty$ if for every $\epsilon > 0$,

$$\lim_{n \to \infty} P(|X_n - X| > \epsilon) = 0. \tag{A.4}$$

The notation $X_n \overset{p}{\to} X$ will be used to indicate this type of convergence. From a functional viewpoint, the requirement of Equation (A.4) is that the function X_n be within an ϵ-band of X over all points except those in a set that has probability 0, as $n \to \infty$ for every $\epsilon > 0$.

In terms of estimation theory, an estimator $\hat{\theta}_n$ of θ is said to be a *consistent* estimator of θ if $\hat{\theta}_n \overset{p}{\to} \theta$ as $n \to \infty$. Chebychev's Theorem is often a useful tool in establishing the consistency of an estimator. For example, suppose that

Figure A.2 *The standard normal distribution function* Φ *(solid line) and the two-term Taylor approximation (dashed line) of* Φ.

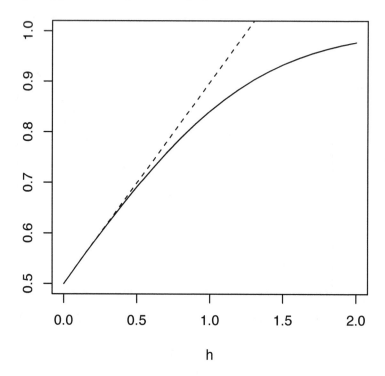

$\hat{\theta}_n$ is unbiased and has a standard error that converges to 0 as $n \to \infty$. Then, Chebychev's Theorem implies that for every $\epsilon > 0$,

$$0 \leq P(|\hat{\theta}_n - \theta| \geq \epsilon) \leq \epsilon^{-2} V(\hat{\theta}_n).$$

Because

$$\lim_{n\to\infty} \epsilon^{-2} V(\hat{\theta}_n) = 0,$$

it follows that

$$\lim_{n\to\infty} P(|\hat{\theta}_n - \theta| > \epsilon) = 0,$$

and therefore $\hat{\theta}_n \xrightarrow{p} \theta$ as $n \to \infty$. Another common consistency result is given by the Weak Law of Large Numbers, which in its simplest form, states that if $\{X_n\}_{n=1}^{\infty}$ is a sequence of independent and identically distributed random variables such that $E(X_n) = \theta < \infty$, then

$$n^{-1} \sum_{i=1}^{n} X_i \xrightarrow{p} \theta, \tag{A.5}$$

as $n \to \infty$. The Weak Law of Large Numbers establishes the consistency of the sample mean under very weak conditions. This result can be used to establish a much broader result for sample moments as shown below.

Consider a sequence of random vectors $\{\mathbf{X}_n\}_{n=1}^\infty$ where $\mathbf{X}_n' = (X_{n1}, \ldots, X_{np})$ where X_{nj} are random variables for $j = 1, \ldots, p$. It can be shown that $\mathbf{X}_n \xrightarrow{p} \mathbf{X}$ as $n \to \infty$, where $\mathbf{X}' = (X_1, \ldots, X_n)$, if $X_{nj} \xrightarrow{p} X_j$ as $n \to \infty$ for every $j = 1, \ldots, p$. That is, convergence in probability of the components of a random vector implies convergence in probability of the corresponding sequence of random vectors. Smooth transformations of random vectors also converge in probability. That is, if $g : \mathbb{R}^p \to \mathbb{R}^q$ is a function that is continuous except perhaps on a set D where $P(\mathbf{X} \in D) = 0$, then $\mathbf{X}_n \xrightarrow{p} \mathbf{X}$ as $n \to \infty$ implies that

$$g(\mathbf{X}_n) \xrightarrow{p} g(\mathbf{X}), \tag{A.6}$$

as $n \to \infty$. The continuity condition is a crucial assumption of this result. See Section 1.7 of Serfling (1980). These result are useful in establishing the consistency of the sample moments.

As an example consider a sequence $X_1, \ldots, X_n$ of independent and identically distributed random variables where $E(X_i) = \mu$ and $V(X_i) = \theta < \infty$. The Weak Law of Large Numbers as stated in Equation (A.5) implies that $\hat{\mu}_n \xrightarrow{p} \mu$ as $n \to \infty$ where

$$\hat{\mu}_n = n^{-1} \sum_{i=1}^n X_i.$$

Similarly, the Weak Law of Large Numbers, along with the smooth transformation result given in Equation (A.6) can be used to show that if $\tau = E(X_i^2)$, then implies that $\hat{\tau}_n \xrightarrow{p} \tau$ as $n \to \infty$ where

$$\hat{\tau}_n = n^{-1} \sum_{i=1}^n X_i^2.$$

Note that it follows then that

$$\begin{bmatrix} \hat{\mu}_n \\ \hat{\tau}_n \end{bmatrix} \xrightarrow{p} \begin{bmatrix} \mu \\ \tau \end{bmatrix},$$

and an additional application of the smooth transformation result of Equation (A.6) implies that

$$\hat{\theta} = \hat{\tau}_n - \hat{\mu}_n^2 \xrightarrow{p} \tau - \mu^2 = \theta$$

as $n \to \infty$. Therefore, the sample variance is a consistent estimator of the population variance. This result can be extended to all sample moments under sufficient conditions. See Theorems 2.2.1A and 2.2.3A of Serfling (1980) and Exercises 4 and 5 of Section A.5.

A stronger mode of convergence is provided by almost sure convergence. Let $\{X_n\}_{n=1}^\infty$ be a sequence of random variables and let X be another random variable. The sequence $\{X_n\}_{n=1}^\infty$ converges almost surely to X as $n \to \infty$ if

$$P(\lim_{n \to \infty} X_n = X) = 1. \tag{A.7}$$

The notation $X_n \xrightarrow{a.s.} X$ will be used to indicate this type of convergence. From a functional viewpoint, Equation (A.7) requires that the function X_n converge at all points in the sample space except perhaps on a set with probability 0. This type of convergence is also called convergence almost everywhere or convergence with probability 1. This mode of convergence is stronger than convergence in probability in the sense that $X_n \xrightarrow{a.s.} X$ implies $X_n \xrightarrow{p} X$ as $n \to \infty$, but that there are sequences that converge in probability that do not converge almost surely. See Theorem 1.3.1 and Example A in Section 1.3.8 of Serfling (1980). Many of the commonly used properties of convergence in probability are also true for almost sure convergence. In particular, random vectors converge almost surely when each of their components converge almost surely, and smooth functions of vectors that converge almost surely also converge almost surely. The Strong Law of Large Numbers, which states that if $\{X_n\}_{n=1}^{\infty}$ is a sequence of independent and identically distributed random variables with finite mean θ, then

$$n^{-1} \sum_{i=1}^{n} X_i \xrightarrow{a.s.} \theta,$$

is the almost sure analog of the Weak Law of Large Numbers.

Weak convergence, or convergence in distribution or law, is another very common mode of convergence used in statistical theory as it is the mode of convergence used in the Central Limit Theorem, summarized in Section A.3. This type of convergence does not require the actual functions represented by the random variables to converge at all. Rather, weak convergence refers to the convergence of the distributions of the random variables. Formally, a sequence of random variables $\{X_n\}_{n=1}^{\infty}$ converges weakly to X if

$$\lim_{n \to \infty} F_n(t) = F(t),$$

for each continuity point of F where $F_n(t) = P(X_n \leq t)$ for $n = 1, \ldots, \infty$ and $F(t) = P(X \leq t)$. The notation $X_n \xrightarrow{w} X$ will be used to denote this mode of convergence. Weak convergence is important to statistical theory and can be defined in very general settings. See Billingsley (1999) and Chapter 7 of Pollard (2002) for further details.

Weak convergence is the weakest mode of convergence considered in this book in the sense that both convergence in probability and almost sure convergence imply weak convergence. It is a simple matter to construct an example to demonstrate that the converse is not always true. Suppose $X \sim N(0,1)$ and $X_n = (-1)^n X$ for $n = 1, \ldots, \infty$. It is clear that $X_n \xrightarrow{w} X$ as $n \to \infty$ but that X_n does not converge in probability to X. Indeed, from a functional viewpoint, the function X_n does not converge at all.

The properties of weak convergence can differ significantly from convergence in probability and almost sure convergence. In particular, componentwise weak convergence of the elements of a random vector does not insure that the corresponding random vectors converge weakly. This is due to the fact that a set of marginal distributions does not uniquely determine a joint distribution. The Cramér-Wold Theorem is required to show the weak convergence of random vectors. See Theorem 1.5.2 of Serfling (1980) for further details. Note that the weak convergence of random vectors can always be established by showing that the corresponding joint distribution functions converge, but it is usually easier to work with the marginal distributions. Smooth functions of random vectors that converge weakly also converge weakly as with convergence in probability and almost sure convergence.

As an example, suppose $X \sim N(0,1)$ and $Y \sim N(0,1)$ are independent random variables and define sequences of random variables $\{X_n\}_{n=1}^{\infty}$ and $\{Y_n\}_{n=1}^{\infty}$ such that $X_n \xrightarrow{w} X$ and $Y_n \xrightarrow{w} Y$ as $n \to \infty$. Does it necessarily follow that $X_n + Y_n \xrightarrow{w} X + Y \sim N(0,2)$? The truth of this result depends on the relationship between X_n and Y_n. Suppose that $X_n = X$ and $Y_n = Y$ for $n = 1, \ldots, \infty$. Then the result is trivially true. If, however, $X_n = X$ and $Y_n = -X$ then $X_n + Y_n$ converges weakly to a degenerate distribution at 0, and the result is not true.

The final property of weak convergence used periodically in this book is the fact that if $X_n \xrightarrow{w} c$, where c is a real constant, then $X_n \xrightarrow{p} c$ as $n \to \infty$. Therefore, in the second construction of the example given above, it follows that $X_n + Y_n \xrightarrow{p} 0$ as $n \to \infty$.

A.3 Central Limit Theorem

In its most basic formulation, the Central Limit Theorem refers to the result that under a wide variety of conditions, a standardized sample mean weakly converges to a normal distribution. The classic result in this context is the Lindeberg-Lévy Theorem, given below.

LINDEBERG-LÉVY THEOREM. Suppose $X_1, \ldots, X_n$ is a set of independent and identically distributed random variables with $\mu = E(X_i)$ and $\sigma^2 = V(X_i) < \infty$. Then $n^{1/2}(\bar{X} - \mu)/\sigma \xrightarrow{w} N(0,1)$ as $n \to \infty$.

For a proof of this Theorem see Section A.5 of Lehmann (1999). The result can be easily extended to the multivariate case given below.

MULTIVARIATE CENTRAL LIMIT THEOREM. Suppose $\mathbf{X}_1, \ldots, \mathbf{X}_n$ is a set of independent and identically distributed d-dimensional random vectors with $\boldsymbol{\mu} = E(\mathbf{X}_i)$ and covariance matrix $V(\mathbf{X}_i) = \boldsymbol{\Sigma}$. If $\boldsymbol{\Sigma}$ has finite elements and is positive definite then $\boldsymbol{\Sigma}^{-1/2}(\bar{\mathbf{X}} - \boldsymbol{\mu}) \xrightarrow{w} N_d(\mathbf{0}, \mathbf{I})$ as $n \to \infty$.

These theorems are special cases of several results with slightly weaker assumptions. See Section 1.9 of Serfling (1980) for an exposition of several of these results.

The Central Limit Theorems form the basis for large sample approximate inference for means when the population is unknown, but it is reasonable to assume that the variances are finite. In this book these results are extended beyond the mean parameter by means of the smooth function model, which extends certain asymptotic results to smooth functions of vector means. The motivating result behind this model is given in the Theorem below.

SMOOTH TRANSFORMATION THEOREM. Suppose $\mathbf{X}_1, \ldots, \mathbf{X}_n$ is a set of independent and identically distributed d-dimensional random vectors with $\boldsymbol{\mu} = E(\mathbf{X}_i)$ and covariance matrix $\boldsymbol{\Sigma}$ that is positive definite with finite elements. Let $g : \mathbb{R}^d \to \mathbb{R}^q$ be a function of $\mathbf{x}' = (x_1, \ldots, x_d)$ such that $g(\mathbf{x}) = (g_1(\mathbf{x}), \ldots, g_q(\mathbf{x}))'$ and $dg_i(\mathbf{x})/d\mathbf{x}$ is non-zero at $\mathbf{x} = \boldsymbol{\mu}$. Let $\mathbf{D}$ be a $q \times d$ matrix with $(i, j)^{\text{th}}$ element

$$D_{ij} = \left. \frac{\partial g_i(\mathbf{x})}{\partial x_j} \right|_{\mathbf{x}=\boldsymbol{\mu}}.$$

Then $n^{1/2}(\mathbf{D}\boldsymbol{\Sigma}\mathbf{D}')^{-1/2}[g(\bar{\mathbf{X}}) - g(\boldsymbol{\mu})] \xrightarrow{w} N_q(\mathbf{0}, \mathbf{I})$ as $n \to \infty$.

For a proof and a more general version of this result see Section 3.3 of Serfling (1980). To see the usefulness of this result, let $X_1, \ldots, X_n$ be a set of independent and identically distributed random variables from a distribution F with mean η and variance θ. Define a new two-dimensional random vector $Y_i = (X_i, X_i^2)'$ for $i = 1, \ldots, n$. In this case, the Multivariate Central Limit Theorem implies that $n^{1/2}\boldsymbol{\Sigma}^{-1/2}(\bar{\mathbf{Y}} - \boldsymbol{\mu}) \xrightarrow{w} N_2(\mathbf{0}, \mathbf{I})$ where $\boldsymbol{\mu}' = (\eta, \theta + \eta^2)$ and

$$\boldsymbol{\Sigma} = \begin{bmatrix} \theta & \gamma - \eta(\theta + \eta^2) \\ \gamma - \eta(\theta + \eta^2) & \kappa - (\theta + \eta^2)^2 \end{bmatrix},$$

where $\gamma = E(X_i^3)$ and $\kappa = E(X_i^4)$. Define a function $g(x_1, x_2) = x_2 - x_1^2$ so that $g(\boldsymbol{\mu}) = \theta$ and $\mathbf{D} = [-2\eta, 1]$. It follows from the Smooth Transformation Theorem that $n^{1/2}(\mathbf{D}\boldsymbol{\Sigma}\mathbf{D}')^{-1/2}[g(\bar{\mathbf{X}}) - \theta] \xrightarrow{w} N(0, 1)$ as $n \to \infty$, where

$$(\mathbf{D}\boldsymbol{\Sigma}\mathbf{D}')^{-1/2} = (3\eta^4 - 4\eta\gamma + 6\eta^2\theta + \kappa - \theta^2)^{-1/2}.$$

Note that

$$g(\bar{\mathbf{X}}) = \frac{1}{n}\sum_{i=1}^{n} X_i^2 - \left(\frac{1}{n}\sum_{i=1}^{n} X_i \right)^2,$$

which is the sample variance. Therefore the Smooth Transformation Theorem has been used to establish an asymptotic normality result for the sample variance. This type of argument is used extensively in this book to establish asymptotic results for smooth functions of means such as correlations, ratios of means and ratios of variances.

A.4 Convergence Rates

The error terms from asymptotic expansions are often complicated functions which may depend on properties of the function being approximated as well as other variables. When using the Taylor expansion the error term depends on the next derivative that was not used in the approximation and how far away from the expansion point the function is being approximated. In asymptotic expansions used primarily in statistics, the key component of the error term is usually the sample size. The usefulness of approximations lies in the fact that they are simpler than the original function and are easier to work with. The gains in simplicity are erased is the error term itself must be accounted for explicitly in each calculation. Typically this is not the case, but some properties of the error term are usually required for the approximation to be useful.

The error terms in question usually have some limiting properties that are of interest. For example, the error term in the Taylor expansion converges to 0 as the point where the function is being approximated approaches the point the expansion is centered around. Similarly, in statistical applications the error terms often converge to 0 either as a deterministic or stochastic sequence as the sample size becomes very large. It is not only important to account for the fact that these error terms converge to 0, but it is also important to keep track of how fast they converge to zero.

As an example, consider two Taylor series approximations of the normal distribution function $\Phi(h)$, expanded around the point $t = 0$. As shown in Section A.1, a two-term Taylor expansion for $\Phi(h)$ is given by $\Phi(h) \simeq \frac{1}{2} + h(2\pi)^{-1/2}$ and a four term Taylor expansion is given by $\Phi(h) \simeq \frac{1}{2} + h(2\pi)^{-1/2} - h^3(2\pi)^{-1/2}/6$. Both of these approximations are plotted in Figure A.3. The error of these approximations is plotted in Figure A.4. It is clear from these plots that both errors converge to 0 as $h \to 0$. It appears that the four term-approximation is superior to the two term approximation. Does this behavior hold as $h \to 0$? To investigate this question the ratio of the absolute errors of the four term approximation to the two-term approximation is plotted in Figure A.5. Two conclusions can be made on the basis of this plot. First, the error from the four-term expansion is uniformly smaller than that of the two-term expansion for the range $h \in (0, 2)$. The second conclusion is a limiting property. Note that even though both errors converge to 0 as $h \to 0$, it is apparent that the error from the four-term approximation converges to 0 at a faster rate than the error for the two-term approximation, because the ratio of the absolute errors converges to 0. This type of argument provides the basis for concluding that one sequence converges at a faster rate to 0 than another.

To formalize this procedure, let $a(t)$ and $b(t)$ be real functions. Then $a(t)$ is of smaller order than $b(t)$ as $t \to L$, denoted as $a(t) = o(b(t))$ if

$$\lim_{t \to L} \frac{a(t)}{b(t)} = 0.$$

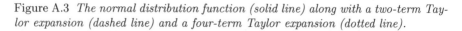

Figure A.3 *The normal distribution function (solid line) along with a two-term Taylor expansion (dashed line) and a four-term Taylor expansion (dotted line).*

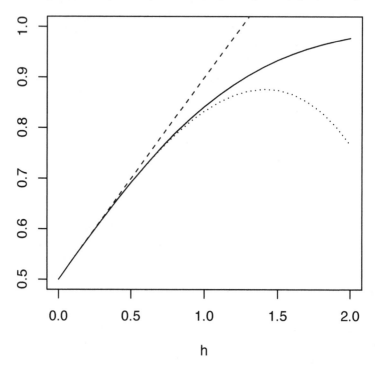

While this notion is defined in terms of continuous functions, the definition will also be applied to sequences of constants where the limit of the index is usually taken to be ∞. In the statistical applications in this book, the function b will often be taken to be of the form $n^{-k/2}$, where k is a specified positive integer. For error terms converging to 0, a sequence $\{a_n\}_{n=1}^{\infty}$ with the property $a_n = o(n^{-k/2})$ will be said to have faster than k^{th}-order convergence to 0.

Errors that are of asymptotically similar sizes are also compared though their ratio, which in this case is assumed to stay bounded in the limit. That is $a(t)$ is of the same order as $b(t)$ as $t \to L$, denoted as $a(t) = O(b(t))$ if $|a(t)/b(t)|$ remains bounded as $t \to L$. As with the definition given above, the function b will often be taken to be of the form $n^{-k/2}$, where k, and for error terms converging to 0, a sequence $\{a_n\}_{n=1}^{\infty}$ with the property $a_n = O(n^{-k/2})$ will be said to have k^{th}-order convergence to 0.

The error terms from Taylor approximations can be written in terms of this

Figure A.4 *Errors in approximating the normal distribution function with a two-term Taylor expansion (dashed line) and a four-term Taylor expansion (dotted line).*

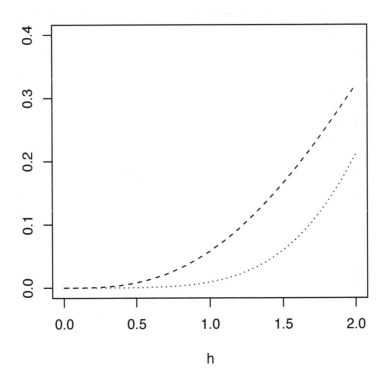

order notation. In particular it can be shown that

$$f(x+h) = \sum_{k=0}^{p} \frac{f^{(k)}(x)}{k!} h^k + o(h^k),$$

and

$$f(x+h) = \sum_{k=0}^{p} \frac{f^{(k)}(x)}{k!} h^k + O(h^{k+1}),$$

as $h \to 0$. This implies that the Taylor expansions for the normal distribution function can be written as $\Phi(h) = \frac{1}{2} + h(2\pi)^{-1/2} + o(h^2)$ and $\Phi(h) = \frac{1}{2} + h(2\pi)^{-1/2} - h^3(2\pi)^{-1/2}/6 + o(h^4)$, where it is noted that the even terms in this expansion are 0.

The errors associated with stochastic approximations can be compared in a similar manner to their deterministic counterparts, though additional technicalities arise due to the random behavior of the sequences. To define these concepts let $\{X_n\}_{n=1}^{\infty}$ be a sequence of random variables and $\{a_n\}_{n=1}^{\infty}$ be a

Figure A.5 *Ratio of the absolute errors in approximating the normal distribution function with a four-term Taylor expansion to a two-term Taylor expansion.*

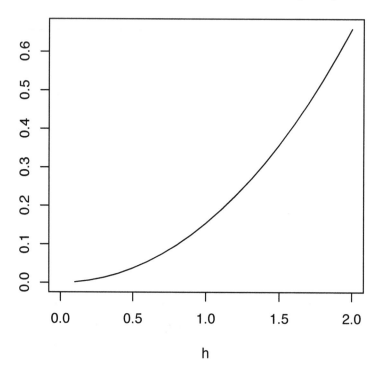

sequence of constants. Then $X_n = o_p(a_n)$ as $n \to \infty$ if $X_n/a_n \overset{p}{\to} 0$ as $n \to \infty$. To define sequences that are of the same order, first define $X_n = O_p(1)$ if the sequence $\{X_n\}_{n=1}^{\infty}$ is bounded in probability. That is, for every $\epsilon > 0$ there exist B_ϵ and n_ϵ such that $P(|X_n| \le B_\epsilon) > 1 - \epsilon$ for every $n > n_\epsilon$. In most applications the fact that $X_n \overset{w}{\to} X$ as $n \to \infty$ implies that $X_n = O_p(1)$ can be used. A sequence of random variables $\{X_n\}_{n=1}^{\infty}$ is then defined to have the property $X_n = O_p(a_n)$ if $X_n/a_n = O_p(1)$ as $n \to \infty$.

The most common application of the results in this book is with respect to the sample moments. In Section A.2 it is established through the Weak Law of Large Numbers that if $\{X_n\}_{n=1}^{\infty}$ is a sequence of independent and identically distributed random variables with finite mean θ, then

$$\bar{X}_n = n^{-1} \sum_{i=1}^{n} X_i \overset{p}{\to} \theta$$

as $n \to \infty$. It follows from this result that $\bar{X}_n - \theta = o_p(1)$, or that $\bar{X}_n = \theta + o_p(1)$ as $n \to \infty$. Similarly, the Central Limit Theorem of Section A.3 shows that if X_n has finite variance $\sigma^2 < \infty$ then it follows that $n^{1/2}(\bar{X}_n - \theta) \overset{w}{\to} Z$, where $Z \sim N(0,1)$, as $n \to \infty$. This indicates that $n^{1/2}(\bar{X}_n - \theta) = O_p(1)$ as $n \to \infty$. Therefore, it follows that $\bar{X}_n - \theta = O_p(n^{-1/2})$, or that $\bar{X}_n = \theta + O_p(n^{-1/2})$. These results can be extended to all of the sample moments under certain conditions. See Exercises 4 and 5 in Section A.5 for further details.

A.5 Exercises

1. Compute one-, two-, and three-term Taylor approximations for the function $\sin(h)$ for values of h near 0. Express the error of these approximations in term of the o and O order operators. Plot the function and the approximations on a single set of axes for values of h between 0 and $\frac{\pi}{2}$. Comment on the role that both h and p play in the accuracy of the approximations.

2. This exercise will explore two possible methods that can be used to approximate the function $\sin^2(h)$ for values of h near 0. As a first approximation, directly compute a two-term Taylor approximation of $\sin^2(h)$. Express the error of this approximation in terms of the o order operator. A second approximation can be derived by squaring a two-term Taylor approximation for $\sin(h)$. Express the error of this approximation, as accurately as possible, in terms of the o order operator. Comment on the difference between the two approximations. The results from Exercise 7 may be required to establish these results.

3. Let $\{X_n\}_{n=1}^{\infty}$ be a sequence of random variables such that $X_n \xrightarrow{w} X$ as $n \to \infty$ where the distribution function of X is degenerate at a real constant c. That is, the distribution function of X is $F(t) = P(X \le t) = \delta(t; [c, \infty))$. Prove that $X_n \xrightarrow{p} c$ as $n \to \infty$.

4. This exercise will establish the consistency of the sample moments as described in Section A.2. Let $X_1, \ldots, X_n$ be a sequence of independent and identically distributed random variables with k^{th} moment defined as $\mu_k = E(X_i^k)$.

   (a) Use the Weak Law of Large Numbers as stated in Equation (A.5) and the smooth transformation result for convergence in probability to establish the consistency of the sample moment

   $$\hat{\mu}_k = n^{-1} \sum_{i=1}^{n} X_i^k,$$

   as an estimator of μ_k. State the conditions on X_i that are required for the result to be justified.

   (b) Use the Multivariate Central Limit Theorem and the Smooth Transformation Theorem to prove that $\hat{\mu}_k$ is asymptotically normal, for *some* matrix $\mathbf{D}$ that need not be explicitly identified. Use this result to conclude that $\hat{\mu}_k = \mu_k + O_p(n^{-k/2})$ as $n \to \infty$.

5. This exercise will establish the consistency of the sample central moments as described in Section A.2. Let $X_1, \ldots, X_n$ be a sequence of independent and identically distributed random variables with k^{th} central moment defined as $\nu_k = E[(X_i - \mu_1)^k]$, where $\mu_1 = E(X_i)$. The kth sample central moment is given by

$$\hat{\nu}_k = n^{-1} \sum_{i=1}^{n} (X_i^k - \hat{\mu}_1)^k.$$

(a) Show that $\hat{\nu}_k$ is a consistent estimator of ν_k using the binomial expansion

$$\nu_k = \sum_{i=0}^{n} \binom{n}{i} \mu_1^i \mu_{k-i}.$$

Use the consistency of the sample moments established in Exercise 4 as a starting point along with the results from Exercises 7–9.

(b) Use the Multivariate Central Limit Theorem and the Smooth Transformation Theorem to prove that $\hat{\nu}_k$ is asymptotically normal, for *some* matrix $\mathbf{D}$ that need not be explicitly identified. Use this result to conclude that $\hat{\nu}_k = \nu_k + O_p(n^{-k/2})$ as $n \to \infty$.

6. Using the univariate version of the Weak Law of Large Numbers as stated in Equation (A.5), establish the multivariate form of the Weak Law of Large Numbers. That is, suppose that $\mathbf{X}_1, \ldots, \mathbf{X}_n$ is a sequence of independent and identically distributed random vectors with $\boldsymbol{\theta} = E(\mathbf{X}_i)$. Show that

$$n^{-1} \sum_{i=1}^{n} \mathbf{X}_i \xrightarrow{P} \boldsymbol{\theta},$$

as $n \to \infty$.

7. Let $\{a_n\}_{n=1}^{\infty}$, $\{b_n\}_{n=1}^{\infty}$, $\{c_n\}_{n=1}^{\infty}$ and $\{d_n\}_{n=1}^{\infty}$ be sequences of constants.

(a) Suppose that $a_n = o(c_n)$ and $b_n = o(d_n)$. Prove that $a_n b_n = o(c_n d_n)$.
(b) Suppose that $a_n = O(c_n)$ and $b_n = O(d_n)$. Prove that $a_n b_n = O(c_n d_n)$.
(c) Suppose that $a_n = o(c_n)$ and $b_n = O(d_n)$. Prove that $a_n b_n = o(c_n d_n)$.

8. Let $\{X_n\}_{n=1}^{\infty}$ and $\{Y_n\}_{n=1}^{\infty}$ be a sequences of random variables and $\{a_n\}_{n=1}^{\infty}$ and $\{b_n\}_{n=1}^{\infty}$ be sequences of constants.

(a) Suppose that $X_n = o_p(a_n)$ and $Y_n = o_p(b_n)$. Prove that $X_n Y_n = o_p(a_n b_n)$.
(b) Suppose that $X_n = O_p(a_n)$ and $Y_n = O_p(b_n)$. Prove that $X_n Y_n = O_p(a_n b_n)$.
(c) Suppose that $X_n = o_p(a_n)$ and $Y_n = O_p(b_n)$. Prove that $X_n Y_n = o_p(a_n b_n)$.

9. Let $\{X_n\}_{n=1}^{\infty}$ be a sequence of random variables and $\{a_n\}_{n=1}^{\infty}$, $\{b_n\}_{n=1}^{\infty}$ and $\{c_n\}_{n=1}^{\infty}$ be sequences of constants.

(a) Suppose that $X_n = o_p(a_n)$ and $b_n = o(c_n)$. Prove that $X_n b_n = o_p(a_n c_n)$.

(b) Suppose that $X_n = O_p(a_n)$ and $b_n = O(c_n)$. Prove that $X_n b_n = O_p(a_n c_n)$.

(c) Suppose that $X_n = o_p(a_n)$ and $b_n = O(c_n)$. Prove that $X_n b_n = o_p(a_n c_n)$.

10. This exercise will prove the collection of results commonly known as Slutsky's Theorem. Let $\{X_n\}_{n=1}^\infty$ and $\{Y_n\}_{n=1}^\infty$ be sequences of random variables such that $X_n \xrightarrow{w} X$ and $Y_n \xrightarrow{p} c$ as $n \to \infty$ where c is a real constant.

(a) Prove that $X_n + Y_n \xrightarrow{w} X + c$ as $n \to \infty$.

(b) Prove that $X_n Y_n \xrightarrow{w} cX$ as $n \to \infty$.

(c) Prove that if $c \neq 0$ then $X_n Y_n \xrightarrow{w} X/c$ as $n \to \infty$.

11. Suppose it is known that $n^{1/2}(\hat{\theta} - \theta)/\sigma \xrightarrow{w} N(0,1)$ as $n \to \infty$ and that $\hat{\sigma}$ is a consistent estimator of the true standard error of $\hat{\theta}$ with the property that $\hat{\sigma} \xrightarrow{p} \sigma$ as $n \to \infty$. Prove that $n^{1/2}(\hat{\theta} - \theta)/\hat{\sigma} \xrightarrow{w} N(0,1)$.

12. Suppose it is known that $n^{1/2}(\hat{\theta} - \theta)/\hat{\sigma} \xrightarrow{w} N(0,1)$ as $n \to \infty$ and that $\hat{\sigma}$ is a biased estimator of the true standard error of $\hat{\theta}$, where $E(\hat{\sigma}) = b_n \sigma$ for a sequence of constants $\{b_n\}_{n=1}^\infty$ with the property that $b_n \to 1$ as $n \to \infty$. Prove that $n^{1/2}(\hat{\theta}-\theta)/\tilde{\sigma} \xrightarrow{w} N(0,1)$ where $\tilde{\sigma} = b_n^{-1}\hat{\sigma}$ where $\hat{\sigma}$ is a consistent estimator of σ.

References

Adkins, L. C. and Hill, R. C. (1990). An improved confidence ellipsoid for linear regression models. *Journal of Statistical Computation and Simulation*, **36**, 9–18.

Agresti, A. and Coull, B. A. (1998). Approximate is better than "exact" for interval estimation of binomial proportions. *The American Statistician*, **52**, 199–126.

Ahrens, L. H. (1965). Observations on the Fe-Si-Mg relationship in chondrites. *Geochimica et Cosmochimica Acta*, **29**, 801–806.

Akritas, M. G. (1986). Bootstrapping the Kaplan-Meier estimator. *Journal of the American Statistical Association*, **81**, 1032–1038.

Apostol, T. M. (1967). *Calculus: Volume I*. John Wiley and Sons, New York.

Azzalini, A. and Bowman, A. W. (1990). A look at some data on the Old Faithful geyser. *Applied Statistics*, **39**, 357–365.

Barber, S. and Jennison, C. (1999). Symmetric tests and confidence intervals for survival probabilities and quantiles of censored survival data. *Biometrics*, **55**, 430–436.

Barndorff-Nielsen, O. E. and Cox, D. R. (1989). *Asymptotic Techniques for Use in Statistics*. Chapman and Hall, London.

Bhattacharya, R. N. and Ghosh, J. K. (1978). On the validity of the formal Edgeworth expansion. *The Annals of Statistics*, **6**, 435–451.

Billingsley, P. (1999). *Convergence of Probability Measures*. John Wiley and Sons, New York.

Booth, J. G. and Sarkar, S. (1998). Monte Carlo approximation of bootstrap variances. *The American Statistician*, **52**, 354–357.

Bowman, A. W. (1984). An alternative method of cross-validation for the smoothing of density estimates. *Biometrika*, **71**, 353–360.

Bowman, A. W. and Azzalini, A. (1997). *Applied Smoothing Techniques for Data Analysis*. Oxford University Press, Oxford.

Boyce, M. S. (1987). Time-series analysis and forecasting of the Aransas/Wood Buffalo Whooping Crane population. *Proceedings of the 1985 Crane Workshop*. J. C. Lewis, Editor. Platte River Whooping Crane Habitat Maintenance Trust, Grand Island, Nebraska, USA, 1–9.

Braun, J. and Hall, P. (2001). Data sharpening for nonparametric inference subject to constraints. *Journal of Computational and Graphical Statistics*, **10**, 786–806.

Brown, L. D., Cai, T. T., and DasGupta, A. (2003). Interval estimation in exponential families. *Statistica Sinica*, **13**, 19–49.

Canadian Wildlife Service and U. S. Fish and Wildlife Service (2005). *Draft International Recovery Plan for the Whooping Crane*. Recovery of Nationally Endangered Wildlife, Ottawa, Canada and U.S. Fish and Wildlife Service, Albuquerque, New Mexico, USA.

Casella, G. and Berger, R. L. (2002). *Statistical Inference*, Second Edition. Duxbury, Pacific Grove, California.

Cheng, M.-Y. and Hall P. (1998). Calibrating the excess mass and dip tests of modality. *Journal of the Royal Statistical Society, Series B*, **60**, 579–589.

Cheung, K. Y. and Lee, S. M. S. (2005). Variance estimation for sample quantiles using the m out of n bootstrap. *Annals of the Institute of Statistical Mathematics*, **57**, 279–290.

Chikae, M., Kagan, K., Nagatani, N., Takamura, Y. and Tamiya, E. (2007). An electrochemical on-field sensor system for the detection of compost maturity. *Analytica Chimica Acta*, **581**, 364–369.

Choi, E. and Hall, P. (1999). Data sharpening as a prelude to density estimation. *Biometrika*, **86**, 941–947.

Chou, Y.-M. (1994). Selecting a better supplier by testing process capability indices. *Quality Engineering*, **18**, 41–52.

Chou, Y.-M., Owen, D. B. and Borrego, A. S. A. (1990). Lower confidence limits on process capability indices. *Journal of Quality Technology*, **22**, 239–229.

Chu, C.-K. and Marron, J. S. (1991). Choosing a kernel regression estimator. *Statistical Science*, **6**, 404–436.

Conover, W. J. (1980). *Practical Nonparametric Statistics*. John Wiley and Sons, New York.

Cox, D. R. (1966). Notes on the analysis of mixed frequency distributions. *British Journal of Mathematical and Statistical Psychology*, **19**, 39–47.

Cramér. H. (1946). *Mathematical Methods of Statistics*. Princeton University Press, Princeton.

Curran-Everett, D. (2000). Multiple comparisons: philosophies and illustrations. *American Journal of Physiology-Regulatory, Integrative and Comparative Physiology*, **279**, R1–R8.

Dalgaard, P. (2002). *Introductory Statistics with R*. Springer, New York.

Davis, C. S. (2002). *Statistical Methods for the Analysis of Repeated Measures*. Springer, New York.

Davison, A. C. and Hinkley, D. V. (1997). *Bootstrap Methods and their Application.* Cambridge, UK: Cambridge University Press.

Dennis, B., Munholland, P. L. and Scott, J. M. (1991). Estimation of growth and extinction parameters for endangered species. *Ecological Monographs,* **61**, 115–143.

Dette, H., Munk, A. and Wagner, T. (1998). Estimating the variance in nonparametric regression-what is a reasonable choice? *Journal of the Royal Statistical Society, Series B,* **60**, 751–764.

Devore, J. L. (2006). *Probability and Statistics for Scientists and Engineers.* Sixth Edition. Thomson, Belmont, CA.

Diaconis, P. and Efron, B. (1983). Computer intensive methods in statistics. *Scientific American,* **248**, 96–108.

Donoho, D. L. (1988). One-sided inference about functionals of a density. *The Annals of Statistics,* **16**, 1390–1420.

Efron, B. (1979). The bootstrap: Another look at the jackknife. *The Annals of Statistics,* **7**, 1–26.

Efron, B. (1981a). Nonparametric standard errors and confidence intervals, (with discussion) *Canadian Journal of Statistics,* **9**, 139–172.

Efron, B. (1981b). Censored data and the bootstrap. *Journal of the American Statistical Association,* **76**, 312–319.

Efron, B. (1987). Better bootstrap confidence intervals. *Journal of the American Statistical Association,* **82**, 171–200.

Efron, B. and Feldman, D. (1991). Compliance as an explanatory variable in clinical trials. *Journal of the American Statistical Association,* **86**, 9–26.

Efron, B. and Gong, G. (1983). A leisurely look at the bootstrap, the jackknife, and cross validation. *The American Statistician,* **37**, 36–48.

Efron, B., Holloran, E., and Holmes, S. (1996). Bootstrap confidence levels for phylogenetic trees. *Proceedings of the National Academy of Sciences of the United States of America,* **93**, 13492–13434.

Efron, B. and Tibshirani, R. J. (1986). Bootstrap methods for standard errors, confidence intervals, and other measures of statistical accuracy. *Statistical Science,* **1**, 54–57.

Efron, B. and Tibshirani, R. J. (1993). *An Introduction to the Bootstrap.* Chapman and Hall, New York.

Efron, B. and Tibshirani, R. J. (1996). The problem of regions. Stanford Technical Report 192.

Efron, B. and Tibshirani, R. J. (1998). The problem of regions. *The Annals of Statistics,* **26**, 1687–1718.

Eicker, F. (1963). Asymptotic normality and consistency of the least squares estimators for families of linear regressions. *The Annals of Mathematical Statistics,* **34**, 447–456.

Epanechnikov, V. A. (1969) Non-parametric estimation of a multivariate probability density. *Theory of Probability and its Applications*, **14**, 153–158.

Ernst, M. D. and Hutson, A. D. (2003). Utilizing a quantile function approach to obtain exact bootstrap solutions. *Statistical Science*, **18**, 231–240.

Falk, M. and Kaufman, E. (1991). Coverage probabilities of bootstrap confidence intervals for quantiles. *The Annals of Statistics*, **19**, 485–495.

Falk, M. and Reiss, R.-D. (1989). Weak convergence of smoothed and nonsmoothed bootstrap quantile estimates. *The Annals of Probability*, **17**, 362–371.

Feigl, P. and Zelen, M. (1965). Estimation of exponential survival probabilities with concomitant information. *Biometrics*, **21**, 826–838.

Felsenstein, J. (1985). Confidence limits on phylogenies: An approach using the bootstrap. *Evolution*, 783–791.

Ferguson, T. S. (1996) *A Course in Large Sample Theory*. Chapman and Hall, London.

Finner, H. and Strassburger, K. (2002). The partitioning principle. *The Annals of Statistics*, **30**, 1194–1213.

Fischer, M. P. and Polansky, A. M. (2006). Influence of flaws on joint spacing and saturation: Results of one-dimensional mechanical modeling. *Journal of Geophysical Research*, **111**, B07403, doi:10.1029/2005JB004115.

Fisher, R. A. (1915). Frequency distribution of the values of the correlation coefficient in samples from an indefinitely large population. *Biometrika*, **10**, 507–521.

Fisher, N. I. and Marron, J. S. (2001). Mode testing via the excess mass estimate. *Biometrika*, **88**, 499–517.

Fix, E. and Hodges, J. L. (1951). Discriminatory analysis-nonparametric discrimination: Consistency properties. *Report No. 4, Project No. 21-29-004*, USAF School of Aviation Medicine, Randolph Field, Texas.

Fix, E. and Hodges, J. L. (1989). Discriminatory analysis-nonparametric discrimination: Consistency properties. *International Statistical Review*, **57**, 238–247.

Flury, B. and Riedwyl, H. (1988). *Multivariate Statistics: A Practical Approach*. Chapman and Hall, London.

Flynn, M. R. (2004). The beta distribution-a physically consistent model for human exposure to airborne contaminants. *Stochastic Environmental Research and Risk Assessment*, **18**, 306–308.

Frangos, C. C., and Schucany, W. R. (1990). Jackknife estimation of the bootstrap acceleration constant. *Computational Statistics and Data Analysis*, **9**, 271–281.

Franklin, L. A. and Wasserman, G. S. (1991). Bootstrap confidence interval estimates of C_{pk}: An introduction. *Communications in Statistics-Simulation and Computation*, **20**, 231–242.

Franklin, L. A. and Wasserman, G. S. (1992a). A note on the conservative nature of the tables of lower confidence limits for C_{pk} with a suggested correction. *Communications in Statistics-Simulation and Computation*, **21**, 1165–1169.

Franklin, L. A. and Wasserman, G. S. (1992b). Bootstrap lower confidence limits for capability indices. *Journal of Quality Technology*, **24**, 196–209.

Freedman, D. A. (1981). Bootstrapping regression models. *The Annals of Statistics*, **9**, 1218–1228.

Freedman, D. A. and Peters, S. C. (1984). Bootstrapping a regression equation: Some empirical results. *Journal of the American Statistical Association*, **79**, 97–106.

Games, P. A. (1971). Multiple comparisons of means. *American Educational Research Journal*, **8**, 531–565.

Garwood, F. (1936). Fiducial limits for the Poisson distribution. *Biometrika*, **28**, 437–442.

Gasser, T., Sroka, L. and Jennen-Steinmetz, C. (1986). Residual variance and residual pattern in nonlinear regression. *Biometrika*, **73**, 625–633.

Gelman, A. B., Carlin, J. S., Stern, H. S., and Rubin, D. B. (1995). *Bayesian Data Analysis*. Chapman and Hall, London.

Gibbons, J. D. and Chakraborti, S. (2003). *Nonparametric Statistical Inference*. Marcel Dekker, New York.

Good, I. J and Gaskins, R. A. (1980). Density estimation and bump-hunting by the penalized likelihood method exemplified by scattering and meteorite data. *Journal of the American Statistical Association*, **75**, 42–56.

Hàjek, J. (1969). *Nonparametric Statistics*. Holden-Day, San Francisco.

Hàjek, J. and Šidák, Z. (1967). *Theory of Rank Tests*. Academic Press, New York.

Hall, P. (1986). On the number of bootstrap simulations required to construct a confidence interval. *The Annals of Statistics*, **14**, 1453–1462.

Hall, P. (1988). Theoretical comparison of bootstrap confidence intervals. *The Annals of Statistics*, **16**, 927–985.

Hall, P. (1989). Unusual properties of bootstrap confidence intervals in regression problems. *Probability Theory and Related Fields*, **81**, 247–273.

Hall, P. (1992a). *The Bootstrap and Edgeworth Expansion*. Springer, New York.

Hall, P. (1992b). On bootstrap confidence intervals in nonparametric regression. *The Annals of Statistics*, **20**, 695–711.

Hall, P., DiCiccio, T. J., and Romano, J. P. (1989). On smoothing and the bootstrap. *The Annals of Statistics*, **17**, 692–704.

Hall, P. and Kang, K.-H. (2005). Unimodal kernel density estimation by data sharpening. *Statistica Sinica*, **15**, 73–98.

Hall, P., Kay, J. W. and Titterington, D. M. (1990). Asymptotically optimal difference-based estimation of variance in nonparametric regression. *Biometrika*, **77**, 521–528.

Hall, P. and Marron, J. S. (1987a). Extent to which least-squares cross-validation minimizes integrated squared error in non-parametric density estimation. *Probability Theory and Related Fields*, **74**, 568–581.

Hall, P. and Marron, J. S. (1987b). Estimation of integrated squared density derivatives. *Statistics and Probability Letters*, **6**, 109–115.

Hall, P. and Marron, J. S. (1991). Local minima in cross-validation functions. *Journal of the Royal Statistical Society-Series B*, **53**, 245–252.

Hall, P. and Martin, M. A. (1987). Exact convergence rate of bootstrap quantile variance estimator. *Probability Theory and Related Fields*, **80**, 261–268.

Hall, P. and Martin, M. A. (1988). On bootstrap resampling and iteration. *Biometrika*, **75**, 661–671.

Hall, P. and Martin, M. A. (1989). A note on the accuracy of bootstrap percentile method confidence intervals for a quantile. *Statistics and Probability Letters*, **8**, 197–200.

Hall, P. and Minnotte, M. C. (2002). High order data sharpening for density estimation. *Journal of the Royal Statistical Society, Series B*, **64**, 141–157.

Hall, P. and Ooi, H. (2004). Attributing a probability to the shape of a probability density. *The Annals of Statistics*, **32**, 2098–2123.

Hall, P. and York, M. On the calibration of Silverman's test for multimodality. *Statistica Sinica*, **11**, 515–536.

Hancock, G. R. and Klockars, A. J. (1996). The quest for α: Developments in multiple comparison procedures in the quarter century since Games (1971). *Review of Educational Research*, **66**, 269–306.

Härdle, W. (1990). *Applied Nonparametric Regression*. Cambridge University Press, Cambridge.

Hartigan, J. A. and Hartigan, P. M. (1985). The DIP test of unimodality. *The Annals of Statistics*, **13**, 70–84.

Heller, G. and Venkatraman, E. S. (1996). Resampling procedures to compare two survival distributions in the presence of right-censored data. *Biometrics*, **52**, 1204–1213.

Hettmansperger, T. P. (1984). *Statistical Inference Based on Ranks*. John Wiley, New York.

Ho, Y. H. S. and Lee, S. M. S. (2005a). Calibrated interpolated confidence intervals for population quantiles. *Biometrika*, **92**, 234–241.

Ho, Y. H. S. and Lee, S. M. S. (2005b). Iterated smoothed bootstrap confidence intervals for population quantiles. *The Annals of Statistics*, **33**, 437–462.

Hochberg, Y. and Tamhane, A. C. (1987). *Multiple Comparison Procedures*. John Wiley and Sons, New York.

Hodges, J. A. (1931). The effect of rainfall and temperature on corn yields in Kansas. *Journal of Farm Economics*, **13**, 305–318.

Hollander, M. and Wolfe, D. A. (1999). *Nonparametric Statistical Methods*. Second Edition. John Wiley and Sons, New York.

Holling, C. S. (1992). Cross-scale morphology, geometry, and dynamics of ecosystems. *Ecological Monographs*, **62**, 447–502.

Jackson, D. A. (1993). Stopping rules in principal component analysis: a comparison of heuristical and statistical approaches. *Ecology*, **74**, 2204–2214.

Jackson, D. A. (1995). Bootstrapped principal components - reply to Mehlman *et. al. Ecology*, **76**, 644–645.

James, L. F. (1997). A study of a class of weighted bootstraps for censored data. *The Annals of Statistics*, **25**, 1595–1621.

Jennrich, R. I. (1969). Asymptotic properties of non-linear least squares estimators. *The Annals of Mathematical Statistics*, **40**, 633–643.

Johnson, R. A. and Wichern, D. W. (1998). *Applied Multivariate Statistical Analysis*. Fourth Edition. Prentice Hall, New Jersey.

Jolicoeur, P., and Mosimann, J. E. (1960). Size and shape variation in the painted turtle: a principal component analysis. *Growth*, **24**, 339–354.

Jones, M. C. and Sheather, S. J. (1991). Using non-stochastic terms to advantage in kernel-based estimation of integrated squared density derivatives. *Statistics and Probability Letters*, **11**, 511–514.

Jørgensen, S. (1985). *Tree Felling With Original Neolithic Flint Axes in Draved Wood*. National Museum of Denmark, Copenhagen.

Kane, V. E. (1986). Process capability indices. *Journal of Quality Technology*, **18**, 41–52.

Kaplan, E. L., and Meier, P. (1958). Nonparametric estimation from incomplete observations. *Journal of the American Statistical Association*, **53**, 457–481.

Konakov, V., and Mammen, E. (1998). The shape of kernel density estimates in higher dimensions. *Mathematical Methods of Statistics*, **6**, 440–464.

Kotz, S. and Johnson, N. L. (1993). *Process Capability Indices*. Chapman and Hall, New York.

Kushler, R. and Hurley, P. (1992). Confidence bounds for capability indices. *Journal of Quality Technology*, **24**, 188–195.

Le Cam, L. (1973). Convergence of estimates under dimensionality restrictions. *The Annals of Statistics*, **1**, 38–53.

Le Cam, L. (1986). *Asymptotic Methods in Statistical Decision Theory*. Springer, Berlin.

Lehmann, E. L. (1999). *Elements of Large Sample Theory*. Springer, New York.

Lehmann, E. L. (2006). *Statistical Methods Based on Ranks*. Springer, New York.

Mallows, C. L. (1973). Some comments on C_p. *Technometrics*, **15**, 661–675.

Mammen E. (1995). On qualitative smoothness of kernel density estimators. *Statistics*, **26**, 253–267.

Mammen, E., Marron, J. S., and Fisher, N. I. (1992). Some asymptotics of multimodality tests based on kernel density estimators. *Probability Theory and Related Fields*, **91**, 115–132.

Manly, B. F. J. (1996). Are there bumps in body-size distributions? *Ecology*, **77**, 81–86.

Mann, H. B.and Whitney, D. R. (1947). On a test of whether one of two random variables is stochastically larger than the other. *Annals of Mathematical Statistics*, **18**, 50–60.

Maritz, J. S. and Jarrett, R. G. (1978). A note on estimating the variance of the sample median. *Journal of the American Statistical Association*, **73**, 194–196

Marron, J. S. (1993). Discussion of: "Practical performance of several data driven bandwidth selectors" by Park and Trulach. *Computational Statistics*, **8**, 17–19.

Marron, J. S., and Wand, M. P. (1992). Exact mean integrated squared error. *The Annals of Statistics*, **20**, 712–736.

Mathieu, J. R. and Meyer, D. A. (1997). Comparing axes heads of stone, bronze and steel: Studies in experimental archaeology. *Journal of Field Archaeology*, **24**, 333–351.

McCullagh, P. and Nelder, J. A. (1989). *Generalized Linear Models*. Second Edition. Chapman and Hall, London.

McMurry, T. and Politis, D. N. (2004). Nonparametric regression with infinite order flat-top kernels. *Journal of Nonparametric Statistics*, **16**, 549–562.

Mehlman, D. W., Shepherd, U. L., and Kelt, D. A. (1995). Bootstrapping principal components-a comment. *Ecology*, **76**, 640–643.

Milan L., and Whittaker, J. (1995). Application of the parametric bootstrap to models that incorporate a singular value decomposition. *Applied Statistics*, **44**, 31–49.

Miller, R. G. (1981). *Simultaneous Statistical Inference.* Second Edition. Springer-Verlag, New York.

Miwa, T. (1997). Controversy over multiple comparisons in agronomic research. *Japanese Journal of Applied Statistics,* **26**, 99–109.

Montgomery, D. C. (1997). *Design and Analysis of Experiments.* Fourth Edition. John Wiley and Sons, New York.

Morrison, D. F. (1990). *Multivariate Statistical Methods.* McGraw Hill, New York.

Muenchow, G. (1986). Ecological use of failure time analysis. *Ecology,* **67**, 246–250.

Müller, H. G. (1988). *Nonparametric Regression Analysis of Longitudinal Data.* Springer-Verlag, New York.

Müller, D. W. and Sawitzki, G. (1991). Excess mass estimates and tests for multimodality. *Journal of the American Statistical Association,* **86**, 738–746.

Müller, H. G. Stadtmüller, U. (1987). Estimation of heteroscedaticity in regression analysis. *The Annals of Statistics,* **15**, 610–635.

Müller, H. G. Stadtmüller, U. (1988). Detecting dependencies in smooth regression models. *Biometrika,* **75**, 639–650.

Nadaraya, E. A. (1964). On estimating regression. *Theory and Probability and its Applications,* **10**, 186–190.

Nelder, J. A. and Wedderburn, R. W. M. (1972). Generalized linear models. *Journal of the Royal Statistical Society, Series A,* **135**, 370–384.

O'Keefe, D. J. (2003). Colloquy: Should familywise alpha be adjusted? Against familywise alpha adjustment. *Human Communication Research,* **29**, 431–447.

Park, B. U. and Marron, J. S. (1990). Comparison of data-driven bandwidth selectors. *Journal of the American Statistical Association,* **85**, 66–72.

Parzen, E. (1962). On the estimation of a probability density function and the mode. *The Annals of Mathematical Statistics,* **33**, 1065–1076.

Pearson G. W., and Qua, F. (1993). High precision ^{14}C measurement of Irish oaks to show the natural ^{14}C variations from AD 1840-5000 BC: A correction. *Radiocarbon,* **35**, 105–123.

Polansky, A. M. (2000). Stabilizing bootstrap-t confidence intervals for small samples. *Canadian Journal of Statistics,* **28**, 501–516.

Polansky, A. M. (2003a). Supplier selection based on bootstrap confidence regions of process capability indices. *International Journal of Reliability, Quality and Safety Engineering,* **10**, 1–14.

Polansky, A. M. (2003b). Selecting the best treatment in designed experiments. *Statistics in Medicine,* **22**, 3461–3471.

Polansky, A. M., and Kirmani, S. N. U. A. (2003). Quantifying the capability of industrial processes. In *Handbook of Statistics, Volume 22*. Edited by B. Khattree and C. R. Rao, Elsevier Science, Amsterdam, The Netherlands. 625–656.

Poliak, C. D. (2007). *Observed Confidence Levels for Regression Models*. Ph. D. Dissertation, Division of Statistics, Northern Illinois University.

Pollard, D. (2002). *A User's Guide to Measure Theoretic Probability*. Cambridge University Press.

Polonik, W. (1995). Measuring mass concentrations and estimating density contour clusters-an excess mass approach. *The Annals of Statistics*, **23**, 855–881.

Potthoff, R. F. and Roy, S. N. (1964). A generalized multivariate analysis of variance model useful especially for growth curve problems. *Biometrika*, **51**, 313–326.

Putter, H. and van Zwet, W. R. (1996). Resampling: consistency of substitution estimators. *The Annals of Statistics*, **24**, 2297–2318.

Randles, R. H. and Wolfe, D. A. (1979). *Introduction to the Theory of Nonparametric Statistics*. John Wiley and Sons, New York.

Reid, N. (1981). Estimating the median survival time. *Biometrika*, **68**, 601–608.

Rice, J. (1984). Bandwidth choice for nonparametric regression. *The Annals of Statistics*, **12**, 1215–1230.

Rosenblatt, M. (1956). Remarks on some nonparametric estimates of a density function. *The Annals of Mathematical Statistics*, **27**, 832–837.

Rudemo, M. (1982). Empirical choice of histograms and kernel density estimators. *Scandinavian Journal of Statistics*, **9**, 65–78.

Ruppert, D., Sheather, S. J. and Wand, M. P. (1995). An effective bandwidth selector for local least squares regression. *Journal of the American Statistical Association*, **90**, 1257–1270.

Ruppert, D. and Wand, M. P. (1994). Multivariate locally weighted least squares regression. *Annals of Statistics*, **22**, 1346–1370.

Saville, D. J. (1990). Multiple comparison procedures: The practical solution. *The American Statistician*, **44**. 174–180.

Schenker, N. (1985). Qualms about bootstrap confidence intervals. *Journal of the American Statistical Association*, **80**, 360–361.

Scott, D. W. (1992). *Multivariate Density Estimation*. John Wiley and Sons, New York.

Scott, D. W. and Terrell, G. R. (1987). Biased and unbiased cross-validation in density estimation. *Journal of the American Statistical Association*, **82**, 1131–1146.

Seifert, B., Gasser, T. and Wolfe, A. (1993). Nonparametric estimation of residual variance revisited. *Biometrika*, **80**, 373–383.

Sen, P. K. and Singer, J. M. (1993). *Large Sample Theory in Statistics*. Chapman and Hall, London.

Serfling, R. L. (1980). *Approximation Theorems of Mathematical Statistics*. John Wiley and Sons, New York.

Shao, J. and Tu, D. (1995). *The Jackknife and Bootstrap*. Springer, New York.

Sheather, S. J. and Jones, M. C. (1991). A reliable data-based bandwidth selection method for kernel density estimation. *Journal of the Royal Statistical Society*, Series B, **53**, 683–690.

Shoemaker, O. J. and Pathak, P. K. (2001). The sequential bootstrap: A comparison with regular bootstrap. *Communications in Statistics: Theory and Methods*, **30**, 1661–1674.

Silverman, B. W. (1981). Using kernel density estimates to investigate multimodality. *Journal of the Royal Statistical Society*, Series B, **43**, 97–99.

Silverman, B. W. (1983). Some properties of a test for multimodality based on kernel density estimates. In *Probability, Statistics and Analysis*. Edited by J. F. C. Kingman, and G. E. H. Reuter. Cambridge, UK, Cambridge University Press. 248–259.

Silverman, B. W. (1986). *Density Estimation for Statistics and Data Analysis*. Chapman and Hall, London.

Simonoff, J. S. *Smoothing Methods in Statistics*. Springer, New York.

Snow, T. P. (1987). *The Dynamic Universe*, Second Edition. West Publishing Company, St. Paul, MN.

Stauffer, D. F., Garton, E. O., and Steinhorst, R. K. (1985). A comparison of principal components from real and random data. *Ecology*, **66**, 1693–1698.

Stefansson, G., Kim, W.-C., and Hsu, J. C. (1988). On confidence sets in multiple comparisons. In *Statistical Decision Theory and Related Topics IV*. Edited by S.S. Gupta and J. O. Berger. New York, Academic Press. **2**, 18–104.

Strawderman, R. L. and Wells, M. T. (1997). Accurate bootstrap confidence limits for the cumulative hazard and survivor functions under random censoring. *Journal of the American Statistical Association*, **92**, 1356–1374.

Stuiver, M., Reimer, P. J. and Braziunas, T. F. (1998). High-precision radiocarbon age calibration for terrestrial and marine samples. *Radiocarbon*, **40**, 1127–1151.

Thomas, H. V. and Simmons, E. (1969). Histamine content in sputum from allergic and nonallergic individuals. *Journal of Applied Physiology*, **26**, 793-797.

Thompson J. R. and Tapia R. A. (1990). *Nonparametric Function Estimation, Modeling, and Simulation*. SIAM, Phildelphia, PA.

Tibshirani, R. (1989). Noninformative priors for one parameter of many. *Biometrika*, **76**, 604–608.

Wand, M. P. and Jones, M. C. (1993). Comparison of smoothing parameterizations in bivariate kernel density estimation. *Journal of the American Statistical Association*, **88**, 529–528.

Wand, M. P. and Jones, M. C. (1995). *Kernel Smoothing*. Chapman and Hall, London.

Watson, G. S. (1964). Smooth regression analysis. *Sankhyā Series A*, **26**, 101–116.

Welch, B. L. and Peers, H. W. (1963). On formulae for confidence points based on integrals of weighted likelihoods. *Journal of the Royal Statistical Society - Series B*, **25**, 318–329.

Westfall, P. H. and Young, S. S. (1993). *Resampling-Based Multiple Testing*. John Wiley and Sons, New York.

Wilcoxon, F. (1945). Individual comparisons by ranking methods. *Biometrics*, **1**, 80–83.

Winterbottom, A. (1979). A note on the derivation of Fisher's transformation of the correlation coefficient. *The American Statistician*, **33**, 142–143.

Withers, C. S. (1983). Expansions for the distributions and quantiles of a regular functional of the empirical distribution with applications to nonparametric confidence intervals. *The Annals of Statistics*, **11**, 577–587.

Woolson, R. F. (1987). *Statistical Methods for the Analysis of Biomedical Data*. John Wiley and Sons, New York.

Wu, C. F. J. (1986). Jackknife, bootstrap and other resampling methods in regression analysis. *The Annals of Statistics*, **14**, 1261–1295.

Zellner, A. (1962). An efficient method for estimating seemingly unrelated regressions and tests for aggregation bias. *Journal of the American Statistical Association*, **57**, 348–368.

Zucconi, F., Pera, A., Forte, M., and De Bertoldi, M. (1981). Evaluating toxicity of immature compost. *Biocycle*, **22**, 54–57.

Author Index

263

Subject Index

Accelerated bias-corrected critical point, 43

Accelerated bias-corrected observed confidence level, 43
  asymptotic accuracy, 44
  asymptotic accuracy of bootstrap estimate, 44
  bootstrap estimate, 43, 62

Accelerated bias-correction, 47

Acceleration constant, 43, 52
  bootstrap estimate, 63

Acceptance region, 214

Almost sure convergence, 239

Attained confidence level, 224

Bandwidth, 143, 170
  asymptotically optimal value, 143
  plug-in estimate, 145

Bayesian confidence levels, 228

BC_a bootstrap confidence interval, 52

Bias-corrected critical point, 42

Bias-corrected observed confidence level, 42
  asymptotic accuracy, 43
  asymptotic accuracy of bootstrap estimate, 43
  bootstrap estimate, 42

Bias-correction, 52

Bioequivalence, 6, 49

Bootstrap, 32, 113, 177
  censored data, 203
  double, 61
  iterated, 61
  nested, 61
  smoothed, 47, 147

  standard error estimate, 61

Borel sigma-field, 24

Capable manufacturing process, 7

Centered residuals, 176

Central Limit Theorem, 31, 240, 242, 246

Chebychev's Theorem, 237, 238

Confidence coefficient, 4

Confidence level, 4

Confidence region
  asymptotically accurate, 217
  definition, 4

Conjugate family, 230

Convergence in probability, 237

Convergence rates, 243

Cornish-Fisher expansion, 39, 178

Correlation parameter
  confidence interval, 26
  normalizing transformation, 26
  observed confidence level, 27

Cramér's continuity condition, 29, 34

Cramér-Wold Theorem, 241

Data
  chondrite meteorites, 18
  growth curves, 14
  heights of NBA players, 57
  histamine levels in smokers, 195
  hormone patch bioequivalence, 7
  hyperactivity treatments, 16
  leukemia survival, 198

Distribution free, 193

Distribution free statistic, 31

Milton Keynes UK
Ingram Content Group UK Ltd.
UKHW040444071024
449327UK00020B/994